Brauer Trees
of Sporadic Groups

Brauer Trees of Sporadic Groups

G. Hiss

and

K. Lux

Lehrstuhl D für Mathematik
Rhein.-Westf. Technische Hochschule
Aachen

CLARENDON PRESS · OXFORD
1989

Oxford University Press, Walton Street, Oxford OX2 6DP
Oxford New York Toronto
Delhi Bombay Calcutta Madras Karachi
Petaling Jaya Singapore Hong Kong Tokyo
Nairobi Dar es Salaam Cape Town
Melbourne Auckland
and associated companies in
Berlin Ibadan

Oxford is a trade mark of Oxford University Press

Published in the United States
by Oxford University Press, New York

British Library Cataloguing in Publication Data
Hiss, G.
Brauer trees of sporadic groups.
1. Group theory
I. Hiss, G. II. Lux, K.
512'.22
ISBN 0-19-853381-0

Library of Congress Cataloging in Publication Data
(Data available)

Typeset from data supplied by the authors.

Printed in Great Britain by
Bookcraft (Bath) Ltd., Midsomer Norton, Avon

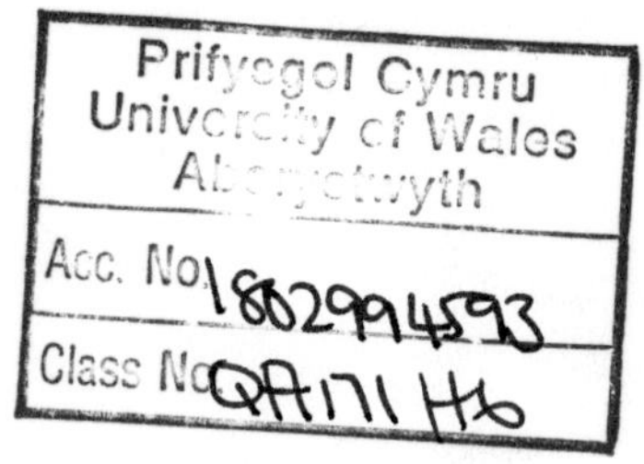

PREFACE

This book compiles the Brauer trees of the sporadic simple groups and their covering groups, as far as they are known. Up to algebraic conjugacy, all trees but 10 are determined. We started to work on this subject when R. Parker first came to Aachen in spring 1984. He introduced us to the problems and to the principal methods.

From this work originated the Ph. D. thesis of the second named author. His dissertation (Lux 1987) has been almost completely included in the present volume. Kapitel 2, 3, and 4 of his thesis make up Chapters 2–4 of our book. All results of Kapitel 5 are contained in our Chapter 6.

We believe that Chapter 4 on Green correspondence will be of some interest to people working on Brauer trees. It describes (partly with proofs) the most powerful methods known to determine the Brauer tree of a block with a cyclic defect group. As far as we know, this method was first used by W. Feit. It was brought to our attention by R. Parker.

Finally we should like to emphasize the fact that large parts of this book have been completely written by computers. We shall discuss this aspect of our work in more detail in the last section of the introduction.

Aachen G. H.
February 1989 K. L.

ACKNOWLEDGEMENTS

In preparing this monograph we were supported by various institutions and persons. We were respectively are financially supported by a grant of the Deutsche Forschungsgemeinschaft (DFG), the first author from April 1984 to August 1987, the second since August 1987. We gratefully acknowledge this financial aid.

About half of the machine calculations were carried out on a MASSCOMP 5300, which was donated to Lehrstuhl D by the DFG. Also most of the programs have been developed on this computer. The other half of the computations, mainly those for the largest groups, were done on a HP 9060, which was donated to the Department of Mathematics by Hewlett Packard.

We wish to thank J. Neubüser for his untiring and unfailing efforts for providing us with great working facilities. The second author is extremely grateful to his Ph. D. supervisor H. Pahlings for his guidance and encouragement.

Above all we are indebted to Richard Parker. He introduced us to the subject when he came to Aachen in May 1984. He supplied his programs which constitute the kernel of the MOC-2 system. It is a great pleasure to thank him for all his assistance, for sharing his ideas with us and for introducing us to the Cambridge style of mathematics. Without him this book would probably never have been written in the present form.

CONTENTS

NOTATION

$\mathbf{N}$	nonnegative integers
$\mathbf{Q}$	field of rational numbers
$\mathbf{Q}_p$	field of p-adic numbers
(K, R, k)	p-modular system
χ	ordinary irreducible character
ϕ	irreducible Brauer character
$\hat{\chi}$	restriction of χ to the p-regular classes
$\bar{\chi}, \bar{\phi}$	complex conjugate character
(χ, ψ)	scalar product of the ordinary characters χ and ψ
ω_χ	central character
ω^*_B	central character of the block B
L	irreducible RG-lattice
$\bar{L}$	reduction modulo p of L
$d_{\chi,\phi}$	multiplicity of the Brauer character ϕ in χ
D_B	decomposition matrix of the block B
Γ	graph
$\mathcal{F}$	family of characters
Γ_B	Brauer tree of a block B
χ_λ	exceptional character
$\circ, \times$	types of a character in a block of cyclic defect
Γ^e_B	canonically embedded Brauer tree of the block B
S_B	real stem of the Brauer tree Γ_B
$M\uparrow^G, M\uparrow^G_H$	induced module
${}^gM$	conjugate module
Tr^G_H	G-trace
$[M, N]$	dimension of $Hom_{kG}(M, N)$
$Hom^1_{\theta G}(M, N)$	projective homorphisms from M to N
$\mathcal{V}(M)$	set of vertices of M
$vtx(M)$	vertex of M
M^{2+}, M^{2-}	symmetric and skew square of M
χ^{2+}, χ^{2-}	symmetric and skew square of χ
$ind(\chi), ind(\phi)$	Frobenius–Schur indicator

$Mod(\theta G)$	set of θG-modules
$Ind(\theta G)$	set of indecomposable θG-modules
$Ind_0(\theta G)$	set of indecomposable non-projective θG-modules
$Ipr(B)$	set of projective indecomposable characters of B
D	defect group of a block B
$N_G(D)$	normalizer of D in G
$f(M), g(N)$	Green correspondents
M^*	dual module
$\Omega(M)$	Heller module of M
b^G	Brauer correspondent of b in G
$T(b)$	inertia subgroup of b
$l(M)$	composition length of the module M
$A, B, C \ldots$	blocks of $N_G(P)$
$1A0, 1B0, 1C0 \ldots$	representatives for the irreducible modules in the blocks of $N_G(P)$
lXm	indecomposable $kN_G(P)$-module with socle $1Xm$ and length l
$\approx$	restriction of a character or module to a block
$\approx_X$	restriction of a character or module to block X
$p^a \top n$	highest power of p dividing n

1

INTRODUCTION

Throughout this book let G be a finite group. General references for the topics discussed in this introduction are Isaacs (1976), Goldschmidt (1980), Feit (1982) and Alperin (1986).

1.1 The Brauer tree

We are going to outline the definition and some of the fundamental properties of the Brauer tree. In later chapters we shall provide more detailed and comprehensive information. There we shall also give the appropriate references for the results we are going to discuss here.

Let (K, R, k) denote a splitting p-modular system for G. As is well known, the structure of the defect group of a block measures the complexity of its representation theory. If for instance the defect group is trivial, the block contains just one ordinary irreducible and exactly one irreducible Brauer character. The next easiest case is when the defect group of a block is cyclic. Here, one of the fundamental results is the following. The number of isomorphism types of indecomposable kG-modules belonging to a block B is finite if and only if the defect group of B is cyclic. This is one of the reasons why the case of the cyclic defect group is much easier to deal with than the general case.

Let B be a block of G with cyclic defect group D of order p^d. Let e be the number of irreducible Brauer characters in B. Then $e \mid p-1$. If D is of order p (i.e. the block has defect 1) and $e = p-1$, there are exactly $p = e+1$ ordinary irreducible characters in B. We denote them by $\{\chi_1, \chi_2, \ldots, \chi_{e+1}\}$. Suppose now that $e \neq p-1$ or that $d > 1$. Then the set of ordinary irreducible characters of B uniquely partitions into two subsets. The elements in one of these are called the *exceptional characters*. They all have the same restriction to the p-regular classes. Their number is $(p^d - 1)/e$. The number of non-exceptional characters is e. Let $\chi_1, \ldots, \chi_e$ denote the non-exceptional characters and χ_λ, $\lambda \in \Lambda$, the exceptionals. Define

$$\chi_{e+1} = \sum_{\lambda \in \Lambda} \chi_\lambda.$$

Then every projective indecomposable character Φ in B has the form

$$\Phi = \chi_i + \chi_j, \quad i, j \in \{1, \ldots, e+1\}, \quad i \neq j.$$

By Brauer reciprocity, all decomposition numbers are 0 or 1, since the columns of the decomposition matrix give the multiplicities of the ordinary characters in the projective indecomposables. We can now define the *labelled Brauer tree* as follows. Its *vertices* (also called *nodes* in the following) are indexed by $\chi_1, \chi_2, \ldots, \chi_{e+1}$. If $e \neq p-1$ or $d > 1$, the node corresponding to χ_{e+1} is called the exceptional node (or vertex). In a drawing of the tree it is often distinguished from the other nodes by a special mark. Two vertices indexed by χ_i and χ_j are joined by an edge if and only if $\chi_i + \chi_j$ is projective. By the definition of a block, there is no rearrangement of rows and columns of the decomposition matrix, such that this has the form

$$\begin{pmatrix} D_1 & 0 \\ 0 & D_2 \end{pmatrix}.$$

This means that the Brauer tree is a connected graph. It has exactly $e+1$ vertices and e edges, hence is a tree in the graph theoretical sense (observe that there are no loops since $\chi_i + \chi_i$ is never projective).

For an ordinary character χ let $\hat{\chi}$ denote its restriction to the p-regular classes. To say that $\chi_i + \chi_j$ is projective is equivalent to saying that $\hat{\chi}_i$ and $\hat{\chi}_j$ have an irreducible Brauer character in common, again by Brauer reciprocity. They can have no more than one in common, since no two columns of the decomposition matrix are identical. Thus the edges of the Brauer tree can and will be labelled with the irreducible Brauer characters in the block. For brevity, the labelled Brauer tree is just called Brauer tree in the following.

Two important features of the Brauer tree should be mentioned. First, let $x \in G$ generate D. Then there is a positive integer z such that $\chi_i(x) = \pm z$ for all $1 \leq i \leq e+1$. (This follows from Lemma 4.4.6 in the case that a Sylow p-subgroup is of order p, but it is true in general.) A vertex corresponding to χ such that $\chi(x)$ is positive is called a *cross*. Otherwise it is a *nought*. The corresponding map from $\{\chi_1, \chi_2, \ldots, \chi_{e+1}\}$ to the set $\{\times, \circ\}$ is called the *type* function. The type of an exceptional character—if there are any—is the type of χ_{e+1}. A nought can only be joined to a cross and a cross only to a nought, i.e. only nodes of different types are joined. The reason is that a projective character has value 0 on all p-singular classes, in particular on x. This observation already allows us to determine the Brauer tree in special cases. For example, suppose that χ_{e+1} is a nought and $\chi_1, \ldots, \chi_e$ are crosses. Then the Brauer tree is a *star* (see Fig. 1.1).

Finally we define the *real stem* of a Brauer tree. This is the subtree obtained by deleting all non-real vertices and all edges incident to them. A vertex is non-real, if the corresponding character (or one of the correponding characters if it is the exceptional vertex) has a non-real value on some p-

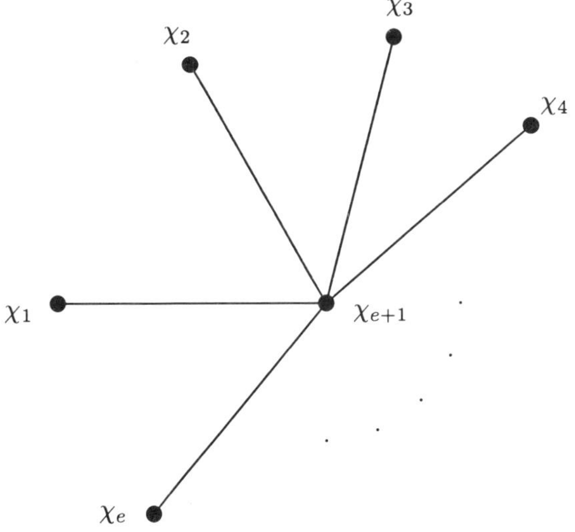

Fig. 1.1 A star

regular element. Of course this can be the empty tree. It is an important fact that the real stem is an open polygon (we also call this a straight line).

If there is at least one real vertex, complex conjugation defines a reflection of the tree with the real stem as reflecting line. A tree coinciding with its real stem is often simply called a real stem in the following.

1.2 Shape of the tree and planar embedded tree

If one deletes the labels from the edges and nodes of a Brauer tree, one obtains the *shape of the tree* (this is not a standard notation, nevertheless quite useful). It is just a tree in the graph theoretical sense with no additional structure. Often it is much easier to determine the shape of the tree than the tree itself. For example, if all ordinary characters in the block are real valued, the Brauer tree is a real stem. In the symmetric groups every ordinary character is real valued, hence the shape of every Brauer tree is a straight line.

The shape of the tree is obtained from the Brauer tree by deleting some information. On the other hand, the *planar embedded tree* contains more information than the labelled Brauer tree. The structure of all indecomposable kG-modules in the block can be derived from it (recall that we have only finitely many indecomposables). Naturally it is more difficult to find the planar embedded tree than the Brauer tree. The planar embedded tree is

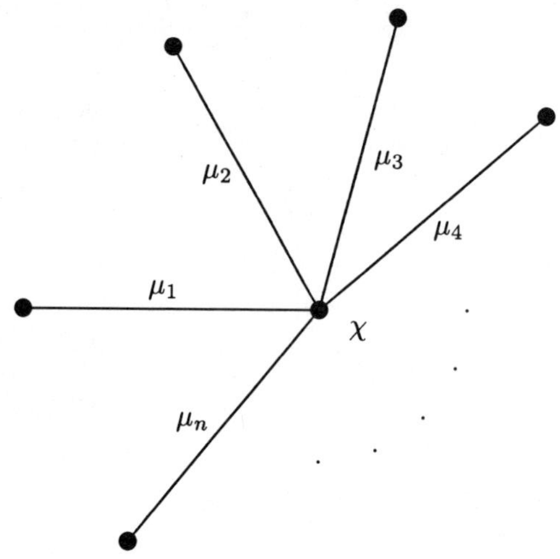

Fig. 1.2 The planar embedding

defined by specifying a planar embedding for every node and all the edges incident to it as follows. Let χ index a vertex of the Brauer tree and let $\mu_1, \mu_2, \ldots, \mu_n$ label all the edges incident to it (Fig 1.2).

Let $M_1, M_2, \ldots, M_n$ be irreducible kG-modules with Brauer characters $\mu_1, \mu_2, \ldots, \mu_n$. From the general theory one knows that there are uniserial kG-modules all with the same Brauer character $\hat{\chi} = \sum_{i=1}^{n} \mu_i$. Since the composition factors of a module are determined by its Brauer character, these modules have composition factors $M_1, \ldots, M_n$. The unique composition series (counting from the bottom) of any one of these determines an ordering of $M_1, \ldots, M_n$. It is a fact that any one of these orderings is obtained from the other by a cyclic permutation. In the planar embedded tree the edges are labelled clockwise according to the ordering of $\mu_1, \ldots, \mu_n$ determined by any one of the uniserial modules with Brauer character $\hat{\chi}$.

If the tree is a real stem, it is automatically planar embedded. On the other extreme, it can be quite a task to determine the planar embedding of a star.

1.3 Possible Brauer trees

In his basic paper Feit (1984a) shows that most trees cannot occur as shapes of Brauer trees. For the moment, let us call a tree *fundamental* if it occurs as the shape of a Brauer tree in the covering group of a finite simple group. Feit shows that every Brauer tree of a finite group is similar to a fundamental

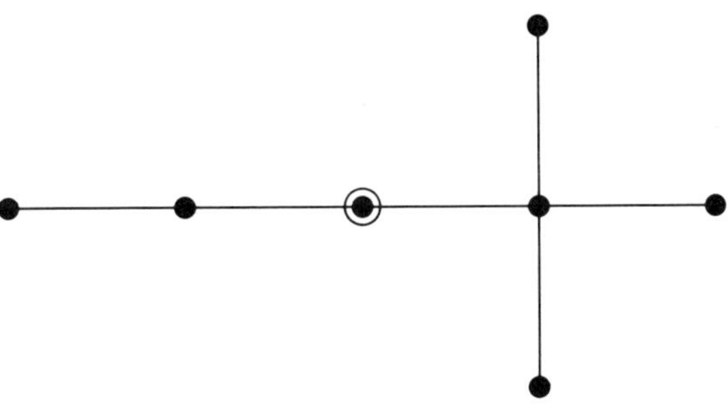

Fig. 1.3 Brauer tree of ${}^2F_4(2)'$ modulo 13

tree, in the sense that they both are *unfoldings* of a single tree around its exceptional vertex. A precise definition of *unfolding* can be found in Feit's paper. The following example may illustrate this phenomenon. The shape of the Brauer tree of the simple Tits group ${}^2F_4(2)'$ in characteristic 13 is shown in Fig. 1.3 (the exceptional node is marked with an extra circle). In the automorphism group ${}^2F_4(2)$ of the Tits group the Brauer tree is a (double) unfolding of the above (Hiss 1986) as shown in Fig. 1.4. The real stem in this example consists of the 5 bottom vertices and the 4 edges joining them.

It should be mentioned that Külshammer (1987) was able to improve on Feit's result by showing that every possible Brauer tree already occurs in the covering group of an automorphism group of a finite simple group.

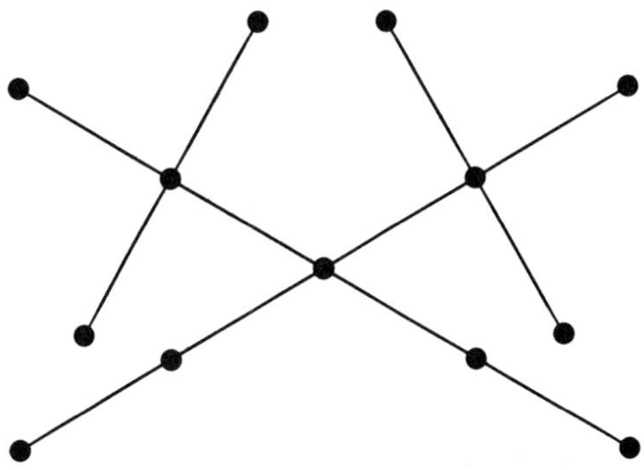

Fig. 1.4 Brauer tree of ${}^2F_4(2)$ modulo 13

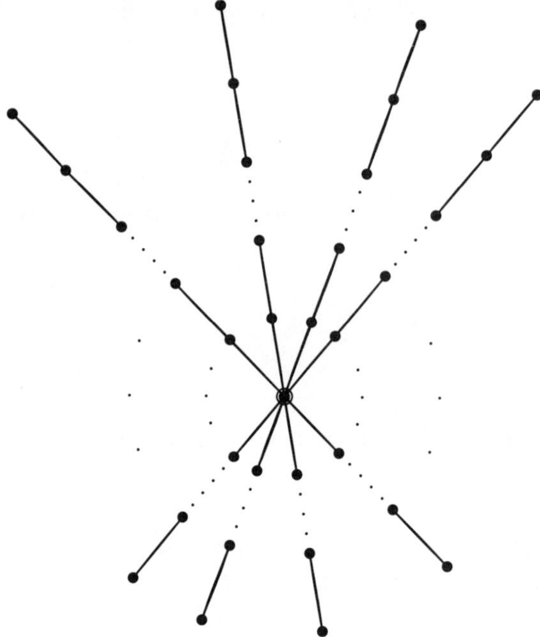

Fig. 1.5 A windmill

Using the classification theorem of finite simple groups, Feit proves that a fundamental tree has at most 248 edges or else is a straight line. The straight lines occur in the classical and in the alternating groups. It is not surprising that one has lines of unlimited length with the exceptional vertex sitting on an arbitrary position. It is a truly amazing fact that 'most' of the Brauer trees are 'windmills', i.e. have the shape shown in Fig. 1.5.

The extra circle around the central node indicates the fact that the exceptional vertex of such a tree (if it exists) is situated in the centre. To find the shapes of the remaining trees one has to consider covering groups of the sporadic simple groups and of the exceptional groups of Lie type.

Of course, Feit's work gives only the shapes of the Brauer trees and not the labelling of the vertices by the ordinary characters in the block. For the classical groups, this problem has been considered by Fong and Srinivasan (1980; 1982; 1984). They completely solved it for the general linear and unitary groups. Fong (1989) has also complete results for the other classical groups. They are as yet unpublished. The Brauer trees for the alternating groups can be derived from the trees for the symmetric groups, which in turn can be determined using some results from James (1978). The Suzuki groups have been examined in Burkhardt (1979), where a complete answer is given. The answer is also known for groups of type $G_2(q)$ (Shamash 1987). The Brauer trees for the Ree groups of type ${}^2G_2(3^{2m+1})$ can easily be calculated using the character tables given in Ward (1966). The first author has recently been able to work out the Brauer trees for the Ree groups of type ${}^2F_4(2^{2m+1})$ using the methods of this book. Finally, it is not too difficult to determine the Brauer trees of the Steinberg triality groups ${}^3D_4(q)$ by making use of the character tables in Deriziotis and Michler (1987). As far as exceptional Chevalley groups are concerned, the question is open for $F_4(q)$, $E_6(q)$, ${}^2E_6(q)$, $E_7(q)$, and $E_8(q)$, because the character tables for these groups are not yet sufficiently known.

In this book, the sporadic simple groups and their covering groups are studied.

1.4 The results

We compile the labelled Brauer trees of all sporadic groups and their covering groups which are known up to now. In those cases where the tree is not completely known, we describe all the possibilities. Some of the trees have been known for quite a while. James (1973) gave a complete list of the Brauer trees of all simple Mathieu groups in his thesis. Humphreys (1982b) and Benson (1985) studied their covering groups. Feit (1984b) considered some other sporadic groups, namely the three smallest Janko groups, the Higman–Sims group, the McLaughlin group and their covering groups. The Higman–Sims group has also been examined in Humphreys (1982a) and the McLaughlin group by Thackray in his thesis (1981). R. Parker has calculated many trees for some of the larger sporadic groups, but he did not publish any of these results. The largest of his examples are trees of the Monster in characteristic 31.

Many of the other trees listed in this book have been calculated by Lux in his Ph. D. thesis (1987). In fact this book contains that thesis as a subset. His results include all trees for the simple Suzuki group, the simple Conway and the simple Fischer groups. At the end of our book we give a list with all the references. It will be clear from this list who calculated which tree for the first time (to the best of our knowledge).

Our drawing of the tree indicates its planar embedding, provided we know it. Of course, a warning is given otherwise. There are only very few cases where this is actually a problem. The embedded trees for the Mathieu groups have been given in Kawata (1987) (for $12M_{22}$ modulo 11 these have also been calculated in Benson (1985)). Feit (1984b) gives the planar embedding of the trees he considers, as well as the Green correspondents of the simple modules.

Even for the trees which appear in published form somewhere in the literature, we give our own proofs in this book (except for the 12-fold cover of M_{22}). First of all, it certainly does no harm to have several different proofs for the same results. Secondly, we have tried to give a complete account of what is known for the sporadic groups, including the proofs. Since we had a computer searching systematically for the trees and the proofs (as will be discussed later) it was not much additional work to deal with the cases already known. Thirdly, copying results from one paper to another is not very interesting and always a source of (typing) errors.

1.5 Open problems

Feit has encountered the problem of *algebraic conjugate* characters. These are characters algebraically conjugate over the rationals $\mathbf{Q}$. If they are not conjugate over the p-adic numbers $\mathbf{Q}_p$, it is enormously difficult to find their position on the Brauer tree. Since in various respects two algebraic conjugate characters are closely related to each other they are hard to distinguish with respect to their location on the tree.

It is often possible to determine a set of nodes of the tree whose elements are occupied by an orbit of algebraic conjugates. But it is impossible to say which character of the orbit goes to which node of this set. We say that the tree is known *up to algebraic conjugacy*, if for every orbit of algebraically conjugate characters we can determine the set of corresponding nodes, but not the correct labelling within this set. In this book we are content with knowing the tree up to algebraic conjugacy. If a tree is only known up to algebraic conjugacy, it is easy to find all its possible labellings. Also, many values of the irreducible Brauer characters are known in this situation, in particular their degrees.

In his paper, Feit (1984b) was able to solve some of the problems arising from algebric conjugate characters. The most difficult of these was the case of J_3 modulo 19, whereas J_3 modulo 17 is still open. Unfortunately, we cannot offer any solution to a problem of this type which has not been settled before, although we tried very hard for some time. However, we found another proof for the J_3 modulo 19 case.

There is also the problem of consistency. It occurs when there are several blocks and characters in distinct blocks whose irrationalities generate the same field. We have situations where it is possible to find the labelled Brauer tree for each individual block by choosing an order of the characters inside the block, but where it is not possible to do this consistently. In other words we cannot give the Brauer trees for all blocks simultaneously after having fixed a p-modular system. Luckily enough, situations like this are not very frequent.

Suppose H is a normal subgroup of G with index some prime number. Then the trees of G are just unfoldings of the trees of H by Feit's theorem, so their shapes are easy to determine. To label the edges, however, can be a very hard problem, even if the labelled trees for H are completely known. The problems that arise are very similar to the problem of algebraic conjugate characters, indeed, are exactly the same in many cases. This is one of the reasons why we do not consider automorphism groups of sporadic groups in this work.

There are, however, other problems arising from the magnitude of the numbers involved. For example, in the O'Nan group, we obtain only projective characters very close to the regular character, which contains every ordinary irreducible character with a multiplicity given by its degree. Of course, the regular projective character contains almost no restriction on the Brauer tree. So even if the tree is an open polygon, there are still too many possible labellings which are consistent with all the projectives we have. If we consider the triple cover of the O'Nan group instead, the situation becomes considerably better. This is due to the fact that in the triple cover there are ordinary characters with much smaller degrees than in the simple group. These yield smaller projectives which are of some use for the factor group. However, there is no such trick available for the Monster. We see no chance at the moment to attack problems of this kind successfully. Fortunately enough, there are only 10 trees which cannot be determined up to algebraic conjugacy. In any case the real stem has been worked out.

1.6 Methods

There is one principal method which is used all over the place. We produce a list of projective characters (mainly by tensoring defect 0 with ordinary irreducible characters). Then we try to find all Brauer trees which are consistent with all the projectives. If there is just one, we are done.

With a very small number of exceptions (e.g. for the largest Janko group J_4, where we have induced characters from the four biggest maximal subgroups by making use of the tables in Fischer (1986)) we have deliberately used only projective characters originating from the character table, such as tensor products and symmetrizations. This makes it easier to check the results by hand. If these projectives suffice to decide the tree, it shows that it is completely determined by the ordinary character table.

If the method described above does not work, we try to use *Green correspondence*. This gives a one-to-one correspondence between the indecomposable modules in the block and certain indecomposables in the normalizer of the defect group. In a certain sense, this correspondence commutes with taking tensor products and symmetric or skew squares. So if one knows what is going on in the normalizer of the defect group (which is of course the case in our examples) one can use this information to reduce the number of possibilities for the Brauer tree via Green correspondence. This method is explained in great detail in Chapter 4.

1.7 Computer proofs

Most of the results in this book have been automatically derived and proved by computers. This includes all the trees where elementary methods were sufficient to find the tree. We used the program systems CAS (see Neubüser *et al.* (1984) for a description) and MOC-2. The latter was developed at Aachen University by R. Parker and the authors. CAS was used mainly as a source for the character tables. In an earlier state of the work it was also applied to produce projective characters and their decomposition into ordinaries. These were then used by us to determine many Brauer trees by hand. Once the MOC-2 system was developed sufficiently, all calculations were done by MOC-2 exclusively. Thus many of the trees have been computed several times by various methods.

The most important program of the MOC-2 system for our purposes is the so-called TRE-program. It was designed and written solely by R. Parker. In a very ingenious way it enumerates all (labelled) trees with a given number of edges. For every tree it checks whether it is consistent with a given set of projective characters. Of course it uses the fact that there is a real stem if the block is invariant under complex conjugation. It also makes use of noughts and crosses, by never joining two noughts or two crosses.

Unfortunately, Parker's program did not give any proofs. (As a man who has invented and written the famous MEAT-AXE, he can for good reasons rely on his programs. But there are people who do not like this sort of computer mathematics; they have good reasons for this, too.) So the authors have added some features to MOC-2 which enable the system to prove its theorems.

Let us shortly describe the course of a typical calculation to see how this works, and thereby discuss some features of the programs. Firstly, a CAS-character table has to be converted into the MOC-2 format. Next, projective characters are computed and expressed in terms of ordinary characters. Every projective character receives a generating number from which its origin can be identified. Usually, a very large number of projectives is generated to make sure that no information is neglected. Preferably, all defect 0 characters are tensored with all ordinary irreducible characters. For reasons of computer disc space this is only possible for groups with up to about 60 conjugacy classes. In the remaining cases, only some of the smallest (by degree) defect 0 characters are tensored with the irreducibles. Next the blocks of defect 1 are calculated. With two exceptions, in a sporadic group a block with a non-trivial cyclic defect group has defect 1. Then the projectives are restricted to the block and pairs of complex conjugate characters are calculated. The TRE-program finds all trees consistent with these projectives. It marks those projectives which were actually used for the proof, i.e. which ruled out a possibility. It is a very remarkable fact that in general out of many hundreds of projectives only very few are actually needed to find the tree. Now another program traces back the origin of the projectives using their generating numbers. All this information is written on to a file in the TeX-input format. Finally it is processed by TeX to produce the pages of this book.

If there are several trees consistent with all projectives the indecomposables which are common to all possibilities are stored on file. They can be used in a second run to produce new projectives e.g. by tensoring them with ordinary characters. Then the procedure starts again with the extended set of projectives. This could in principle be iterated several times but it turns out that it is never useful to go beyond a second iteration.

We feel that the most important aspect of these computer proofs is the fact that they can be checked fairly easily by hand. We shall give some examples of how this can be done in a later chapter. In a sense the computer is used only as a gigantic sieve. In many cases it would have been very hard, if not impossible, to find the tree by hand, but once this has been achieved by the computer it is comparatively easy to check the result. From a huge pile of information (namely the large set of projective characters) it draws the right conclusions and in the process of doing so it throws away most of the redundant information.

It should be mentioned that the computer does not always find the best possible proof, i.e. the proof which succeeds with the smallest number of projective characters. There are examples where one could delete one or two projectives from the pile given by the computer, and still derive the same conclusion. One of the reasons for this is that the algorithm used by TRE is totally different from any algorithm a human beeing would use. Another cause lies in the ordering of the projective characters. Suppose we have two projective characters such that the information of the first (the computer gets) is contained properly in the information of the second. Then the computer has to use both of them. We have taught the computer two little tricks. It knows that the trivial character always is irreducible and that the character with the largest degree in the block (if there are characters of different degrees) is always on a node which is incident to at least two edges.

We decided not to improve the computer proofs by hand, for the following reasons. Firstly, it can be harder to check the correctness and uniqueness of a given tree if the numbers of projectives is too small. With some superfluous information (of course not too much of this) one can sometimes draw the conclusions more easily. Some examples of this phenomenon will be given later. The second reason is that every editing of the files produced by the computer bears the danger of misprints and errors.

2

THE BRAUER TREE OF A BLOCK WITH CYCLIC DEFECT

In this chapter we shall describe the decomposition matrix of a block with cyclic defect. Using the decomposition matrix we shall introduce the Brauer tree of a such a block. Furthermore we shall describe the method which we use to calculate the blocks and the defect groups from the ordinary character table of G. We shall end this chapter with a remark about self-dual blocks of cyclic defect, which will enable us to define a subgraph of the Brauer tree called the real stem. Finally we will introduce a natural embedding of the Brauer tree into the real plane. This embedding will be used in the fourth chapter, where Green correspondence will be applied to determine Brauer trees.

Throughout this book let (K, R, k) be a splitting p-modular system for the finite group G. The maximal ideal of R is generated by the element π.

2.1 Basic Definitions

Definition 2.1.1. A set $\mathcal{F}$ of ordinary irreducible characters is called a *(p-)family*, if all $\chi \in \mathcal{F}$ restrict in the same way to the p-regular classes and if each $\chi \in Irr(G)$, which restricts to the p-regular classes in the same way as the characters in $\mathcal{F}$, is already contained in $\mathcal{F}$.

Definition 2.1.2. The *Brauer graph* Γ_B of a block B of positive defect is defined as follows:

a) If there is more than one family of characters in B, the nodes of Γ_B are in one-to-one correspondence with the families of characters in B and are labelled by a chosen character of the corresponding family. Two nodes are joined by an edge if they are different and if there exists an irreducible Brauer character ϕ, which is a common constituent of the restriction to the p'-classes of the two ordinary characters χ and ψ corresponding to the nodes, i.e. if $d_{\chi,\phi} \neq 0$ and $d_{\psi,\phi} \neq 0$ holds.

b) If there is only one family of characters in B, there is obviously only one irreducible Brauer character ϕ in B. In this case the Brauer graph contains only one edge corresponding to ϕ. The nodes adjacent to this edge are labelled by two different, arbitrarily chosen characters of the family $\mathcal{F}$.

Remark 2.1.3. If ϕ is an irreducible Brauer character of G, χ is an ordinary irreducible character of G and $d_{\chi,\phi} \neq 0$, we say that ϕ is a constituent of the reduction modulo p of χ. If L is an RG-lattice with character χ, the kG-module $\bar{L} := L/\pi L$ is called the reduction modulo p of L. If M is an irreducible kG-module, which is a composition factor of $\bar{L}$, then M is also called an irreducible constituent of $\bar{L}$.

The Brauer graph Γ_B may be relatively complicated in general.

Example 2.1.4. Let $G = A_5$, $p = 2$ and B be the principal block. The decomposition matrix D_B has the following entries, see for example Landrock (1983), page 75–76:

	1	$\mathbf{2_1}$	$\mathbf{2_2}$
1	1	0	0
3_1	1	1	0
3_2	1	0	1
5	1	1	1

Hence the Brauer graph Γ_B is

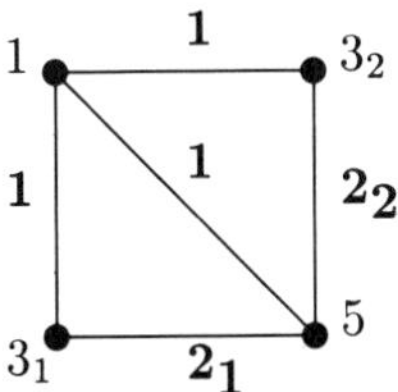

The following theorem shows that the Brauer graph of a block of cyclic defect has a very nice structure. This theorem was originally proved by Brauer (1941) for blocks of defect 1, and was later generalized by Dade (1966) to blocks with an arbitrary cyclic defect group.

Theorem 2.1.5. *Let B be a block with cyclic defect group Q_B of order p^d. Then there exist $e \mid (p-1)$ irreducible Brauer characters $\phi_1, \ldots, \phi_e$ in B. The number of ordinary irreducible characters in B is $e + \frac{p^d-1}{e}$. These characters will be denoted by*

$$\chi_1, \ldots, \chi_e, \chi_\lambda, \quad \lambda \in \Lambda,$$

Λ an index set, $\mid \Lambda \mid = \frac{p^d-1}{e}$.

The $\chi_\lambda, \lambda \in \Lambda$, are called the exceptional characters in B and they all restrict in the same way to the p-regular classes. We define:

$$\chi_{e+1} := \sum_{\lambda \in \Lambda} \chi_\lambda$$

and denote $\{\chi_1, \ldots, \chi_{e+1}\}$ *by* $Irr_0(B)$. *The decomposition matrix* D_B *can be described as follows:*

Let Φ_i *be the ordinary character of the projective indecomposable module* P_i *corresponding to the Brauer character* ϕ_i. *Then it follows that*

$$\Phi_i = \chi_{i(1)} + \chi_{i(2)}, \quad i(1) \neq i(2), \quad i(1), i(2) \in \{1, \ldots, e+1\}.$$

This statement is equivalent to: each Brauer character $\phi_i, i = 1, \ldots, e$, *is an irreducible constituent of* $\chi_{i(1)}, \chi_{i(2)}$ *only (with multiplicity* 1*).*

Remark 2.1.6. a) Since Γ_B is connected and has e edges and $e+1$ nodes, it follows that Γ_B is a tree. In the sequel we will identify the nodes with the corresponding ordinary characters and the edges with the corresponding Brauer characters. If a node belongs to one of the exceptional characters, we mark this node on the tree with a circle, in case there is more than one exceptional character. If ϕ is a constituent of χ and χ', we say that ϕ links χ and χ' or that ϕ is a common edge of χ and χ'.

b) Due to the duality betweeen the set of irreducible Brauer characters and the characters of the projective indecomposable kG-modules, the Brauer tree Γ_B admits a second interpretation: We identify the nodes of the tree with the characters $\chi \in Irr_0(B)$ and the edges with the characters $\Phi \in Ipr(B)$.

c) If ϕ is an irreducible Brauer character in B and χ is a node adjacent to ϕ, it follows that ϕ is the restriction of χ to the p-regular classes, if and only if χ is adjacent to one edge only, namely the edge corresponding to ϕ. In this case we say that χ is a *leaf* of Γ_B.

In order to determine the Brauer trees for blocks of cyclic defect in a group G, we first have to calculate the distribution of the ordinary irreducible characters of G into the blocks.

Definition 2.1.7. Let χ be an ordinary irreducible character of G, and let $Cl(g)$ denote the cardinality of the conjugacy class of $g \in G$. Then the class function ω_χ defined by

$$\omega_\chi(g) = \frac{\chi(g)Cl(g)}{\chi(1)}, \; g \in G,$$

is called the *central character* of χ.

Theorem 2.1.8. *a) Two characters* χ *and* $\chi' \in Irr(G)$ *belong to the same block if and only if the following holds for all p-regular elements* g *of* G:

$$\omega_\chi(g) \equiv \omega_{\chi'}(g) \mod \pi.$$

b) If $p^a \top |G|$ *and if* p^d *is the order of* Q_B, *a defect group of* B, *it follows that* p^{a-d} *is the highest p-power which divides* $\chi(1)$ *for all* $\chi \in Irr(B)$.

Proof. See Theorem 4.2 and Theorem 4.5, Feit (1982), page 150. □

Definition 2.1.9. Let B be a block of G. The *central character* ω_B^* is defined to be the reduction modπ of a central character ω_χ with $\chi \in Irr(B)$. The preceding theorem shows that this definition is independent of the choice of χ.

The following theorem tells us how to determine the isomorphism type of the defect groups, once we know the distribution of ordinary irreducible characters into the blocks of G.

Theorem 2.1.10. *Let B be a block of G.*

a) Suppose

$$\omega_B^*(g) \neq 0$$

for a p-regular element $g \in G$. Then the defect group Q_B of B is conjugate to a subgroup of a Sylow p-subgroup of $C_G(g)$.

b) There exists a p-regular element $g \in G$ with

$$\omega_B^*(g) \neq 0$$

and Q_B is conjugate to a Sylow p-subgroup of $C_G(g)$.

Proof. See Theorem 8.5, Goldschmidt (1980). □

The next theorem indicates how the nodes of a tree Γ_B for a block of cyclic defect can be distributed into two disjoint sets:

Theorem 2.1.11. *Let B be a block with cyclic defect group of order p^d, $p > 2$, and $p^a \top |G|$. Then:*

a) $p^{a-d} \top \chi(1)$ for $\chi \in Irr(B)$.

b) If $\chi, \psi \in Irr_0(B)$, $\chi + \psi \in Ipr(B)$ and if f_χ and f_ψ are the p' parts of $\chi(1)$ and $\psi(1)$, then

$$f_\chi + f_\psi \equiv 0 \bmod p^d.$$

Proof. *a*) See Feit (1982), Theorem 2.16, page 278.

b) Since $\chi + \psi$ is projective, it follows that $p^a \mid \chi(1) + \psi(1)$. Part a) completes the proof. □

Definition 2.1.12. By Theorem 2.1.11 and 2.1.5 it follows that the nodes of a Brauer tree can be divided into two disjoint sets N_1 and N_2: all the characters whose degrees have p' parts which are congruent modulo p^d belong to one of the sets. In order to distinguish between these two sets on the Brauer tree the nodes are indexed by $\circ$ or $\times$, and a node χ is said to be of *type* $\circ$ or $\times$. We say that the trivial character is always of type $\times$. Obviously, only nodes of opposite type can have a common edge. In Section 4 we shall give a representation theoretic interpretation of this distinction.

The following lemma deals with the Brauer trees, which can be determined using only the type of the characters in the block.

Lemma 2.1.13. *Let B be a block of cyclic defect and assume that the number of irreducible Brauer characters in B is more than 1. Then the following statement holds: Γ_B is a star if and only if there is exactly one node of either type $\times$ or $\circ$ in B.*

Proof. If Γ_B is a star, it follows that all nodes besides one of them are leaves. The unique node not being a leaf is linked to all the others and hence is the only node of its type on the tree. If, on the other hand, there exists a unique node of a certain type, the remaining nodes must be linked to this node by Theorem 2.1.11. □

Example 2.1.14. Let $G = J_4$, the fourth Janko group, $p = 5$. The sixth block (in CAS-notation) contains the following irreducible ordinary characters:

Chr.nr.	12	13	33	34	50
type	$\times$	$\times$	$\times$	$\times$	$\circ$

Hence the Brauer tree Γ_B is

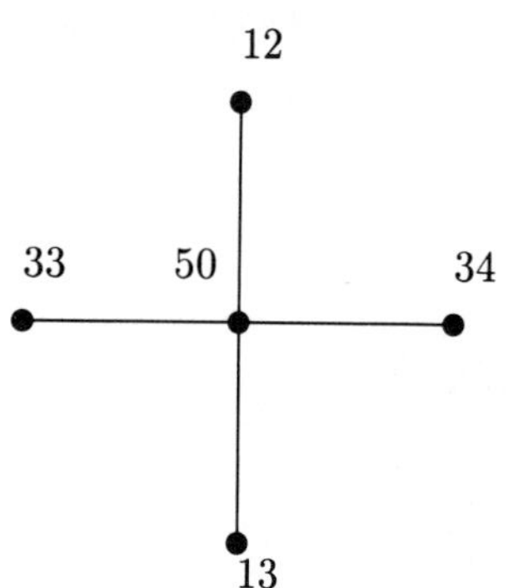

Definition 2.1.15. Let B be a block of G. The block B is called *self-dual*, if there exists $\chi \in Irr(B)$ such that $\bar{\chi} \in Irr(B)$.

Corollary 2.1.16. *Let B be a block of G. The following statements are equivalent:*

a) B is self-dual.

b) If $\chi \in Irr(B)$, then $\bar{\chi} \in Irr(B)$.

c) There exists $\phi \in IBr(B)$ such that $\bar{\phi} \in IBr(B)$.

d) If $\phi \in IBr(B)$, then $\bar{\phi} \in IBr(B)$.

Proof. *a)* $\Rightarrow$ *b)*: This follows easily, if we use that the central character of a block is independent of the choice of $\chi \in Irr(B)$.

b) $\Rightarrow$ *c)*: If ϕ is a constituent of the reduction modulo p of the ordinary irreducible character χ, it follows that $\bar{\phi}$ is a constituent of the reduction modulo p of $\bar{\chi}$. Since $\bar{\chi} \in Irr(B)$, we conclude that $\bar{\phi} \in IBr(B)$.

c) $\Rightarrow$ *d)*: Let $\phi, \psi \in IBr(B)$ with $\bar{\phi} \in IBr(B)$. Then there exists a sequence of ordinary irreducible characters $\chi_1, \ldots, \chi_n$ such that χ_i and χ_{i+1} $(1 \leq i \leq n-1)$ have a common constituent, if we reduce the characters modulo p, and ϕ is a constituent of χ_1 and ψ is a constituent of χ_n. Since $\bar{\phi} \in IBr(B)$, it follows that $\bar{\chi}_1 \in Irr(B)$. Since $\bar{\chi}_1 \in Irr(B)$, it follows that each constituent of the reduction modulo p of $\bar{\chi}_1$ is contained in B, etc. Finally we get that $\bar{\psi} \in IBr(B)$.

d) $\Rightarrow$ *a)*: If $\chi \in Irr(B)$ and $\phi \in IBr(B)$ is a constituent of the reduction modulo p of χ, it follows that $\bar{\chi} \in Irr(B)$ because $\bar{\phi} \in IBr(B)$. □

Lemma 2.1.17. *Let B be a block of cyclic defect and $p > 2$. If B is self-dual, then there exists a real valued character $\chi \in Irr_0(B)$.*

Proof. Using Theorem 2.1.5 we conclude that either the number of edges or the number of nodes in Γ_B is odd. Since B is self-dual, Lemma 2.1.16 shows that complex conjugation acts on the sets $Irr_0(B)$ and $IBr(B)$.

It follows that there exists either a real valued character in $Irr_0(B)$ or a real valued Brauer character ϕ in B. So let us assume that there exists a real valued Brauer character ϕ. Looking at the nodes χ and χ' adjacent to ϕ we see that χ and χ' are either fixed or swapped under complex conjugation. By Theorem 2.1.11 χ and χ' are of the opposite type but $\bar{\chi}$ and $\bar{\chi}'$ are of the same type as χ resp. χ'. It follows that $\chi = \bar{\chi}$ resp. $\chi' = \bar{\chi}'$. □

If B is a self-dual block of cyclic defect, we can restrict the shape of the Brauer tree Γ_B considerably. The following theorem of Brauer gives a description of the reduction modulo p of certain RG-lattices in blocks of cyclic defect.

Theorem 2.1.18. *Let B be a block of cyclic defect, and let $\phi \in IBr(B)$ and $\chi \in Irr_0(B)$ such that $d_{\chi,\phi} \neq 0$.*

a) There exists an RG-lattice L with character χ such that $\bar{L}$, the reduction modulo p, is uniserial and the irreducible kG-module with Brauer character ϕ is the head of $\bar{L}$.

b) If L' is an RG-lattice with character χ, whose reduction modulo p, $\bar{L}'$ is uniserial, the composition series of $\bar{L}'$ is obtained by a cyclic permutation of the composition factors of $\bar{L}$.

Proof. See Brauer (1941) or Theorem 11.15, Goldschmidt (1980). □

Definition 2.1.19. Let B be a block of cyclic defect and let Γ_B be the Brauer tree of B. We define the *planar embedded Brauer tree* Γ_B^e by ordering the edges around each node via the sequence described in Theorem 2.1.18, if we walk around the tree clockwise. This means that the edge ϕ_2 is the successor of the edge ϕ_1 at node χ, if there exists an RG-lattice L with character χ, whose reduction modulo p $\bar{L}$ is uniserial and ϕ_1, ϕ_2 are successive composition factors in the ascending composition series of $\bar{L}$.

Theorem 2.1.20. *Let B be a self-dual block of cyclic defect, $p > 2$.*

a) If χ is a node of Γ_B and $(\phi_1, \dots, \phi_n)$ is the ordering of the edges at χ, then the ordering at $\bar{\chi}$ is $(\bar{\phi}_1, \bar{\phi}_n, \dots, \bar{\phi}_2)$

b) The real valued ordinary irreducible characters and the real valued irreducible Brauer characters form a connected subgraph S_B of Γ_B, which is an open polygon i.e. a straight line. S_B is called the real stem of Γ_B.

c) Complex conjugation induces a graph automorphism of the planar embedded Brauer tree Γ_B^e which is a reflection along the real stem.

Proof. *a*) By Theorem 2.1.18 there exists an RG-lattice L with character χ such that $\bar{L}$, the reduction mod p is uniserial and such that $\bar{L}$ has the composition series $(\phi_1, \dots, \phi_n)$. The dual lattice L^* is a lattice with character $\bar{\chi}$, whose reduction mod p is again uniserial. It follows that the composition series of $\bar{L}^*$ is $(\bar{\phi}_n, \dots, \bar{\phi}_1)$. Since the ordering of the edges around a node is defined only up to a cyclic permutation of the edges, the result follows easily.

b) The nodes adjacent to a real valued edge ϕ belong to real valued ordinary irreducible characters as was shown above. If there exists only one real valued character in $Irr(B)$, every Brauer character is non-real and the claim is trivially true. So let us assume that there exists more than one real valued ordinary irreducible character in B. We first show that the real stem S_B is connected.

Let χ and χ' be two different real valued ordinary characters in B, and let W be a path connecting χ and χ' on the tree. Let W run through the nodes $\chi_1 = \chi, \dots, \chi_n = \chi'$ and the edges $\phi_1, \dots, \phi_{n-1}$. It follows that the path W' which runs through the nodes $\bar{\chi}_1 = \chi, \dots, \bar{\chi}_n = \chi'$ and the edges $\bar{\phi}_1, \dots, \bar{\phi}_{n-1}$ is also a path from χ to χ'. Since Γ_B is a tree, there is a unique path connecting two nodes, and therefore W and W' must be the same, i.e. $\chi_i = \bar{\chi}_i$, $(1 \le i \le n)$, and $\phi_i = \bar{\phi}_i$ $(1 \le i < n)$. So all the edges and the nodes on the path connecting χ and χ' are real valued and it follows easily that S_B is connected.

It remains to be shown that each real valued ordinary character $\chi \in Irr_0(B)$ has at most two real valued constituents on reduction mod p.

From the preceding we know that χ has a real valued constituent ϕ on reduction mod p. By Theorem 2.1.18 there exists an RG-lattice L with character χ, such that $\bar{L}$, the reduction mod p, is uniserial and such that the head of $\bar{L}$ is the irreducible kG-module M with Brauer character ϕ. Since χ is real valued, the dual lattice L^* is again a lattice with character χ. If n is the length of the composition series of $\bar{L}$ and if $\phi = \phi_1, \dots, \phi_n$ is the sequence of the Brauer characters belonging to successive composition factors of $\bar{L}$, it follows that $\bar{\phi}_n, \ldots, \bar{\phi}_1 = \phi$ is the sequence of the Brauer characters of the successive composition factors of $\bar{L}^*$. By Theorem 2.1.18 we conclude that the cyclic permutation $\rho = (\phi_1, \bar{\phi}_n, \bar{\phi}_{n-1}, \ldots, \bar{\phi}_2)$ transforms the sequence of Brauer characters of the composition series of $\bar{L}$ into the sequence of Brauer characters of the composition series of $\bar{L}^*$. If $\phi_i, i \neq 1$, is a real valued constituent of χ, it follows that $n - i + 2 = i$, hence $i = n/2 + 1$. Finally we get that χ has at most two real valued constituents.

c) Using the following arguments it is easily seen that complex conjugation induces a graph automorphism of the tree Γ_B:

$$\chi \in Irr_0(B) \Rightarrow \bar{\chi} \in Irr_0(B)$$

resp.

$$\phi \in IBr(B) \Rightarrow \bar{\phi} \in IBr(B)$$

and

$$d_{\chi,\phi} \neq 0 \Rightarrow d_{\bar{\chi},\bar{\phi}} \neq 0.$$

By part *a)* above, complex conjugation also preserves the ordering of the edges around the nodes. □

The following theorem is a useful generalization of Theorem 2.1.20.

Theorem 2.1.21. *Let B be a block of cyclic defect and let α be an automorphism of G such that B is invariant under the composition of α and complex conjugation.*

a) The nodes belonging to the invariant ordinary irreducible characters and the edges belonging to the invariant irreducible Brauer characters form a connected subgraph of the Brauer tree Γ_B, which is an open polygon.

b) The composition of α and complex conjugation induces a graph automorphism of the planar embedded Brauer tree Γ_B^e.

Proof. Analogous to Theorem 2.1.20. □

A useful tool for determining the Brauer trees of a group is the following lemma.

Lemma 2.1.22. *Let B be a block of cyclic defect for G. Let χ be a node of Γ_B and let $\phi_1, \ldots, \phi_n$ $(n \geq 1)$ be the edges adjacent to χ. Then the following statement can be easily derived from the definition of the Brauer tree:*

$$\sum_{i=1}^{n} \phi_i(1) = \chi(1),$$

hence,

$$\phi_i(1) \leq \chi(1).$$

In particular, if the number of edges of Γ_B is bigger than 1 and χ is a node of highest degree in B, then χ is not a leaf.

3

ELEMENTARY METHODS FOR DETERMINING BRAUER TREES

This chapter deals with the various elementary methods for determining Brauer trees. The elementary methods can roughly be described as the methods, which only use the theory of Brauer characters and which do not use the module structure of the indecomposable kG-modules. Among the elementary methods a vital role is played by the generation of projective characters. Most of the trees given in this book have been determined completely by the elementary methods.

3.1 The methods

In the sequel θ will denote either the complete discrete valuation ring R or the field k of characteristic p.

Notation:

If M and N are θG-modules and M is isomorphic to a direct summand of N we abbreviate this fact by $M \mid N$. If N is a θH-module for a subgroup H of G we denote the induced module $\theta G \otimes_{\theta H} N$ by $N\uparrow^G$. If $g \in G$, then $g \otimes N$ is a θgHg^{-1}-submodule of $N\uparrow^G$, which we denote by ${}^gN$.

Definition 3.1.1. Let H be a subgroup of G. A θG-module M is called H*-projective* if there exists a θH-module N with $M \mid N\uparrow^G$

Lemma 3.1.2. *Let M be a θG-module. The following statements are equivalent:*

a) M is H-projective.

b) $M \mid M\downarrow_H\uparrow^G$.

c) If $G = \bigcup_{i=1}^n g_iH$ is a decomposition of G into disjoint left cosets of H in G, then there is a θH-endomorphism α of M such that $\sum_{i=1}^n g_i\alpha g_i^{-1} = 1_M$.

Proof. See for example Theorem 3.8, Feit (1982), page 89. □

Definition 3.1.3. Let M,N be θG-modules and let H be a subgroup of G. Let g_i, $i = 1, \ldots, n$, be a system of representatives for the left cosets of H in G.

If α is a θH-homomorphism of M to N, then we call the θG-homomorphism:

$$Tr_H^G(\alpha) = \sum_{i=1}^{n} g_i \alpha g_i^{-1}$$

the *G-trace* of α. The G-trace of α is independent of the choice of the system of representatives for the left cosets. The map:

$$Tr_H^G : Hom_{\theta H}(M, N) \to Hom_{\theta G}(M, N)$$

is a group homomorphism and is called the *trace homomorphism* from H to G. The image of Tr_H^G will be denoted by $Hom_{\theta G}(M, N)_H$ and if $\alpha \in Hom_{\theta G}(M, N)_H$ then α is called an *H-projective homomorphism* of M to N. In case $H = \{1\}$, α is called a *projective homomorphism.*

Lemma 3.1.4. *Let M, N, X and Y be θG-modules, and let H be a subgroup of G. Let α be an H-projective homomorphism of M to N. Then the following statements hold:*

a) If $\beta \in Hom_{\theta G}(X, M)$, then

$$Tr_H^G(\alpha\beta) = Tr_H^G(\alpha)\beta.$$

b) If $\gamma \in Hom_{\theta G}(N, Y)$, then

$$Tr_H^G(\gamma\alpha) = \gamma Tr_H^G(\alpha).$$

c) $Hom_{\theta G}(M, M)_H$ is an ideal of $Hom_{\theta G}(M, M)$.

d) We introduce the following notation:

$$Hom_{\theta G}^1(M, N) := Hom_{\theta G}(M, N)/Hom_{\theta G}(M, N)_1,$$

where $1 = \{1\} < G$.

Proof. *a*) and *b*) are easily proved using the definition of a projective homomorphism, see for example Lemma 3.7, Feit (1982), page 89, and c) is an immediate consequence of *a*) and *b*). □

Lemma 3.1.5. *Let M be a θG-module, and let g be an element of G. Let P be a Sylow p-subgroup of G.*

a) M is H-projective $\iff$ M is gHg^{-1}-projective.

b) Every θG-module M is P-projective.

c) Let $U \leq H \leq G$ and let N be a U-projective θH-module. If $M \mid N\uparrow^G$, then M is a U-projective module.

Proof. *a*) Since

$$N \uparrow_H^G \cong {}^gN \uparrow_{gHg^{-1}}^G$$

a) follows immediately.

b) Since $f = \frac{1}{|G:P|} 1_M \in End_{\theta P}(M)$ for a θG-module M and $Tr_P^G(f) = 1_M$, *b*) follows using Lemma 3.1.2.

c) Let S be a θU-module with $N \mid S \uparrow^H$. Since $M \mid N \uparrow^G$, it follows that $M \mid S \uparrow^G$. Hence M is a U-projective θG-module. □

Definition 3.1.6. Let M be a θG-module. The set of subgroups V of G, for which M is V-projective, will be denoted by $\mathcal{V}(M)$. We define a partial ordering on the set $\mathcal{V}(M)$ as follows: Let $V_1 \leq V_2$ for V_1 and V_2 in $\mathcal{V}(M)$, if there exists $g \in G$ such that $gV_1g^{-1} \subseteq V_2$. In this case we call V_1 *subconjugate* to V_2. A subgroup $V \in \mathcal{V}(M)$ is called a *vertex* of M if V is a minimal element with respect to this partial ordering of $\mathcal{V}(M)$. The set of vertices of M is denoted by $vtx(M)$.

The following theorem of Mackey will play a crucial role in the chapter about Green correspondence.

Theorem 3.1.7. *Let H and K be subgroups of G, let D be a system of representatives of the double cosets for H, K in G and let M be a θH-module. Then the following statement holds:*

$$M \uparrow^G \downarrow_K \cong \bigoplus_{d \in D} {}^dM \downarrow_{{}^dH \cap K} \uparrow^K .$$

Proof. See for example Lemma 7, Alperin (1986), page 61. □

Lemma 3.1.8. *Let M be a θG-module and let V be a vertex of M. Then:*

a) V is a p-subgroup of G.

b) V is uniquely determined up to conjugacy with elements in G.

Proof. *a*) follows from Lemma 3.1.5 part *b*), and *b*) is a consequence of Lemma 3.1.5 part *a*) and Mackey's theorem. □

Lemma 3.1.9. *Let M be an indecomposable θG-module and let V be a vertex of M. Since M is V-projective, there exists an indecomposable θV-module S with $M \mid S \uparrow_V^G$. (Obviously V has to be a vertex of S.) A θV-module with the property described above is called a source of M. If S is a source of M, it follows easily that ${}^nS$ with $n \in N_G(V)$ is again a source of M. Furthermore: S is uniquely determined up to conjugacy with elements in $N_G(V)$.*

Proof. The statements above can easily be deduced using Mackey's theorem, see for example Lemma 4.5, Feit (1982), page 113. □

Definition 3.1.10. Let M be a θG-module and let V be a vertex of M. M is called a *trivial source module* if 1_V is a source of M.

By definition a trivial source module is a direct summand of a permutation module. But even more is true:

Lemma 3.1.11. *Let M be a θG-module. Then M is a trivial source module, if and only if M is a direct summand of a permutation module.*

Proof. See Lemma 12.5, Landrock (1983), page 174. □

Lemma 3.1.12. *Let M be a kG-module. If M is a trivial source module, then there exists an RG-lattice L such that M is isomorphic to the reduction mod p of L. In this case we call M a liftable kG-module and L a lift of M.*

Proof. See Theorem 12.4, Landrock (1983), page 173. □

Lemma 3.1.13. *Let M be a θG-module. If M is a trivial source module, it follows that the dual module M^* is a trivial source module.*

Proof. Let $M \mid 1_H \uparrow^G$ where H is a subgroup of G. It follows that $M^* \mid 1_H^* \uparrow^G$. Hence, since $1_H = 1_H^*$:

$$M^* \mid 1_H \uparrow^G .$$

This proves that M^* is a trivial source module. □

Remark 3.1.14. It is clear from the Definition 3.1.1 that the 1-projective θG-modules are exactly the projective θG-modules.

The following two lemmas are used for generating projective characters.

Lemma 3.1.15. *Let H be a subgroup of G, and let U be a subgroup of H. If N is a U-projective θH-module, then $N \uparrow^G$ is a U-projective θG-module.*

In particular, if N is projective, then $N \uparrow^G$ is projective.

Proof. The statement given above follows directly from part *c*) of Lemma 3.1.5. □

Lemma 3.1.16. *Let M and N be indecomposable θG-modules with $Q \in vtx(M)$ and $Q' \in vtx(N)$. It follows that every indecomposable direct summand of $M \otimes N$ is ${}^gQ \cap Q'$-projective for some $g \in G$. In case M or N is projective, it follows that $M \otimes N$ is projective.*

Proof. See Lemma 2.7, Feit (1982), page 83. □

Definition 3.1.17. Let M be a θG-module and let $p > 2$. Let $M^{2+} = \langle m \otimes n + n \otimes m \mid m, n \in M \rangle$ and $M^{2-} = \langle m \otimes n - n \otimes m \mid m, n \in M \rangle$. It follows that M^{2+} and M^{2-} are θG-submodules of $M \otimes M$ and furthermore:

$$M \otimes M = M^{2+} \oplus M^{2-}.$$

In particular, if M is projective, then M^{2+} and M^{2-} are projective. In case M is a projective RG-lattice and Ψ is the ordinary character of M, then Ψ^{2+} and Ψ^{2-}, which are given by:

$$\Psi^{2+}(g) = \frac{\Psi(g)^2 + \Psi(g^2)}{2}$$

resp.

$$\Psi^{2-}(g) = \frac{\Psi(g)^2 - \Psi(g^2)}{2}$$

for $g \in G$, are projective characters.

Lemma 3.1.18. *a) If Ψ is a defect zero character of G and χ is an ordinary character of G, then $\Psi \otimes \chi$ is projective.*

b) If Ψ is a defect zero character of G and $p > 2$, then Ψ^{2+} and Ψ^{2-} are projective characters.

Remark 3.1.19. In order to determine the Brauer tree of a block B of cyclic defect, we use the projective characters, which we have generated using Lemma 3.1.15, Lemma 3.1.18 and Definition 3.1.17. We first decompose a projective character Ψ into the sum of the projective characters of the blocks of G. If B is a block of G, the part of Ψ lying in B, expressed as a sum of ordinary irreducible characters of B, is just the sum of the ordinary irreducible characters of B, which are constituents of Ψ. Only the knowledge of the ordinary character table of G is needed for applying the methods of generating and decomposing projective characters. Nevertheless the examples calculated in the last chapter show that these methods are very strong and usually sufficient to determine the Brauer trees of a sporadic simpe group. In this last chapter the projective characters we use are collected into a table. The columns of this table correspond to the characters in $Irr(B)$, where we already have chosen a representative for the exceptional characters. Above this table another table gives us information about the ordinary characters in the block, their degrees etc.

Another useful information about the ordinary irreducible characters is the Frobenius–Schur indicator.

Definition 3.1.20. The *Frobenius–Schur indicator* of an ordinary irreducible character $\chi \in Irr(G)$ is defined by:

$$ind(\chi) = \frac{1}{|G|} \sum_{g \in G} \chi(g^2).$$

Lemma 3.1.21. *If $\chi \in Irr(G)$, then $ind(\chi)$ is either 0, 1 or -1.*

a) If $\chi \neq \bar{\chi}$, then $ind(\chi) = 0$. This statement holds if and only if 1_G is not a constituent of $\chi \otimes \chi$.

b) If $\chi = \bar{\chi}$ and 1_G is a constituent of χ^{2+}, then $ind(\chi) = 1$.

c) If $\chi = \bar{\chi}$ and 1_G is a constiutent of χ^{2-}, then $ind(\chi) = -1$.

The Frobenius–Schur indicator admits a generalization to the irreducible Brauer characters. But there does not exist an analogous formula to the one given above in the general case. In the following definition the Frobenius–Schur indicator of an irreducible Brauer character is introduced using the Frobenius–Schur indicator of a specific ordinary irreducible character.

Definition 3.1.22. The *Frobenius–Schur indicator* of $\phi \in IBr(G)$ is defined as follows:

a) If $\phi \neq \bar{\phi}$, then $ind(\phi) = 0$.
b) If $\phi = \bar{\phi}$, i.e. ϕ is real valued, there exists $\chi \in Irr(G)$ with $\chi = \bar{\chi}$ and $d_{\chi,\phi} \not\equiv 0 \bmod 2$. Given such χ, the Frobenius–Schur indicator of ϕ is defined to be equal to the Frobenius–Schur indicator of χ.

Remark 3.1.23. The existence of an irreducible ordinary character as needed in part *b*) in the definiton above, is proved in Willems (1976). He also proves that $ind(\phi)$ is independent of the choice of χ with the needed properties. Similar results have been achieved in Thompson (1986), using different methods.

The application to blocks of cyclic defect is contained in the following theorem.

Theorem 3.1.24. *Let B be a self-dual block of cyclic defect. Suppose that there exist ordinary irreducible characters χ and ψ in B $ind(\chi) = 1$ and $ind(\psi) = -1$. Then the following statements hold.*

a) The family of exceptional characters in B contains at least two characters. For each exceptional character χ_λ we get: $\chi_\lambda \neq \bar{\chi}_\lambda$.
b) The real stem S_B has the following shape: The exceptional node separates the nodes belonging to the characters with Frobenius–Schur indicator 1 from those nodes belonging to the characters with Frobenius–Schur indicator -1.

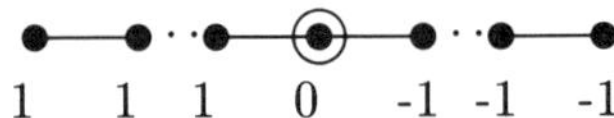

Proof. By the definition of the modular Frobenius–Schur indicator a real valued irreducible Brauer character ϕ in B cannot be a constituent of two ordinary irreducible characters χ and ψ with $ind(\chi) = 1$ and $ind(\psi) = -1$. Since the real stem is a straight line, there must exist a node on the real stem, where the corresponding character has Frobenius–Schur indicator 0. But this can only be true if this node was the exceptional node. □

Example 3.1.25. Let $G = MCL$, the McLaughlin simple group, $p = 7$. The second block (CAS-notation) contains the following ordinary irreducible characters:

Cas.Nr.	2	10	13	18	19
Type	$\times$	$\circ$	$\circ$	$\times$	$\times$
Indicator	1	1	-1	0	0

The characters with CAS-numbers 18 and 19 form the family of exceptional characters. Since the block contains only real valued characters (when restricted to the p-regular classes), the Brauer tree Γ_B is a real stem. The node corresponding to the character with CAS-number 18 cannot be a leaf, since this character has the highest degree of all the ordinary irreducible characters in the block. It follows therefore that the node corresponding to the character with CAS-number 2 has to be a leaf. Hence we are left with the following two possible Brauer trees:

a)

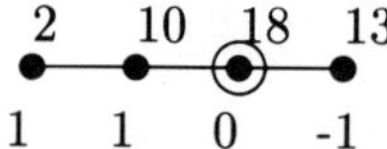

or b)

2 13 18 10

1 -1 0 1

The theorem about the modular Frobenius–Schur indicator disproves case *b*). This shows that *a*) is the correct Brauer tree.

4

NONELEMENTARY METHODS FOR DETERMINING BRAUER TREES

4.1 The Green correspondence

In this chapter we derive a one-to-one correspondence between the indecomposable nonprojective θG-modules and the indecomposable nonprojective θK-modules for certain subgroups K using Green correspondence. In order to get this correspondence we have to assume that a Sylow p-subgroup D of G has order p. We shall only need to apply these results if this assumption is satisfied. It follows under this assumption that D is a trivial intersection set for G, i.e. $D \cap gDg^{-1} = \{1\}$ for $g \notin N_G(D)$, and it is due to this fact that the Green correspondence supplies us with the very simple bijection between the indecomposable nonprojective θG-modules and the indecomposable nonprojective θK-modules for $K \geq N_G(D)$. In this chapter the following notation will be used:

$Mod(\theta G)$ is the set of θG-modules,

$Ind(\theta G)$ is the set of indecomposable θG-modules,

$Ind_0(\theta G)$ is the set of indecomposable nonprojective θG-modules (all up to isomorphism),

H is $N_G(D)$, the normalizer of D in G.

Lemma 4.1.1. *There exists a bijection between $Ind_0(\theta G)$ and $Ind_0(\theta H)$. This bijection is easily defined by the following three statements:*

a) Let $N \in Ind_0(\theta H)$, then:

$$N \uparrow^G = M \oplus P, \ M \in Ind_0(\theta G), \quad P \text{ projective.}$$

We denote M, the unique indecomposable nonprojective summand, by $g(N)$.

b) Let $M \in Ind_0(\theta G)$, then:

$$M \downarrow_H = N \oplus P', \ N \in Ind_0(\theta H), \quad P' \text{ projective.}$$

We denote N, the unique indecomposable nonprojective summand, by $f(M)$.

c) Let $M \in Ind_0(\theta G)$ and $N \in Ind_0(\theta H)$, then:

$$g(f(M)) = M \quad \text{and} \quad f(g(N)) = N.$$

f(M) is called the Green correspondent of M and g(N) is called the Green correspondent of N.

Proof. See for example Theorem 1, Alperin (1986), page 71.

a) If $N \in Ind_0(\theta H)$ and if T denotes a system of representatives for the double cosets for H in G, then the following statement holds using Mackey's theorem:

$$N \uparrow^G \downarrow_H \cong \bigoplus_{r \in T} ({}^rN) \uparrow^H_{{}^rH \cap H} .$$

Since a Sylow p-subgroup of G is a t.i. set, it follows that ${}^rH \cap H$ is a p'-subgroup of G for $r \notin H$. Hence $N \uparrow^H_{{}^rH \cap H}$ is projective for $r \notin H$.

Thus:

$$N \uparrow^G \downarrow_H \cong N \oplus P, \quad P \text{ projective}.$$

Let

$$N \uparrow^G = \bigoplus_{i=1}^{n} M_i, \quad M_i \in Ind(\theta G).$$

be the decomposition of $N \uparrow^G$ into indecomposable direct summands. In Chapter 3 we have shown that M_i is projective, if and only if $M_i \downarrow_H$ is projective, since p does not divide the index $|G : H|$. Hence $N \mid M_{i_0} \downarrow_H$ holds for exactly one i_0, $1 \leq i_0 \leq n$. M_{i_0} is not projective, since N is not projective. The Theorem of Krull–Schmidt tells us that M_i is projective for $i \neq i_0$.

b) Let $M \in Ind_0(\theta G)$. Since M is H-projective, there exists an indecomposable θH-module N with $M \mid N \uparrow^G$. N is not projective, since M is not projective. From part *a*) we get:

$$M \downarrow_H = N \oplus Q, \quad Q \text{ projective}.$$

c) Follows easily from the definiton of f and g. □

An easy consequence of the preceding lemma is:

Lemma 4.1.2. *Let K be a subgroup of G with $K \geq H = N_G(D)$. There exists a bijection between the indecomposable nonprojective θK-modules and the indecomposable nonprojective θG-modules, analogously to Lemma 4.1.1:*

a) If $N \in Ind_0(\theta K)$, then:

$$N \uparrow^G = M \oplus P, \ M \in Ind_0(\theta G), \quad P \ \textit{projective}.$$

We denote M by $g(N)$.

b) If $M \in Ind_0(\theta G)$, then:

$$M \downarrow_K = N \oplus P', \ N \in Ind_0(\theta K), \quad P' \ \textit{projective}.$$

We denote N by $f(M)$.

c) If $M \in Ind_0(\theta G)$ and $N \in Ind_0(\theta K)$, then :

$$g(f(M)) = M \quad \text{and} \quad f(g(N)) = N.$$

$f(M)$ is called the Green correspondent of M (with respect to K and G) and $g(N)$ is called the Green correspondent of N (with respect to K and G).

Proof. The statements of this lemma are easily proved by applying the Green correspondence to the pairs $(G, N_G(D))$ and $(K, N_K(D) = N_G(D))$ using Lemma 4.1.1. □

Lemma 4.1.3. *Let K be a subgroup of G with $K \geq N_G(D)$. Let M be an indecomposable nonprojective θG-module and let N be an indecomposable nonprojective θK-module. Then the following statements hold.*

a) If M is a trivial source module, then $f(M)$ is a trivial source module.

b) If N is a trivial source module, then $g(N)$ is a trivial source module.

Proof. *a*) Since M is a trivial source module, it follows that:

$$M \mid 1_D \uparrow^G,$$

if D is a vertex of M. Hence, using Lemma 4.1.2, we get:

$$f(M) \mid 1_D \uparrow^G \downarrow_K \cong \bigoplus_{r \in T} (1_{{}^rD \cap K}) \uparrow^K,$$

where T is a system of representatives for the double cosets for D, K in G. Since D is also a vertex of $f(M)$, we get:

$$f(M) \mid 1_D \uparrow^K .$$

b) Since N is a trivial source module, it follows that:

$$N \mid 1_D \uparrow^K .$$

Hence, we get using Lemma 4.1.2, since $g(N) \mid N \uparrow^G$:

$$g(N) \mid 1_D \uparrow^G . \ \square$$

Lemma 4.1.4. *Let K be a subgroup of G, and let $K \geq N_G(D)$. We define the maps f and g on the sets $Mod(\theta G)$ and $Mod(\theta K)$ as follows.*

a) If P is a projective indecomposable θG-module we define $f(P) := 0$. Analogously $g(Q) := 0$ for a projective indecomposable θK-module Q.

b) Let

$$f : Mod(\theta G) \to Mod(\theta K)$$

$$g : Mod(\theta K) \to Mod(\theta G)$$

be now defined as follows: For $M \in Mod(\theta G)$ we define

$$f(M) = \bigoplus_{i=1}^{n} f(M_i), \quad if\ M \cong \bigoplus_{i=1}^{n} M_i, M_i \in Ind(\theta G).$$

Analogously we define $g(N)$ for $N \in Mod(\theta K)$:

$$g(N) = \bigoplus_{i=1}^{m} g(N_i), \quad if\ N \cong \bigoplus_{i=1}^{m} N_i, N_i \in Ind(\theta K).$$

Obviously $f(M)$ and $g(N)$ are projective-free, i.e. they do not contain an indecomposable projective direct summand. Furthermore:

$$M \downarrow_K = f(M) \oplus P, \quad P\ projective,$$

resp.

$$N \uparrow^G = g(N) \oplus Q, \quad Q\ projective.$$

The next lemma has turned out to be quite useful for determining Brauer trees.

Lemma 4.1.5. *Let M and N be θG-modules and let K be a subgroup of G with $K \geq N_G(D)$. It follows that:*

a)

$$f(M) \otimes f(N) = f(M \otimes N) \oplus P, \quad P\ projective.$$

b) If $p > 2$:

$$f(M)^{2+} = f(M^{2+}) \oplus Q, \quad Q\ projective.$$

$$f(M)^{2-} = f(M^{2-}) \oplus Q', \quad Q'\ projective.$$

Proof. *a*) Let $M \downarrow_K = f(M) \oplus P_1$ with P_1 projective and $N \downarrow_K = f(N) \oplus P_2$ with P_2 projective. On the one hand, we get, using Lemma 4.1.4:

$$(M \otimes N) \downarrow_K = f(M \otimes N) \oplus Q, \quad Q \text{ projective.}$$

On the other hand,

$$(M \otimes N) \downarrow_K = (f(M) \oplus P_1) \otimes (f(N) \oplus P_2).$$

Hence:

$$(M \otimes N) \downarrow_K = f(M) \otimes f(N) \oplus Q_1, \quad Q_1 \text{ projective.}$$

Since $f(M \otimes N)$ is projective-free, the Theorem of Krull–Schmidt tells us that each indecomposable direct summand of $f(M \otimes N)$ is a direct summand of $f(M) \otimes f(N)$. Hence the statement follows.

b) Let $M \downarrow_K = f(M) \oplus P$, P projective. Using Lemma 3.1.17 we get:

$$M \otimes M = M^{2+} \oplus M^{2-},$$

i.e.

$$(M \otimes M) \downarrow_K = M^{2+} \downarrow_K \oplus M^{2-} \downarrow_K .$$

On the other hand,

$$(M \otimes M) \downarrow_K = (f(M) \oplus P) \otimes (f(M) \oplus P)$$

$$= (f(M) \otimes f(M)) \oplus (f(M) \otimes P) \oplus (P \otimes f(M)) \oplus (P \otimes P)$$

Hence:

$$(M \otimes M) \downarrow_K = f(M)^{2+} \oplus f(M)^{2-} \oplus P', \quad P' \text{ projective.} \tag{4.1}$$

$f(M)^{2+}$ is obviously a θK- submodule of M^{2+}, since $f(M)$ is a direct summand of $M \downarrow_K$, and it follows from part *a*) that $f(M)^{2+}$ is a direct summand of $(M \otimes M) \downarrow_K$. Hence it follows that $f(M)^{2+}$ is a direct summand of $M^{2+} \downarrow_K$. Using a similar argument we get that $f(M)^{2-}$ is a direct summand of $M^{2-} \downarrow_K$. Equation 4.1 tells us that

$$f(M)^{2+} \oplus P'' = (M^{2+}) \downarrow_K, \quad P'' \text{ projective.}$$

On the other hand,

$$M^{2+} \downarrow_K = f(M^{2+}) \oplus Q_1, \quad Q_1 \text{ projective.}$$

Since $f(M^{2+})$ is projective-free, it follows using the Theorem of Krull–Schmidt that

$$f(M)^{2+} = f(M^{2+}) \oplus Q, \quad Q \text{ projective.}$$

In the same way we derive

$$f(M)^{2-} = f(M^{2-}) \oplus Q', \quad Q' \text{ projective. } \square$$

Lemma 4.1.6. *Let $M \in Mod(\theta G)$, and let M^* denote the dual module of M. Then the following statement holds:*

$$f(M^*) = f(M)^*.$$

In particular, if M is self-dual, then $f(M)$ is self-dual.

Proof. Let $M \downarrow_K = f(M) \oplus P$, P projective. Then the following statement is true:

$$(M^*) \downarrow_K = (M \downarrow_K)^* = f(M)^* \oplus P^*.$$

Since a module is projective, if and only if the dual module is projective, the statement follows. $\square$

Lemma 4.1.7. *Let M and N be θG-modules. Using the notation of Chapter 3 we get:*

$$Hom^1_{\theta G}(M, N) \cong Hom^1_{\theta K}(f(M), f(N)),$$

i.e. the space of θG-homomorphisms from M to N modulo the projective homomorphisms is isomorphic to the space of θK-homomorphisms of the Green correspondents modulo the projective homomorphisms.

Proof. See Theorem 3, Alperin (1986), page 74. □

Definition 4.1.8. Let M be a θG-module and let P_M be the projective cover of M. The kernel of the natural epimorphism from P_M onto M is called the *Heller module* $\Omega(M)$ of M. The map:

$$\Omega : Ind_0(\theta G) \longrightarrow Ind_0(\theta G)$$

with

$$M \longmapsto \Omega(M)$$

is a bijection. The inverse map is denoted by Ω^{-1}.

Lemma 4.1.9. *Let M be an indecomposable nonprojective θG-module. Then the following statements hold:*

a)

$$M \cong \Omega(\Omega(M)^*)^*.$$

b) If $V \in vtx(M)$, then $V \in vtx(\Omega(M))$.

c) If S is a source of M, then $\Omega(S)$ is a source of $\Omega(M)$.

d) If there exists an RG-lattice L such that $L/\pi L \cong M$, i.e. such that M is liftable, then $\Omega(M)$ is liftable, too.

e) The Green correspondence commutes with the map Ω:

$$f(\Omega(M)) \cong \Omega(f(M)).$$

Proof. See for example Corollary 17.6b, Green (1974a), page 151. □

Definition 4.1.10. Let M be an indecomposable θG-module with vertex $V \neq 1$. M is called a *cotrivial source module* if $\Omega(1_V)$ is a source of M.

Lemma 4.1.11. *Let M be an indecomposable nonprojective θG-module. The following statements are equivalent:*

a) M is a cotrivial source module.

b) M is a direct summand of $\Omega(1_H \uparrow^G)$ for a subgroup H of G.

Proof. $a) \Longrightarrow b)$: By Schanuel's Lemma (see for example Feit (1982), page 9):

$$(\Omega(1_V)) \uparrow^G \cong \Omega(1_V \uparrow^G) \oplus P, \quad P \text{ projective.}$$

Since M is an indecomposable nonprojective direct summand of $\Omega(1_V) \uparrow^G$, the statement follows from the Theorem of Krull–Schmidt.

$b) \Longrightarrow a)$: If M is a direct summand of $\Omega(1_H \uparrow^G)$, there exists an indecomposable nonprojective direct summand N of $1_H \uparrow^G$ with $M = \Omega(N)$. Since the module N is a direct summand of a permutation module, it is a trivial source module. Hence the result follows directly from Lemma 4.1.9 part c). □

Lemma 4.1.12. *Let M be an indecomposable cotrivial source kG-module. Then M is liftable, i.e. there exists an RG-lattice L such that $L/\pi L \cong M$.*

Proof. By Lemma 4.1.11 there exists a subgroup U of G with:

$$M \mid \Omega(1_U \uparrow^G).$$

Hence:

$$M = \Omega(N),$$

where N is an indecomposable direct summand of $(1_U) \uparrow^G$. N is a trivial source module, and hence liftable, and therefore Lemma 4.1.9 tells us that M is liftable. □

Lemma 4.1.13. *Let M and N be θG-modules. Then the following statement holds:*

$$Hom^1_{\theta G}(M, N) \cong Hom^1_{\theta G}(\Omega(M), \Omega(N))$$

Proof. See for example Lemma 5.10, Feit (1982), page 119. □

Definition 4.1.14. Let M and N be θG-modules, and let $\Omega(M)$ be the Heller-module of M. We define:

$$Ext^1_{\theta G}(M, N) = Hom^1_{\theta G}(\Omega(M), N).$$

$Ext^1_{\theta G}(M, N)$ classifies the extensions of M by N up to cohomological equivalence. If for example $Ext^1_{kG}(M, N) \cong k$, there exists a unique nonsplit extension of M by N, see for example Appendix I, Landrock (1983), page 257.

Lemma 4.1.15. *Let M and N be indecomposable θG-modules. Then the following statement is true:*

There exists a nonsplit extension of M by N, if and only if there exists a nonsplit extension of $f(M)$ by $f(N)$.

Proof. The statement is an easy consequence of Lemma 4.1.7 and the Definition 4.1.14. □

The Green correspondence defines a one-to-one correspondence betweeen the indecomposable nonprojective θK-modules of a subgroup K with $K \geq N_G(D)$ and the indecomposable nonprojective θG-modules. In certain examples it is useful to remove the condition $K \geq N_G(D)$ and replace it by the more general one $K \geq D$. Under this assumption we cannot expect to get a one-to-one correspondence between the indecomposable nonprojective kK-modules and the indecomposable nonprojective kG-modules but we still can describe quite explicitly the structure of the modules induced from K to G.

Theorem 4.1.16. *Let K be a subgroup of G with $K \geq D$. Let U be an indecomposable nonprojective kK-module with Green correspondent $f(U)$ in $N_K(D)$. Then the following statement holds:*

If

$$f(U)\uparrow^{N_G(D)} \cong \bigoplus_{i=1}^{n} U_i \oplus P,$$

where the U_i are indecomposable nonprojective $N_G(D)$-modules and P is a projective $N_G(D)$-module, then

$$U\uparrow^G \cong \bigoplus_{i=1}^{n} g(U_i) \oplus Q, \quad Q \text{ projective.}$$

Proof. See also Burry (1979). By Lemma 4.1.4:

$$f(U)\uparrow^K = U \oplus P_1, \quad P_1 \text{ projective.}$$

Hence:

$$f(U)\uparrow^G = U\uparrow^G \oplus P_2, \quad P_2 \text{ projective.}$$

On the other hand, it follows by the assumptions above:

$$f(U)\uparrow^G = \bigoplus_{i=1}^{n} (U_i\uparrow^G) \oplus P_3, \quad P_3 \text{ projective.}$$

By Lemma 4.1.4:

$$U_i\uparrow^G = g(U_i) \oplus Q', \quad Q' \text{ projective.}$$

Hence:

$$f(U)\uparrow^G = \bigoplus_{i=1}^{n} g(U_i) \oplus P, \quad P \text{ projective.}$$

Since each indecomposable summand $g(U_i)$ is nonprojective, the result follows from the Theorem of Krull–Schmidt. □

It remains to determine the structure of the modules $U\uparrow_{N_K(D)}^{N_G(D)}$. We shall give an answer to this question at the end of the following section, once we have shown that all indecomposable $kN_G(D)$-modules are uniserial.

4.2 The blocks of $N_G(P)$

In order to be able to apply Green correspondence to the blocks of defect 1, we first have to determine the blocks in the normalizer N of a Sylow p-subgroup of G. We shall also give a classification of the indecomposable kN-modules. Since all indecomposable kN-modules are uniserial, the homomorphisms and the projective homomorphisms between them can be easily found.

So let P be a Sylow p-subgroup of G of order p. Let N denote the normalizer of P in G and let C denote the centralizer of P in G. It follows from the Theorem of Schur–Zassenhaus applied to C, that there exists a complement Q to P in C, i.e. $C = Q \times P$.

Let $\{\lambda_0 = 1_P, \lambda_1, \ldots, \lambda_{p-1}\}$ be the ordinary irreducible characters of P. It is well known that the ordinary irreducible characters of C can be described in terms of the ordinary irreducible characters of P and Q as $\lambda_i \otimes \chi$ where $0 \le i \le p-1$ and χ runs through $Irr(Q)$. Moreover the inflation of a character χ of Q is defined to be the character $1_P \otimes \chi$ of C.

There exists a group homomorphism

$$\beta : N \to Aut(P)$$

which is defined by conjugation:

$$\beta(n) : x \longmapsto n^{-1}xn \quad (n \in N, x \in P)$$

and for which $kernel\beta = C$. From β we derive the k-linear representation α of N as follows: If $\beta(n)(x) = n^{-1}xn = x^{\bar{\alpha}(n)}$ for $n \in N$ and $x \in P$ with $\bar{\alpha}(n) \in \mathbf{N}$, then let $\alpha(n) = \bar{\alpha}(n)1_k$. We do not distinguish between the linear representation α and the Brauer character α in the sequel. Since N/C is isomorphic to a subgroup of $Aut(P)$, it follows that N/C is cyclic and the order of N/C is a divisor of $p-1 = |Aut(P)|$.

We first determine the blocks of C in order to derive from this result the blocks of N.

Theorem 4.2.1. *The blocks of C are in bijection with the ordinary irreducible characters of Q. Each block B contains a unique ordinary irreducible character χ_0, whose kernel contains P. Hence χ_0 is the inflation of an ordinary irreducible character of Q, i.e. $\chi_0 = 1_P \otimes \hat{\chi}_0$, $\hat{\chi}_0 \in Irr(Q)$. χ_0 is called the canonical character of B. The ordinary irreducible characters of B are the characters $\chi_i = \lambda_i \otimes \hat{\chi}_0$ for $\lambda_i \in Irr(P)$, $0 \le i \le p-1$. Obviously there exists exactly one irreducible Brauer character ϕ_0 in B and it follows that ϕ_0 is the restriction of χ_0 onto the p-regular classes of C. Hence the Brauer tree Γ_B is:*

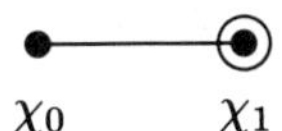

Proof. We first determine the irreducible Brauer characters of C.

Since P is a normal p-subgroup in C, P lies in the kernel of every irreducible Brauer character of C. Hence every Brauer character ϕ of C is the inflation of a Brauer character of the factor group Q. Since Q is a p'-group, the ordinary irreducible characters of Q are the irreducible Brauer characters. Therefore we do not distinguish between these two sets. The irreducible Brauer characters of C are of the form $1_P \otimes \chi$, $\chi \in IBr(Q)$. The ordinary characters $\lambda_i \otimes \chi$ all restrict to the Brauer character $1_P \otimes \chi$. Hence the decomposition matrix D of C is determined and the distribution of the ordinary irreducible characters of C follows immediately. □

In order to deal with the blocks of N, we have to use the theorems which describe the connection between the blocks of a normal subgroup H of G and the blocks of G.

Definition 4.2.2. Let H be a subgroup of G, let b be a block of H and let ω_b be the central character of b. We define the class function ω_b^G of G as follows: If x is a conjugacy class of G and $\hat{x}$ is its class sum, then let

$$\omega_b^G(\hat{x}) = \omega_b(\widehat{x \cap H}).$$

In case ω_b^G is the central character of a block B of G, we say that b^G is defined and denote this by $b^G = B$. The map $b \mapsto b^G$ is called the *Brauer correspondence.*

Lemma 4.2.3. *Let H be a subgroup of G, let b be a block of H and let χ be an ordinary irreducible character in b. Then the following statement holds:*

If $\chi\uparrow^G$ is irreducible, then b^G is defined and $\chi\uparrow^G$ lies in in b^G.

Proof. See Lemma 1.2, Feit (1982), page 193. □

Lemma 4.2.4. *Let Q be a p-subgroup of G, and let $C_G(Q)Q \leq H \leq N_G(Q)$. Then b^G is defined for all blocks b of H. If B is a block of G, then $B = b^G$ for a block b of H, if and only if Q is subconjugate to a defect group Q_B of B.*

Proof. See Theorem 9.6, Goldschmidt (1980). □

Brauer's first main Theorem shows that the blocks of G with nontrivial defect group can already be described locally:

Theorem 4.2.5. *[Brauer] Let Q be a p-subgroup of G and let $N_G(Q)$ be the normalizer of Q in G. Then the Brauer correpondence induces a bijection between the blocks of $N_G(Q)$ with defect group Q and the blocks of G with defect group Q.*

Proof. See Theorem 9.8, Goldschmidt (1980) or Theorem 9.7, Feit (1982), page 137. □

Definition 4.2.6. Let H be a normal subgroup of G, let b be a block of H and let B be a block of G.

a) We say that B covers b if there exists an ordinary irreducible character χ in B, which, by restriction to H, contains an ordinary irreducible constituent χ' in b.

b) G acts on the blocks b of H by conjugation: If $\chi \in Irr(b)$, we define b^g by $\chi^g \in Irr(b^g)$, see for example Feit (1982), page 195. The subgroup $T(b) = \{g \in G \mid b^g = b\}$ is called the *inertia subgroup* of b. Obviously $T(b)$ contains the inertia subgroups of the ordinary irreducible characters and the inertia subgroups of the irreducible Brauer characters in the block b.

Lemma 4.2.7. *Let H be a normal subgroup of G, let B be a block of G and let b be a block of H, which is covered by B. Then the following statements hold:*

a) Every ordinary character χ in B contains a constituent in b by restriction to H.

b) Every Brauer character ϕ in B contains a Brauer character of b by restriction to H.

Proof. See Lemma 4.10, Feit (1982), page 153. □

Lemma 4.2.8. *let H be a normal subgroup of G, let b be a block of H and let B be a block of G. Then we get:*

a) There is a block of G which covers b.

b) The blocks of H covered by B form an orbit under conjugation in G.

Proof. See Lemma 4.10, Feit (1982), page 153. □

Lemma 4.2.9. *Let Q be a normal p-subgroup of G and let $C_G(Q) \leq H \triangleleft G$. Then every block of H is covered by a unique block of G.*

Proof. See Lemma 3.10, Feit (1982), page 200. □

The following theorem shows that the blocks of N can be determined once we know the blocks of the inertia subgroups of the blocks of C.

Theorem 4.2.10. *Let b be a block of H, where H is a normal subgroup of G. It follows that the Brauer correspondence $\hat{B} \mapsto \hat{B}^G$ defines a bijection between the blocks $\hat{B}$ of $T(b)$, which cover b , and the blocks of G, which cover b. Moreover, if $\hat{B}$ is a block of $T(b)$, which covers b, the following statements hold:*

a) The induction map $\chi \mapsto \chi\uparrow^G$ defines a bijection between the ordinary irreducible characters in $\hat{B}$ and the ordinary irreducible characters in $\hat{B}^G$.

b) The induction map $\phi \mapsto \phi\uparrow^G$ defines a bijection between the irreducible Brauer characters of $\hat{B}$ and $\hat{B}^G$.

c) $\hat{B}$ and $\hat{B}^G$ have the same decomposition matrix.

d) $\hat{B}$ and $\hat{B}^G$ have a common defect group.

e) b is the unique block of H, which is covered by $\hat{B}$.

Proof. See Theorem 2.5, Feit (1982), page 197. The statements follow if we use the preceding lemmas and Clifford's Theorem. □

Now we come back to our original setting. Using the theorems above we can describe the blocks of N.

Remark 4.2.11. Let B be a block of N, then P, the Sylow p-subgroup, is the defect group for B, since P is a normal subgroup of N. Lemma 4.2.8 tells us that B covers an orbit of blocks of C and B is the only block which covers the blocks in this orbit by Lemma 4.2.9. Moreover there is one-to-one correspondence between the blocks of N and the orbits of N on the blocks of C. Using the structure of the blocks of C we see that the blocks of N are in one-to-one correspondence with the orbits of N on the ordinary irreducible characters of Q. This follows, since if b is a block of C with canonical character χ_0, then the inertia subgroup $T(b)$ is the inertia subgroup of χ_0.

Theorem 4.2.10 tells us that there exists a unique block $\hat{B}$ in $T(b)$, which covers b. The Brauer tree of the block $\hat{B}$ and B look the same, since the decomposition matrices of B and $\hat{B}$ are the same by Theorem 4.2.10. The nodes and the edges of Γ_B are the induced nodes and edges of $\Gamma_{\hat{B}}$.

It remains to determine the block $\hat{B}$ of $T(b)$. We first calculate the ordinary, irreducible characters χ of $T(b)$, which contain a character of b by restriction to C. χ must be in $\hat{B}$, since it has to be in a block of $T(b)$ by Definition 4.2.6, which covers b, and $\hat{B}$ is the only block of $T(b)$, which covers b. By Clifford's Theorem a character χ of $T(b)$, which restricted to C contains the canonical character χ_0, has to be an extension of χ_0 to $T(b)$. Let α be the linear character defined in the beginning of this section, and let $\alpha' = \alpha\downarrow_{T(b)}$ be the restriction of α to $T(b)$. If ζ is an extension of χ_0 to $T(b)$, all extensions of χ_0 to $T(b)$ can be described as follows:

$$\zeta_j = \zeta \otimes (\alpha')^j, \quad 0 \le j < e = |T(b)/C|.$$

(Remember: N/C is cyclic)

The ordinary irreducible characters of $T(b)$, which contain a character χ_i $i \neq 0$ by restriction onto C, can be calculated using Clifford's Theorem:

$T(b)/C$ acts regularly on the characters $\chi_i = \lambda_i \otimes \chi_0$ $(i \neq 0)$, since N/C acts regularly on $P\backslash\{1\}$ and hence it acts regularly on $Irr(P)\backslash\{1_P\}$. Therefore it follows, if $\chi_{i_1}, \dots, \chi_{i_{\frac{p-1}{e}}}$ are the orbit representatives of $T(b)/C$ on $\{\chi_i \mid i \neq 0\}$, that

$$\zeta_{\lambda_1} = \chi_{i_1}\uparrow^{T(b)}, \dots, \zeta_{\lambda_{\frac{p-1}{e}}} = \chi_{i_{\frac{p-1}{e}}}\uparrow^{T(b)}$$

are the ordinary irreducible characters of $T(b)$, whose restriction to C contain a character χ_i with $i \neq 0$.

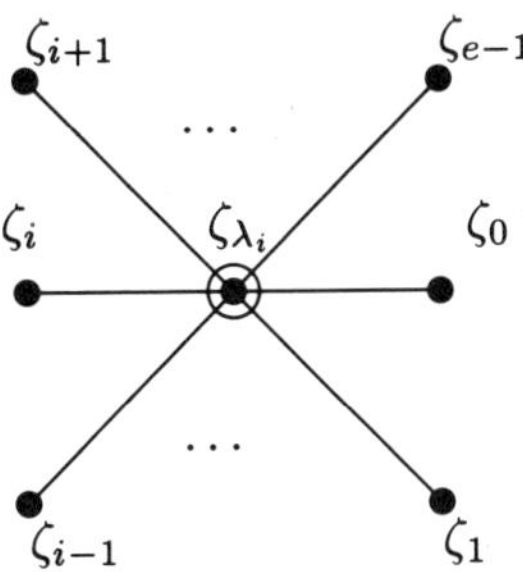

Fig. 4.1

The irreducible Brauer characters can be described analogously: Since $\hat{\chi}_0$ is the only Brauer character in b, which is invariant under the operation of $T(b)/C$, the extensions of $\hat{\chi}_0$ are exactly the irreducible Brauer characters, which by restriction to C contain $\hat{\chi}_0$. These extension are obviously the restriction to the p-regular classes of $T(b)$ of the extensions of χ_0. If we consider $\chi_i \uparrow^{T(b)}$ with $i \neq 0$ restricted to the p-regular classes, then it follows that this Brauer character is the sum of the extensions of $\hat{\chi}_0$. Hence we conclude that all ordinary irreducible characters and all irreducible Brauer characters of $T(b)$, which contain a constituent in b by restriction to C, must belong to the block $\hat{B}$. To summarize: the non-exceptional ordinary characters in $\hat{B}$ are

$$\zeta_j, \quad 0 \leq j \leq e-1,$$

the exceptional characters in $\hat{B}$ are:

$$\zeta_{\lambda_i}, \quad 1 \leq i \leq \frac{p-1}{e}.$$

The Brauer tree $\Gamma_{\hat{B}}$ is shown in Fig. 4.1.

Using Theorem 4.2.10 we get the description of the block B of N. The non-exceptional characters in B are:

$$\zeta_j \uparrow_{T(b)}^{N}, \quad 0 \leq j \leq e-1.$$

If ζ_0 denotes an arbitrarly chosen extension of χ_0 onto $T(b)$, it follows that:

$$\zeta_j \uparrow^N = (\zeta_0 \otimes (\alpha^j \downarrow_{T(b)})) \uparrow^N .$$

Hence:

$$\zeta_j \uparrow^N = \zeta_0 \uparrow^N \otimes \alpha^j.$$

Since $\zeta_0 \uparrow^N$ is an ordinary irreducible character of N, we can describe the situation as follows: The non-exceptional irreducible characters in B are the tensor products of an arbitrary non-exceptional character in B with α^j. The

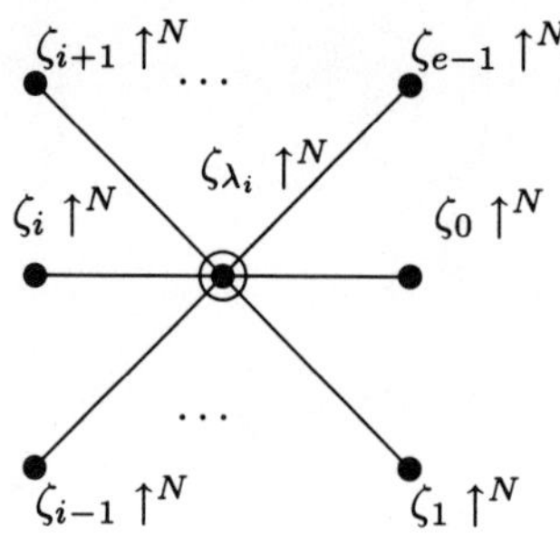

Fig. 4.2

exceptional characters in B are obtained by inducing up the exceptional characters of $\hat{B}$. Since B has the same decomposition matrix as $\hat{B}$, the Brauer tree Γ_B is again a star with the exceptional node in the centre. The irreducible Brauer characters in B can be described in an analogous way to the non-exceptional characters: If ϕ_0 is the reduction mod p of an arbitrary non-exceptional character, it follows that:

$$\phi_j = \phi_0 \otimes \alpha^j, \quad 0 \leq j \leq e-1.$$

The Brauer tree Γ_B is shown in Fig. 4.2.

From the description of the blocks we get:

Lemma 4.2.12. *The irreducible trivial source lattices in $N_G(P)$ are exactly the irreducible $RN_G(P)$-lattices, which belong to the leaves of the Brauer trees of $N_G(P)$ and the cotrivial source lattices with character $\chi \in Irr_0(N_G(P))$ have the sum of the exceptional characters as their character. Moreover the number of irreducible trivial source lattices is equal to the number of irreducible Brauer characters.*

In order to be able to apply Green correspondence to the pair (G, N), we need the structure of the indecomposable modules in a block of N.

Lemma 4.2.13. *Let B be a block of N with defect group P and let e be the number of irreducible Brauer characters in B. Let P_i be the projective indecomposable kN-module corresponding to the simple module with Brauer character ϕ_i. Then it follows that:*

P_i is uniserial of length p and the composition factors of the ascending composition series of P_i have the Brauer characters $\phi_i, \phi_{i+1}, \ldots, \phi_{i+p-2}, \phi_i$, where i has to interpreted modulo e.

Proof. See Theorem 3.5, Feit (1982), page 282. □

Lemma 4.2.14. *Let B be a block of N. Then every indecomposable kN-module U in B is isomorphic to a factor module of a projective indecomposable kN-module and hence is uniserial. $l(U)$ will denote the length of the composition series of U.*

Proof. The result follows easily using the preceding theorem and a theorem of Nakayama about uniserial algebras. See for example Theorem 3.5, Feit (1982), page 282. □

Finally we need the following two lemmas to describe the extensions of kG-modules.

Lemma 4.2.15. *Let U be an indecomposable kN-module in B. Then:*

a) If $l(U) \leq e$, then $Hom_{kN}(U,U) \cong k$.

b) If $l(U) \leq e$ and U_1 is an arbitrary indecomposable kN-module in B, then $Hom_{kN}(U,U_1) \cong k$ or 0 resp. $Hom_{kN}(U_1,U) \cong k$ or 0.

Proof. See Peacock (1975). □

Lemma 4.2.16. *Let U_1 and U_2 be indecomposable kN-modules in the block B whose length of a composition series are l_1 and l_2.*

a) If $l_1 + l_2 \leq p$, then

$$Hom^1_{kN}(U_1, U_2) \cong Hom_{kN}(U_1, U_2).$$

b) If $l_1 + l_2 > p$, then

$$Hom^1_{kN}(U_1, U_2) \cong Hom_{kN}(\Omega(U_1), \Omega(U_2)).$$

Proof. See Peacock (1975). □

We now investigate the problem of the possible Green correspondents of the simple kG-modules. Brauer's first main Theorem tells us that there is a bijection between the blocks of G with defect group P and the blocks of $N_G(P)$ with defect group P. The next lemma shows that this bijection commutes with Green correspondence.

Lemma 4.2.17. *Let B be a block of G with defect group P and let b be the Brauer correspondent in $N_G(P)$, i.e. $b^G = B$. If U is an indecomposable θG-module in B, then $f(U)$, the Green correspondent of U, lies in b.*

Proof. See Theorem 7.8, Feit (1982), page 131. □

In the sequel we shall fix a block B of G with defect group P and its Brauer correspondent b in $N_G(P)$. All θG-modules will be in B, all θN-modules will be in b and e should denote the number of irreducible Brauer characters in b. We choose an irreducible Brauer character in b and denote it by ϕ_0. It follows from the description of the irreducible Brauer characters in $N_G(P)$, that all irreducible Brauer characters of b can be represented in the form:

$$\phi_i = \phi_0 \otimes \alpha^i, \quad 0 \leq i < e.$$

Lemma 4.2.18. *Let S be a simple kG-module and let $l(f(S))$ be the length of the composition series of its Green correspondent. Then*

$$l(f(S)) \leq e \quad \text{or} \quad l(f(S)) \geq p - e.$$

Proof. See Theorem 2.7, Feit (1982), page 276. □

Lemma 4.2.19. *Let S be a simple kG-module and let $f(S)$ be the Green correspondent with length l and socle ϕ_i. Let T be a uniserial kN-module in b with length m and socle ϕ_j. Then it follows that:*

a) If $l + m \leq p$, then there exists a unique nontrivial extension E of T by $f(S)$, if and only if $j \equiv i + l$ modulo e. E is again uniserial, has length $l + m$ and socle ϕ_i. E is uniquely determined up to isomorphism.

b) If $l + m > p$ there exists a unique nontrivial extension E of T by $f(S)$ if and only if $j + m - 1 \equiv i$ modulo e. E is the direct sum of the projective cover of the simple module with Brauer character ϕ_i and a uniserial module with length $l + m - p$ and with socle with Brauer character ϕ_j. E is uniquely determined up to isomorphism.

Proof. See Peacock (1975). □

We now come back to the question from the previous section. We had been given a subgroup K of G with $K \geq P$, hence $P \leq N_K(P) \leq N_G(P)$. The problem, which has been left, was to determine how the induced module $U_{N_K(P)} \uparrow^{N_G(P)}$ for $U \in Ind_0(N_K(P))$ splits into indecomposable direct summands. A first step towards the description is:

Lemma 4.2.20. *Let X be a subgroup of $N_G(P)$ with $X \geq P$. Then it follows that if L is a simple kX-module, then $L \uparrow^{N_G(P)}$ is semisimple.*

Proof. See Theorem 11.2, Landrock (1983), page 161. □

Lemma 4.2.21. *Let X be a subgroup of $N_G(P)$ with $X \geq P$. Then, if M is a kX-module, the socle series of the module $M \downarrow_P$ is the restriction of the socle series of M.*

Proof. See Corollary 9.6, Landrock (1983), page 31. □

Theorem 4.2.22. *Let K be a subgroup of G with $K \geq P$. Let U be a uniserial $kN_K(P)$-module with socle L and length m. Let $L \uparrow^{N_G(P)} = \oplus_{i=1}^{n} L_i$ be the decomposition of $L \uparrow^{N_G(P)}$ into simple direct summands. Then the following statement is true:*

$$U \uparrow^{N_G(P)} = \bigoplus_{i=1}^{n} U_i,$$

where U_i is the uniserial $kN_G(P)$-module with socle L_i and length m.

Proof. Let T be an indecomposable direct summand of $U\uparrow^{N_G(P)}$. Let s be the length of T. We claim that $m = s$. By Mackey's Theorem:

$$U\uparrow^{N_G(P)}\downarrow_P \cong \bigoplus {}^rU\downarrow_P .$$

Since U is unserial of length m, so is ${}^rU$. By Lemma 4.2.21 ${}^rU\downarrow_P$ splits into a direct sum of uniserial kP-modules of length m. Again by Lemma 4.2.21 $T\downarrow_P$ splits into a direct sum of uniserial kP-modules of length s. Since $T \mid U\uparrow^{N_G(P)}$ and hence $T\downarrow_P \mid U\uparrow^{N_G(P)}\downarrow_P$, it follows that $m = s$. Since

$$Soc(U\uparrow^{N_G(P)}) \supseteq \bigoplus_{i=1}^{n} L_i$$

there must exist a uniserial $kN_G(P)$-module U_i with socle L_i and length m, which is a direct summand of $U\uparrow^{N_G(P)}$. Since all simple $kN_K(P)$-modules have the same dimension and since all composition factors of a uniserial $kN_G(P)$-module have the same dimension, it follows by comparing the dimensions of $U\uparrow^{N_G(P)}$ and $\bigoplus_{i=1}^{n} U_i$ that

$$U\uparrow^{N_G(P)} \cong \bigoplus_{i=1}^{n} U_i. \ \square$$

4.3 Tensor products of indecomposable $N_G(P)$-modules

This chapter deals with the description of the tensor products of indecomposable kN-modules. We already know that an indecomposable kN-module is uniserial, since it is a factor module of a projective indecomposable kN-module. It follows that the indecomposable kN-modules are uniquely characterized by the socle and the length of the composition series.

Remark 4.3.1. We introduce the following notation for the blocks of N: The blocks of N with defect group P are denoted by the capital letters A, B, ... ,Z. The principal block of N will always be denoted by A. Moreover we choose an irreducible kN-module S of X and denote it by $1X0$. If M is the indecomposable kN-module in X with length l and socle $S \otimes \alpha^j$, then we denote M by lXj (where j has to interpreted modulo $p-1$). We get the following general formula , see Lemma 2.1, Feit (1982), page 341:

$$nXj = 1X0 \otimes nAj$$

and the following formula for the tensor product of two uniserial kN-modules:

$$nXj \otimes mYs = 1X0 \otimes nAj \otimes 1Y0 \otimes mAs,$$

hence:

$$nXj \otimes mYs = (1X0 \otimes 1Y0) \otimes (nAj \otimes mAs)$$

Since P is contained in the kernel of $1X0$ and $1Y0$, the tensor product $1X0 \otimes 1Y0$ can be derived from the ordinary character table of N/P. Hence it remains to determine the tensor product:

$$nAj \otimes mAs.$$

We can rewrite this tensor product as follows:

$$nAj \otimes mAs = \alpha^j \otimes \alpha^s \otimes nA0 \otimes mA0.$$

Hence:

$$nAj \otimes mAs = \alpha^{j+s} \otimes (nA0 \otimes mA0) = 1A(j+s) \otimes (nA0 \otimes mA0). \tag{4.2}$$

So the final task is to decompose the tensor product $nA0 \otimes mA0$ into the direct sum of indecomposable kN-modules. Before we determine this decomposition, we introduce the following kN-modules: $M = \Omega(1_N) = (p-1)A0$ and $W = \Omega^{-1}(1_N) = (p-1)A1$. Benson and Parker (1984) tells us:

$$M \otimes W = 1_N \oplus P, \quad P \text{ projective.} \tag{4.3}$$

Using the general tensor product formula given in Theorem 2.7, Feit (1982), page 344, and using the notation $A \equiv B$, if the kN-modules A and B are isomorphic up to projective summands, it follows that:

$$nAa \otimes M \equiv (p-n)A(a+n-1) \tag{4.4}$$

and

$$nAa \otimes W \equiv (p-n)A(a+n). \tag{4.5}$$

Since we want to apply the tensor products using Green correspondence, we only want to know the indecomposable nonprojective summands of the tensor product $nAa \otimes mAb$. Using the equations 4.3, 4.4 and 4.5 we may assume that the sum of the lengths of the composition series $n+m \leq p$. If $n \leq m$, we get the following formula from the general tensor product formula given in Feit (1982), page 344:

$$nA0 \otimes mA0 = \bigoplus_{i=1}^{n} (n+m+1-2i)A(i-1) = \tag{4.6}$$

$$(n+m-1)A0 \oplus (n+m-3)A1 \oplus \cdots \oplus (m-n+1)A(n-1).$$

Hence it follows by equation 4.2 that:

$$nAa \otimes mAb = \bigoplus_{i=1}^{n} (n+m+1-2i)A(a+b+i-1) \tag{4.7}$$

for $n \leq m$. The general equation 4.6 can be derived from the following equation using induction, see for example Lemma 2.4, Feit (1982), page 343:

$$2A0 \otimes mA0 = (m-1)A1 \oplus (m+1)A0, \quad m \leq p-1.$$

By tensoring with the modules M and W an arbitrary tensor product $xAa \otimes yAb$ (w.l.o.g. $x \leq y$), as already mentioned above, can be reduced to a tensor product of the following form: $nAa \otimes mAb$ with $n \leq m$ and $n + m \leq p$.

a) If $x > p/2$ and $y > p/2$, then $p - x \leq p/2$ and $p - y \leq p/2$, hence $(p - x) + (p - y) \leq p$. So we get using equation 4.3:

$$xAa \otimes yAb \equiv$$

$$M \otimes xAa \otimes W \otimes yAb \equiv (p - x)A(a + x - 1) \otimes (p - y)A(y + b).$$

The tensor product on the right-hand side can be determined using equation 4.7.

b) If $x \leq p/2$ and $y > p/2$, then $x + (p - y) \leq p$ and

$$M \otimes xAa \otimes yAb \equiv xAa \otimes (p - y)A(a + y - 1).$$

The tensor product on the right-hand side can again be determined by equation 4.7. The final result is obtained by tensoring with W.

$$xAa \otimes yAb \equiv W \otimes (xAa \otimes (p - y)A(a + y - 1)).$$

Lemma 4.3.2. *For the indecomposable kN-modules of length 1 and length $p - 1$ we get the following tensor product formulae:*

a)

$$1Aa \otimes 1Ab \equiv 1A(a + b).$$

b)

$$1Aa \otimes (p - 1)Ab \equiv (p - 1)A(a + b).$$

c)

$$(p - 1)Aa \otimes (p - 1)Ab \equiv 1A(a + b - 1).$$

Proof. These formulae are easily derived from the general equation 4.7. □

Remark 4.3.3. In the next section we shall show that the indecomposable kN-modules of length 1 and length $p-1$ are exactly the Green correspondents of the uniserial kG-modules, which are the reduction mod p of RG-lattices with character $\chi \in Irr_0(B)$.

In order to determine the structure of the symmetric and the skew square of an indecomposable module, we need the the following lemma:

Lemma 4.3.4. *Let G be a group, $p \neq 2$, and let A and B be θG-modules. Then*

$$(A \otimes B)^{2+} \cong A^{2+} \otimes B^{2+} \oplus A^{2-} \otimes B^{2-}$$

resp.

$$(A \otimes B)^{2-} \cong A^{2+} \otimes B^{2-} \oplus A^{2-} \otimes B^{2+}.$$

Proof. See also Lindsey (1974). Let

$$c : A \otimes B \otimes A \otimes B \longrightarrow A \otimes A \otimes B \otimes B$$

be the kG-isomorphism

$$c : a_1 \otimes b_1 \otimes a_2 \otimes b_2 \longmapsto a_1 \otimes a_2 \otimes b_1 \otimes b_2.$$

$A^{2+} \mid A \otimes A$ and $B^{2+} \mid B \otimes B$. Hence $A^{2+} \otimes B^{2+} \mid A \otimes A \otimes B \otimes B$. Thus it follows that: $c^{-1}(A^{2+} \otimes B^{2+}) \mid A \otimes B \otimes A \otimes B$. It remains to be shown that $c^{-1}(A^{2+} \otimes B^{2+}) \leq (A \otimes B)^{2+}$. So let

$$x = (a_1 \otimes a_2 + a_2 \otimes a_1) \otimes (b_1 \otimes b_2 + b_2 \otimes b_1) \in A^{2+} \otimes B^{2+}.$$

It follows that:

$$x = a_1 \otimes a_2 \otimes b_1 \otimes b_2 + a_1 \otimes a_2 \otimes b_2 \otimes b_1 + a_2 \otimes a_1 \otimes b_1 \otimes b_2 + a_2 \otimes a_1 \otimes b_2 \otimes b_1.$$

Hence:

$$c^{-1}(x) = a_1 \otimes b_1 \otimes a_2 \otimes b_2 + a_1 \otimes b_2 \otimes a_2 \otimes b_1 + a_2 \otimes b_1 \otimes a_1 \otimes b_2 + a_2 \otimes b_2 \otimes a_1 \otimes b_1$$

$$= (a_1 \otimes b_1 \otimes a_2 \otimes b_2 + a_2 \otimes b_2 \otimes a_1 \otimes b_1) + (a_1 \otimes b_2 \otimes a_2 \otimes b_1 + a_2 \otimes b_1 \otimes a_1 \otimes b_2).$$

Hence it follows that: $c^{-1}(x) \in (A \otimes B)^{2+}$. In an analogous way we can show that the statement for the skew square holds. A comparision of the dimension of the modules involved proves the result. □

Lemma 4.3.5. *Let B be a block of $N_G(P)$ and let mXa be an indecomposable $kN_G(P)$-module in B, then it follows from the preceding lemma that:*

$$mXa^{2+} \cong 1X0^{2+} \otimes mAa^{2+} \oplus 1X0^{2-} \otimes mAa^{2-}$$

resp.

$$mXa^{2-} \cong 1X0^{2+} \otimes mAa^{2-} \oplus 1X0^{2-} \otimes mAa^{2+}.$$

$1X0^{2+}$ and $1X0^{2-}$ can easily be obtained from the ordinary character table of $N_G(P)/P$. The following holds for mAa^{2+} and mAa^{2-}.

a) If $m \leq (p-1)/2$:

$$mAa^{2+} \equiv \bigoplus_{\substack{i=1 \\ i \text{ odd}}}^{m} (2m+1-2i)A(2a+i-1)$$

and

$$mAa^{2-} \equiv \bigoplus_{\substack{i=1 \\ i \text{ even}}}^{m} (2m+1-2i)A(2a+i-1)$$

b) If $m > (p-1)/2$, and $n = p - m$:

$$mAa^{2+} \equiv \bigoplus_{\substack{i=1 \\ i \text{ even}}}^{n} (2n+1-2i)A(2a+2m+i-2).$$

and

$$mAa^{2-} \equiv \bigoplus_{\substack{i=1 \\ i \text{ odd}}}^{n} (2n+1-2i)A(2a+2m+i-2).$$

In particular,

c) for the module $1Xa$:

$$1Xa^{2+} \equiv 1X0^{2+} \otimes 1A2a$$

and

$$1Xa^{2-} \equiv 1X0^{2-} \otimes 1A2a,$$

d) for the module $(p-1)Xa$:

$$(p-1)Xa^{2+} \equiv 1X0^{2-} \otimes 1A(2a-1)$$

and

$$(p-1)Xa^{2-} \equiv 1X0^{2+} \otimes 1A(2a-1).$$

Proof. See Theorem 2.9, Feit (1982), page 346 and Lindsey (1974). □

Lemma 4.3.6. *Using the notation from above we get the following statement: If mAi is the uniserial kN-module of length m and with socle α^i, then the dual module is:*

$$(mAi)^* = mA(-m+1-i).$$

In particular if $m = 1$:

$$1Ai^* = 1A(-i).$$

Proof. Trivial. □

4.4 The Green correspondents of the simple kG-modules

Before we determine the Green correspondents of the simple kG-modules, we investigate the Green correspondents of the RG-lattices with $\chi \in Irr_0(B)$. In this section B will denote a block of G with defect group P and b will denote the Brauer correspondent of B in N. We recall that the Green correspondence commutes with the Brauer correspondence.

Lemma 4.4.1. *The number of irreducible Brauer characters in B is equal to the number of irreducible Brauer characters in b.*

Proof. See Feit (1982), page 275. □

Lemma 4.4.2. *Let L be an irreducible RG-lattice in B. Then it follows that the Green correspondence commutes with the reduction mod p:*

$$f(L/\pi L) \cong f(L)/(\pi f(L)).$$

Proof. See Feit (1982), page 119. □

Lemma 4.4.3. *Let $\chi \in Irr_0(B)$ and let L be a RG-lattice in B with character $\chi \in Irr_0(B)$, whose reduction mod p is uniserial. Then the following statement holds for the Green correspondent $f(L)$:*

$f(L)$ is either an irreducible RN-lattice in b or $\Omega(f(L))$ is an irreducible RN-lattice in b. This means: Either $\overline{f(L)}$ or $\overline{\Omega(f(L))}$ it an irreducible kN-module. If $\overline{f(L)}$ is irreducible, then L is a trivial source lattice. If $\overline{\Omega(f(L))}$ is irreducible, then L is a cotrivial source lattice, since it follows by Lemma 4.2.12 that the irreducible RN-lattices with character χ, whose kernel contains P, are the trivial source lattices of N in b.

Proof. See Lemma 5.8, Feit (1982), page 292. □

Lemma 4.4.4. *Let $\chi \in Irr_0(B)$ and let L be an RG-lattice in B with character χ, whose reduction mod p is uniserial. Then it follows that: L is uniquely determined by $\overline{f(L)}$.*

Proof. The statement is an easy consequence of the preceding two lemmas, because, since $\overline{f(L)}$ is an irreducible kN-module, $f(L)$ is an irreducible RN-lattice. Since the Green correspondence is bijective, L is uniquely determined by $\overline{f(L)}$. If $\Omega(f(L))$ is an irreducible RN-lattice, we use a similar argument to prove the result. □

Lemma 4.4.5. *The Green correspondence induces a bijection between the RG-lattices with character $\chi \in Irr_0(B)$, whose reduction mod p is uniserial, and the RN-lattices with $\chi' \in Irr_0(b)$, whose reduction mod p is uniserial.*

Proof. If e is the number of irreducible Brauer characters in B, we get that there are exactly $2e$ uniserial lattices as described above using the preceding lemma, because every Brauer character ϕ is a constituent of exactly two characters χ_1 and χ_2 in $Irr_0(B)$. By Theorem 2.1.18 and the preceding lemma there exists exactly one RG-lattice with character χ_1 resp. χ_2, whose reduction mod p is uniserial and has the simple module corresponding to ϕ in the head. Since the number of irreducible Brauer characters is the same for B and b, the result follows. □

Lemma 4.4.6. *Let L be an RG-lattice with character $\chi \in Irr_0(B)$, whose reduction mod p is uniserial, and let χ_0 be an arbitrary irreducible character of b whose kernel contains P. Then the values of χ on the p-singular elements are given by the following statements:*

a) If $f(L)$ is an irreducible RG-lattice:

$$\chi(xy) = \chi_0(y), \quad \langle x \rangle = P, \quad y \text{ a } p'\text{-element}, \ y \in C_G(x).$$

b) If $\Omega(f(L))$ is an irreducible RG-lattice:

$$\chi(xy) = -\chi_0(y) \quad \langle x \rangle = P, \quad y \text{ a } p'\text{-element}, \ y \in C_G(x).$$

Proof. If L is an RG-lattice in B with character χ, $\chi \in Irr_0(B)$, then the values of the character χ on the p-singular elements are given by the values of the Green correspondent $f(L)$ on these classes, since:

$$L\downarrow_N = f(L) \oplus Q, \quad Q \text{ projective}.$$

Observe that the character of Q is zero on p-singular classes, since Q is projective. Hence Lemma 4.4.3 completes the proof. □

We will now give a representation theoretic interpretation of the distinction of the ordinary irreducible characters of B into the types $\times$ and $\circ$.

Definition 4.4.7. Let $\chi \in Irr_0(B)$ and let L be an irreducible RG-lattice with character χ, whose reduction mod p is uniserial. We say that χ is of type $\times$, if the Green correspondent $f(L)$ is an irreducible RN-lattice. If $\Omega(f(L))$ is an irreducible RN-lattice, χ is said to be of type $\circ$. Since the character of a projective indecomposable RG-lattice in B is the sum of two characters χ and χ' in B, it follows that only characters of different type can be linked on the Brauer tree Γ_B.

Remark 4.4.8. Lemma 4.4.6 shows that we can derive the values of irreducible characters in b on the p-regular classes lying in $C_G(P)$ from the ordinary characters of B. So knowing the characters of B enables us to determine the characters of b, if we know the character table of N/P.

We now start to determine the Green correspondents of the simple kG-modules. The length of the Green correspondents can be described as follows:

Lemma 4.4.9. *Let S be a simple kG-module in B . Then*

$$l(f(S)) \leq e \quad \text{or} \quad l(f(S)) \geq p - e.$$

Proof. If $e < l(f(S)) \leq p/2$, then the head of $f(S)$ must appear at least one more time in the composition series of $f(S)$, and we get a nontrivial endomorphism of $f(S)$. By Lemma 4.2.16:

$$End_{kN}(f(S)) \cong End^1_{kN}(f(S)).$$

Hence $dim_k(End^1_{kN}(f(S))) > 1$. By Lemma 4.1.7:

$$End^1_{kN}(f(S)) \cong End^1_{kG}(S).$$

Hence $dim_k(End_{kG}(S)) > 1$. This is a contradiction, since S is simple and the Lemma of Schur tells us $End_{kG}(S) \cong k$. If $p/2 < l(f(S)) < p - e$, then $l(\Omega(f(S))) \leq p/2$. By Lemma 4.1.13:

$$End_{kN}(\Omega(f(S))) = End^1_{kN}(\Omega(f(S))) \cong End^1_{kN}(f(S))$$

$$\cong End^1_{kG}(S)$$

hence $l(\Omega(f(S))) \leq e$, and this implies that $l(f(S)) \geq p - e$. □

Furthermore:

Lemma 4.4.10. *The socles of the Green correspondents of the simple kG-modules are all different.*

Proof. We assume that there exist S_1 and S_2 two different, simple kG-modules in B with

$$Soc(f(S_1)) \cong Soc(f(S_2)).$$

Since S_1 and S_2 are different:

$$Hom_{kG}(S_1, S_2) = (0).$$

This implies using Lemma 4.1.7

$$Hom^1_{kN}(f(S_1), f(S_2)) = (0).$$

We distinguish the following two cases:

a) If $l(f(S_1)) + l(f(S_2)) \leq p$, then it follows by Lemma 4.2.16 that:

$$Hom_{kN}(f(S_1), f(S_2)) = (0).$$

Let $l(f(S_1)) \leq l(f(S_2))$ w.l.o.g. Then there exists a nontrivial homomorphism from $f(S_1)$ to $f(S_2)$. Contradiction.

b) If $l(f(S_1)) + l(f(S_2)) > p$, then $l(\Omega(f(S_1))) + l(\Omega(f(S_2))) \leq p$. Hence :

$$Hom_{kN}(\Omega(f(S_1)), \Omega(f(S_2))) \cong Hom^1_{kN}(\Omega(f(S_1)), \Omega(f(S_2)))$$

$$\cong Hom^1_{kN}(f(S_1), f(S_2)) = (0).$$

Let $l(\Omega(f(S_1))) \leq l(\Omega(f(S_2)))$ w.l.o.g. It follows that there exists a nontrivial homomorphism from $\Omega(f(S_2))$ to $\Omega(f(S_1))$, since both have the same head. Contradiction. □

Let χ_0 be the exceptional node of B, if the number of exceptional characters is greater than 1 and let χ_0 otherwise be any node of Γ_B. If χ is a node of Γ_B, let $d(\chi)$ be the distance of the node χ from χ_0, i.e. the number of edges of the path connecting χ and χ_0. If ϕ is an edge of Γ_B, let $n(\phi)$ be the number of nodes of Γ_B, which are not connected to χ_0 after the removal of the edge ϕ. Then the following statement holds:

Lemma 4.4.11. *Let χ be the node adjacent to ϕ, which is not connected to χ_0 after the removal of ϕ.*

a) If χ is of type $\times$, then $n(\phi)$ is the length of the Green correspondent $l(f(\phi))$.

b) If χ is of type $\circ$, then $p - n(\phi)$ is the length of the Green correspondent $l(f(\phi))$.

Proof. See Lemma 9.3, Feit (1982), page 307. □

We may rephrase this lemma in a way which is easier to apply in the calculated examples.

Lemma 4.4.12. *Let ϕ be an edge of Γ_B. The length of the Green correspondent of ϕ is the number of nodes on the part of the tree, to which the node adjacent to ϕ of type $\times$ belongs after removal of ϕ. This part of the Brauer tree will be called the $\times$-part of the edge ϕ. Observe that we have to count the exceptional node $(p-e)$-times.*

Proof. Let ϕ be an edge of Γ_B and let χ be the node adjacent to ϕ, which is not connected to the exceptional node after removal of ϕ. If we count the exceptional node $p - e$ times the number of nodes on the tree is p. If χ is of type $\times$, our statement follows directly from the preceding lemma. If χ is of type $\circ$, then $l(f(\phi)) = p - n(\phi)$ and this is equal to the number of nodes on the $\times$-side of ϕ, if we count the exceptional node $p - e$ times. □

Lemma 4.4.13. *Let χ be a node of Γ_B and let ϕ_1, ϕ_2 be two edges adjacent to χ. It follows that:*

a) If χ is of type $\times$, then

$$l(f(\phi_1)) + l(f(\phi_2)) > p.$$

b) If χ of type $\circ$, then

$$l(f(\phi_1)) + l(f(\phi_2)) < p.$$

Proof. *a*) Every node of Γ_B has to be on the $\times$-side of either ϕ_1 or ϕ_2. χ, however, is on the $\times$-side of both edges. Hence the result follows.

b) A node of Γ_B is either on the $\times$-side of ϕ_1 and not on the $\times$-side of ϕ_2, or vice versa. The node χ does not lie on the $\times$-side of ϕ_1 nor on the $\times$-side of ϕ_2. We obtain the result by using the preceding lemma. □

To complete the determination of the Green correspondents of the simple kG-modules, we have to describe the socles of the Green correspondents. In the sequel we show how we can derive the socles of all the Green correspondents given the socle of one of them and given the embedded Brauer tree Γ_B^e.

Lemma 4.4.14. *Let ϕ be an irreducible Brauer character in B, and let us assume that the Green correspondent of ϕ has the socle ϕ_i and length m. Then: If ϕ' is the (clockwise) successor of ϕ at the node χ of type $\times$ and ϕ'' is the (clockwise) successor of ϕ' at the node χ' of type $\circ$, then the Green correspondent of ϕ'' has the socle $\phi_{i+1} = \phi_i \otimes \alpha$.*

Proof. We have to consider the following part of the Brauer tree Γ_B:

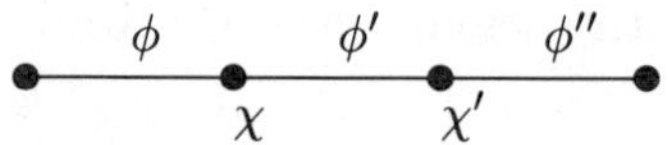

The fact that ϕ and ϕ' resp. ϕ' and ϕ'' are neighbours implies that there exists a nontrivial extension between the corresponding simple kG-modules. By Lemma 4.1.15 it follows that there exists a nontrivial extension between the Green correspondents. Lemma 4.2.19 tells us what this extension can look like.

We distinguish the following three cases:

Case 1: $\phi = \phi'$. Hence χ is a leaf. If χ is not the exceptional node, the Green correspondent of ϕ is of length 1 by Lemma 4.4.12. By Lemma 4.4.13 we get that $l(f(\phi)) + l(f(\phi'')) < p$. Hence it follows that the socle of $f(\phi'')$ is $i + 1$ using Lemma 4.2.19. If χ is the exceptional node, it follows that $l(f(\phi)) = p - e$. Again by Lemma 4.4.13 we get that $l(f(\phi)) + l(f(\phi'')) < p$. Since $p - e \equiv 1 \bmod e$, Lemma 4.2.19 completes the proof.

Case 2: $\phi' = \phi''$. Hence χ' is a leaf. It follows that $f(\phi')$ has length $p - 1$ or e. By Lemma 4.4.13 we get that $l(f(\phi)) + l(f(\phi')) > p$. Lemma 4.2.19 completes the proof again.

Case 3: $\phi \neq \phi' \neq \phi''$. It follows by Lemma 4.4.13 that: $l(f(\phi))+l(f(\phi')) > p$. Let ϕ_i be the socle of $f(\phi)$ and $m = l(f(\phi))$, and let ϕ_j be the socle of $f(\phi')$ and $n = l(f(\phi'))$. Lemma 4.2.19 tells us that:

$$j + n - 1 \equiv i \bmod e.$$

If ϕ_a is the socle of $f(\phi'')$ and $r = l(f(\phi''))$, we conclude again using Lemma 4.2.19:

$$a \equiv j + n \bmod e.$$

Hence:

$$i + 1 \equiv j + n \equiv a \bmod e. \ \square$$

Remark 4.4.15. The preceding lemma can be used to give a proof of the following algorithm, which derives the socles of all the Green correspondents of the simple kG-modules in B given the socle of the Green correspondent of one of them, say L. In order to determine the socles we wander around the embedded Brauer tree Γ_B^e as follows:

We start the walk at the edge ϕ belonging to L and mark this edge by its socle ϕ_i or simply i. We next leave out the successing edge at the $\times$-side ϕ' and go on to the edge ϕ'' successing ϕ' at the $\circ$-side of ϕ'. This edge is marked by its socle ϕ_{i+1} etc. In this way we need exactly $2e$ steps before we return to ϕ. In the principal block we choose the trivial edge to be our starting edge since we know that its Green correspondent is 1_N.

Lemma 4.4.16. *Let $\chi \in Irr_0(B)$ and let ϕ be an edge of Γ_B adjacent to χ. If ϕ_i is the socle of the Green correspondent of ϕ, then there exists exactly one irreducible RG-lattice L with character χ (whose reduction modulo p is uniserial) such that:*

$$soc(f(\bar{L})) \cong \phi_i.$$

If χ is of type $\times$, then $f(\bar{L}) \cong \phi_i$ and if χ is of type $\circ$, then $f(\bar{L})$ is isomorphic to the radical of the projective cover of ϕ_i.

Proof. This is easily proved using the preceding lemmas. □

Definition 4.4.17. Let us denote the irreducible RN-lattice L, whose reduction modulo p has the form $1Xi$ or $(p-1)Xi$, by $\widehat{1Xi}$ and $\widehat{(p-1)Xi}$.

Lemma 4.4.18. *For the RN-lattice $\widehat{1Ai}$ and $\widehat{(p-1)Ai}$ we get analogous formulae as in Lemma4.3.2:*

$$\widehat{1Ai} \otimes \widehat{1Aj} \equiv 1\widehat{A(i+j)}$$

resp.

$$\widehat{1Ai} \otimes \widehat{(p-1)Aj} \equiv \widehat{(p-1)A(i+j)}$$

and

$$\widehat{(p-1)Ai} \otimes \widehat{(p-1)Aj} \equiv 1A(\widehat{i+j}-1).$$

If L_1 and L_2 are RN-lattices in the blocks X and Y, whose reductions modulo p are unserial of length 1 or $(p-1)$, we can derive this tensor product from the following in an analogous way, as was done for the tensor products of indecomposable kN-modules, $\widehat{1X0}$ and $\widehat{1Y0}$ since:

$$\widehat{1Xi} \otimes \widehat{1Yj} \equiv \widehat{1X0} \otimes \widehat{1Y0} \otimes 1\widehat{A(i+j)}$$

resp.

$$\widehat{1Xi} \otimes \widehat{(p-1)Yj} \equiv \widehat{1X0} \otimes \widehat{1Y0} \otimes \widehat{(p-1)A(i+j)}$$

and

$$\widehat{(p-1)Xi} \otimes \widehat{(p-1)Yj} \equiv \widehat{1X0} \otimes \widehat{1Y0} \otimes 1A(\widehat{i+j}-1).$$

For the symmetric and the skew square we get:

$$1\widehat{Xi^{2+}} \equiv 1\widehat{X0^{2+}} \otimes \widehat{1A2i}$$

and

$$\widehat{1Xi^{2-}} \equiv \widehat{1X0^{2-}} \otimes \widehat{1A2i}.$$

For the lattice $(p-\widehat{1)Xj}$ *we get:*

$$(p-\widehat{1)Xj^{2+}} \equiv \widehat{1X0^{2-}} \otimes 1A(\widehat{2j}-1)$$

and

$$(p-\widehat{1)Xj^{2-}} \equiv \widehat{1X0^{2+}} \otimes 1A(\widehat{2j}-1).$$

Proof. All the given statements follow from the preceding lemmas, if we use the formula given in Section 3. □

5

HOW TO READ THE TABLES

5.1 Block information and Brauer tree

In the present chapter we explain the information contained in the tables of the following chapter and how to read them. To have the information for a particular block readily available, we have attempted not to break pages between two tables belonging to the same block. Where this is not possible, we have tried to break the pages in a way that the results are on one page and the proof on the next. A single page, however, can contain the tables for more than one block of a group. We print a rule between the tables for a fixed prime number belonging to different blocks of a group. If the prime changes on one page, a double rule is printed between the tables for the different primes.

We are going to explain the tables by demonstrating several examples. The information for a particular block starts with a headline giving the name of the group, the prime number and the number of the block. For the Mathieu group M23, the prime 23 and the block with number 1 this line looks like this:

Group: M23 Prime: 23 Block: 1

The name of a group is its CAS-name. This is very often different from the ATLAS-name (cf. Conway *et al.* (1985)). At the end of our monograph the reader will find a table with the two names of a group juxtaposed. This simplifies the identification of the groups.

The numbering of the blocks corresponds to the order of the ordinary characters on the CAS-character table, which we have used as input to our programs. Only blocks of nonzero defect are numbered. A block B has number n if and only if the ordinary characters of non-zero defect preceding the first character in B—according to the order on the CAS-table—lie in $n-1$ distinct blocks. Since the trivial character is always the first character on our input tables, the principal blocks are exactly the blocks with number 1.

The next table contains what we call the *block information.* For the principal 23-block of M23 this is displayed in Table 5.1.

The block information table usually has 5 columns. The first column with the heading 'Nr.' contains the current numbers of the ordinary characters in the block. Every character of the exceptional family receives the same number,

Table 5.1. The block information for the principal 23-block of M23

Nr.	CAS-Nr.	Degree	CC	N&C
1	1	1	r	×
2	2	22	r	∘
3	3	45	4	∘
4	4	45	3	∘
5	6	231	r	×
6	7	231	8	×
7	8	231	7	×
8	10	770	11	∘
8	11	770	10	∘
9	12	896	13	∘
10	13	896	12	∘
11	14	990	15	×
12	15	990	14	×

which is 8 in our example. So the numbers in this first column correspond to the nodes of the Brauer tree. The nodes are labelled with these numbers and not—as is done sometimes—with the degrees of the characters. If one looks for instance at the tables for the monster group, the reason is clear: the trees would occupy too much space if their vertices were labelled with the degrees of the ordinary characters.

The second column—headed with 'CAS-Nr.'—contains the numbers of the characters according to their ordering on the CAS-character table. Since this order may differ from the one in the ATLAS, we shall provide a list at the end of this book which allows identification of characters from different tables. Of course most of the time a character is uniquely determined by its degree.

The third column—headed with 'Degree'—lists the degrees of the characters. From these degrees and the Brauer tree one can easily derive the degrees of the irreducible Brauer characters as follows. The degree of a Brauer character corresponding to a leaf is equal to the degree of the ordinary character corresponding to that leaf. Remove the leaf from the tree and subtract its degree from the degree of the node it was originally joined to. Continue until all edges have been removed from the Brauer tree.

The heading 'CC' of the fourth column stands for *complex conjugation.* An entry in this column gives the CAS-number of the character complex conjugate to the one corresponding to the row. It contains an 'r' if the character is real valued. Complex conjugation refers to ordinary and not to Brauer characters. Thus in our example the two exceptional characters of degree 770 with CAS-

Table 5.2. The block information for the second 17-block of 3J3

Nr.	CAS-Nr.	Degree	CC	AC	N&C
1	22	18	23	24	×
2	24	18	25	22	×
3	30	171	31	ar	×
4	32	171	33	34	×
5	34	171	35	32	×
6	36	324	37	ar	×
7	38	1215	39	ar	∘
7	40	1215	41	ar	∘
8	46	2736	47	ar	∘
9	54	3078	55	ar	×

numbers 10 and 11 are complex conjugate as ordinary characters, although they are real valued and identical as 23-modular characters.

Finally the fifth column—headed with 'N&C'—contains a '×' if the character is a cross and a '∘' if the character is a nought, i.e. the type of the character. Please recall the definition of noughts and crosses (see page 2 of the introduction).

In some cases there is an additional column headed with 'AC'. The occurrence of this means that the block is not invariant under complex conjugation, but that there is an outer automorphism of the group which maps the complex conjugate block to the original one. In other words, the block is invariant under the composition of complex conjugation with the outer automorphism. Characters which are invariant under this composition are marked with 'ar' in this column. They can be thought of as constituting the 'real stem' of the tree. For a character, which is not invariant, the CAS-number of its image under the composition is given. Such a character and its image can be thought of as being a 'complex conjugate' pair. Table 5.2 shows an example where such a column appears.

If space allows, the Brauer tree is given next to the table of block information. If there is too little space, the tree is printed below the table. The exceptional node—if it exists—is marked with an extra circle like this ⦿ . The nodes are labelled with the current numbers of the characters as given in the first column of the block information table. If we were not able to find the tree, all possible trees are described. Unless stated otherwise, the tree is drawn according to its planar embedding as defined on page 4 of the introduction. If the planar embedding is not obvious, a proof is given. The labelled planar embedded trees for our two examples are shown in Figs. 5.1 and 5.2.

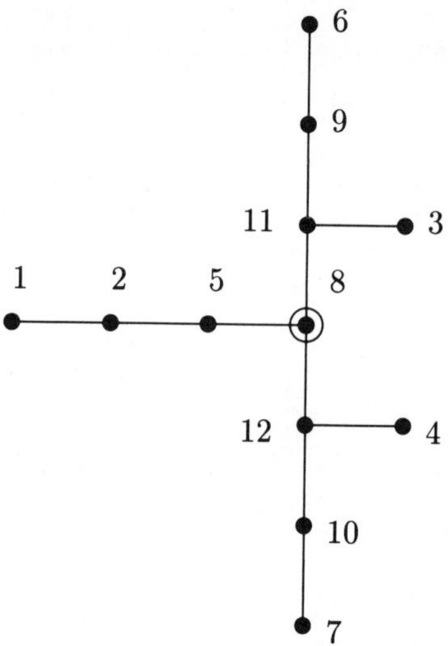

Fig. 5.1 The planar embedded tree for the principal 23-block of M23

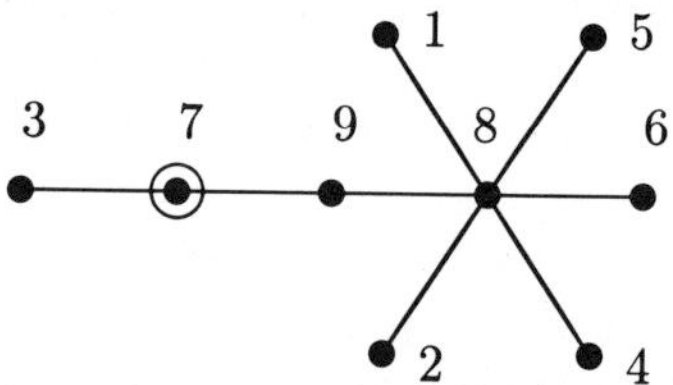

Fig. 5.2 The planar embedded tree for the second 17-block of 3J3

5.2 How to read the proofs

When it is obvious that the given tree is correct, e.g. by Corollary 5.3.2 in the next section, no proof is given. Otherwise the proof can be derived from the table headed '**PROJECTIVES:**' plus some additional remarks, if needed. Above all, supplementary information is necessary if the proof requires Green correspondence. Table 5.3 shows a typical proof for the principal 23-block of M23. The columns of the projective table correspond to the nodes of the tree. Its first three rows are the transpose of the first, fourth and fifth column of the block information table, except that in the projective table only one column is given for the exceptional node. To have this information available in the present form makes it easier to check the trees by hand. An example of how to do this will be given in the next section.

The rows following the first three give the decomposition of certain pro-

Table 5.3 The proof for the principal 23-block of M23

PROJECTIVES:

Nr.:	1	2	3	4	5	6	7	8	9	10	11	12
CC:	r	r	4	3	r	7	6	r	10	9	12	11
N&C:	×	∘	∘	∘	×	×	×	∘	∘	∘	×	×
χ_5^{2+}	1	2			1	1	1		2	2	1	1
$\chi_3 \otimes (\chi_1 + \chi_2)$			1								1	
$\chi_2 \otimes \chi_{16}$					1	1	1	1	2	2	1	1
$\chi_3 \otimes \chi_5$								1	1	1	1	2

Projective χ_5^{2+} shows in a first run that node 1 is joined to node 2, i.e. that $\chi_1 + \chi_2$ is projective. A second run then produces projective Φ_2 which proves that node 3 is joined to node 11 (and hence node 4 to node 12), solving the problem of algebraic conjugacy. The two pairs of nodes $\{9, 10\}$ and $\{6, 7\}$ have irrationalities independent of each other, and independent of the pairs $\{3, 4\}$ and $\{11, 12\}$.
The planar embedding of the tree has been calculated in Kawata (1987). It easily follows from the symmetrization $\chi_3^{2-} \approx \chi_{15}$, using the methods outlined in the chapter on Green correspondence. We omit the details, since there will be many more interesting examples of how this is done.

---

jectives in the block. The origin of these projectives is recorded in the first column. In the projective table the letter χ always refers to an ordinary character, which of course can also be a projective. The ordinary characters can be taken from the group or from a subgroup. Projective characters are denoted by Φ. The symbol Φ_n can have two distinct meanings. It can denote the nth projective character of the projective table. It can also refer to the indecomposable belonging to the irreducible Brauer character of degree n. We hope that it will always be clear which convention we are using.

If a projective is obtained as a tensor product of a Brauer character with a projective, the second factor in the projective table always denotes the projective. There may be some extra rows outside the table, as in the principal 29-block of the Rudvalis group. They explain the origin of some projectives, used in the tensor products. It means that the present tables have been obtained only after an iteration.

If some additional information is given following the projective table, then the ordinary characters are denoted by their CAS-numbers. Since this practice could confuse the reader, the corresponding nodes are always given in brackets.

Sometimes the edges of the trees are labelled with numbers. In this case the numbers can denote the degrees of the irreducible Brauer characters corresponding to the edges. Brauer characters are then denoted by their degrees. The labels on the edges can also refer to the Green correspondents of certain indecomposable modules associated with the tree. This will be explained later on.

The entries of the projective table are obtained as follows: Let Φ be a projective character referred to in the first column of a certain row. Then there is a positive integer d such that d times the entry in the column corresponding to the character χ is the scalar product (Φ, χ). Missing entries have to be interpreted as zeros. The integer d is the same for a fixed projective, but may vary from projective to projective. Usually d equals 1. We do not know d, but it is clear how to calculate it, if necessary: Compute the scalar product of Φ with a character in the block which has a nonzero entry in the row corresponding to Φ. The quotient of the scalar product and this entry equals d. Of course, the entries in the row give the multiplicities of the ordinary characters in the block in the projective Φ/d. So the rows are nonnegative integral linear combinations of the columns of the decomposition matrix. As shown later, one can derive the tree from these. To simplify the usage of the tables we have ordered the projectives lexicographically.

Suppose there are one or more non-trivial orbits of algebraically conjugate characters in a block. Consider a permutation of the characters fixing each orbit. If every such permutation is achieved by a Galois automorphism, then interchanging the position of any two algebraically conjugate characters on the Brauer tree amounts to choosing a different p-modular system. This is what we call the case of *independent irrationalities.* It occurs for example if all non-trivial orbits have length two and the characters in any two distinct orbits generate disjoint fields, i.e. fields intersecting in the rational numbers. Since we do not prescribe a particular p-modular system, we are free to put any character of one fixed orbit in the first possible position. Most of the time the character with the smallest CAS-number is then located left of its conjugates on the tree. We shall always explain the choice we have made.

We complete this section with a word on the proofs requiring Green correspondence. The main tools are tensor products of trivial or cotrivial source modules, the labelling of the edges of the Brauer tree according to Green correspondence and Lemmas 4.3.2 and 4.3.5. Since a tensor product usually has components in several blocks, the symbol $\approx_X$ is used to denote the restriction of a character or module to Block X. We shall be a little sloppy in our language by not distinguishing between an indecomposable trivial (resp. cotrivial) source module, its unique lift to a trivial (resp. cotrivial) source lattice and its (Brauer) character of type cross (resp. nought). Also '*leaf*', originally defined to be a node of the tree incident to exactly one edge, is used to denote the ordinary or modular character of this node as well as the irreducible lattice or module corresponding to the unique edge incident to it. We do not think that this practice will cause confusion.

5.3 Reconstructing the Brauer tree

When trying to reconstruct the Brauer tree from the projective table, the following lemmas are quite useful.

Lemma 5.3.1. *Let Φ be a projective character of a block with cyclic defect group. Suppose that in the decomposition of Φ into ordinary characters there is exactly one, χ say, of type $\times$ or of type $\circ$ (the exceptionals counted only once). Then every character occurring with positive multiplicity in Φ is joined to the node corresponding to χ in the Brauer tree.*

Proof. The projective character Φ is the sum of indecomposables of the form $\vartheta + \psi$, where ϑ and ψ have different type. □

Corollary 5.3.2. *The Brauer tree in a block with cyclic defect group is a star, if and only if there is exactly one ordinary character of type $\times$ or of type $\circ$ (the exceptionals counted only once) in the block.*

Proof. Consider the regular projective and use the lemma above. □

Whenever the Brauer tree is a star, it is trivial to find its labelling by this corollary. We do not give any further comments in our tables, when the corollary applies.

Finally the next lemma turns out to be quite useful in many examples.

Lemma 5.3.3. *Let Φ be a projective character in a block with a cyclic defect group. Let χ label a node of the Brauer tree and let $\psi_1, \ldots, \psi_n$ label all nodes incident to it. Then $\sum_{i=1}^{n}(\Phi, \psi_i) \geq (\Phi, \chi)$. If equality holds, then $\Phi - (\Phi, \chi)\chi - \sum_{i=1}^{n}(\Phi, \psi_i)\psi_i$ is a projective character. If $(\Phi, \psi_i) = 0$ for $1 \leq i < n$, then $\Phi - (\Phi, \chi)(\chi + \psi_n)$ is projective.*

Proof. We have $\Phi = \sum_{i=1}^{n} m_i(\chi + \psi_i) + \Phi_0$, where Φ_0 is a projective not involving χ. The result now follows from $(\Phi, \chi) = \sum_{i=1}^{n} m_i$. □

We are now going to present a nontrivial example of a reconstruction of the Brauer tree from the set of projectives the computer has used. The authors acknowledge with pleasure that this example has been worked out by R. Parker.

Consider the third block of the monster group in characteristic 13. The Brauer tree and the projective table are given in Table 5.4. Remember our convention to number the projectives in the order of their appearence in the projective table.

The nodes of the Brauer tree are denoted by their current numbers in the following. First note that the tree is a real stem. Looking at Φ_1 we see that 3 is joined to 2 and to one of 5 or 6, i.e. either 2—3—5 or 2—3—6 is part of the Brauer tree. Here we have applied Lemma 5.3.3. In any case 1 is not joined to 3, hence it must be joined to 4 using Φ_1. Again by applying Lemma 5.3.3

Table 5.4 The Brauer tree of a 13-block of the monster

Group: M   Prime: 13   Block: 3

Nr.	CAS-Nr.	Degree	CC	N&C
1	6	19360062527	r	×
2	9	36173193327999	r	×
3	33	31569817307122699605	r	∘
4	58	2382987417506242421875	r	∘
5	62	6566555764392010419123	r	×
6	63	7226910362631220625000	r	×
7	109	1037605886984697481755304	r	∘
8	118	3619209050774375426792424	r	∘
9	151	41762322738385820195625000	r	×
10	168	88943820620288343261672393	r	×
11	172	124058385593021471188320256	r	∘
12	185	173865305251972140447265625	r	×
13	186	175867626988794162227008203	r	∘

1 — 4 — 5 — 7 — 10 — 11 — 12 — 13 — 9 — 8 — 6 — 3 — 2

PROJECTIVES:

Nr.	1	2	3	4	5	6	7	8	9	10	11	12	13
CC	r	r	r	r	r	r	r	r	r	r	r	r	r
N&C	×	×	∘	∘	×	×	∘	∘	×	×	∘	×	∘
$\chi_3 \otimes \chi_{18}$	1	4	6	3	2	2							
$\chi_2 \otimes \chi_{23}$		1	2	1	1	1							
$\chi_2 \otimes \chi_{68}$			1	5	7	7	3	10	4	2	1		
$\chi_2 \otimes \chi_{32}$			1	2	2	1							
$\chi_2 \otimes \chi_{73}$							3	6	7	6	3		1
$\chi_2 \otimes \chi_{88}$							1		3	8	8	5	7
$\chi_2 \otimes \chi_{74}$										1	2	4	3

to projective Φ_4, we conclude that 4 is joined to 5. Hence we either have 1—4—5 and 2—3—6 or 1—4—5—3—2. The latter, however, is not consistent with Φ_1. Now since 2 does not occur at all in Φ_3, from 2—3—6 we see that 6 is left with multiplicity 6 in Φ_3 by applying the last part of Lemma 5.3.3. We conclude that 6 is joined to 8, hence 2—3—6—8. This leaves 8 with multiplicity 4 in Φ_3, hence it is joined either to 5 or to 9. If it is joined to 5, we have with the above 2—3—6—8—5—4—1, contradicting Φ_3. We thus have 2—3—6—8—9, leaving 8 with zero multiplicity in Φ_3. Starting from 1—4—5 and using Φ_3, we conclude that 5 is joined to 7, hence 7—5—4—1. This leaves 7 with multiplicity 1 in Φ_3, hence 7 is joined to 10 which in turn must be joined to 11. So far, only using projectives Φ_1, Φ_3 and Φ_4, we have 2—3—6—8—9 and 11—10—7—5—4—1.

Using the second branch and Φ_5, we conclude that 9 is joined to 13, hence 2—3—6—8—9—13 and 11—10—7—5—4—1. Again using the second branch and Φ_7, we conclude that 12 is joined to 13 and to 11. To summarize, we have 2—3—6—8—9—13—12—11—10—7—5—4—1 which is the tree we were looking for.

Please note that we have not used the projectives Φ_2 and Φ_6 at all, whereas the program had to use them to determine the tree. There are two reasons for this. First of all, the algorithm the program uses is totally different from the method applied here. Secondly, the projectives come in a different order from the one in the projective table. Suppose we had seen Φ_2 before knowing of Φ_1. Then, in a later state of the reconstruction, we would have been forced to use another projective in addition to Φ_2, to see where 1 could possibly be joined to.

6

THE TREES

6.1 The Mathieu Group M11

Group: M11 Prime: 5 Block: 1

Nr.	CAS-Nr.	Degree	CC	N&C
1	1	1	r	×
2	5	11	r	×
3	6	16	7	×
4	7	16	6	×
5	8	44	r	∘

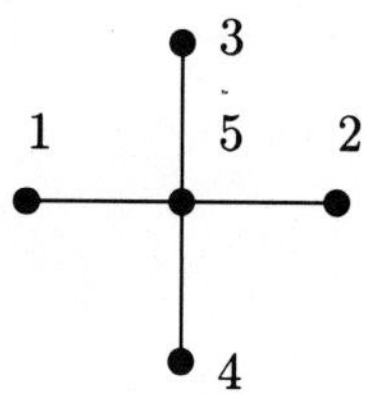

Group: M11 Prime: 11 Block: 1

Nr.	CAS-Nr.	Degree	CC	N&C
1	1	1	r	×
2	2	10	r	○
3	3	10	4	○
4	4	10	3	○
5	6	16	r	○
5	7	16	r	○
6	9	45	r	×

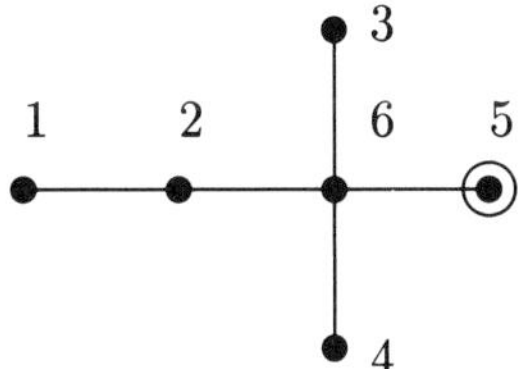

PROJECTIVES:

Nr.:	1	2	3	4	5	6
CC:	r	r	4	3	r	r
N&C:	×	○	○	○	○	×
$\chi_5 \otimes \chi_5$	1	1				
$\chi_3 \otimes \chi_5$				1		1

6.2 The Double Cover of M12

Group: 2M12 Prime: 3 Block: 2

Nr.	CAS-Nr.	Degree	CC	N&C
1	6	45	r	×
2	12	99	r	×
3	14	144	r	∘

1 3 2

Group: 2M12 Prime: 5 Block: 1

Nr.	CAS-Nr.	Degree	CC	N&C
1	1	1	r	×
2	11	66	r	×
3	12	99	r	∘
4	14	144	r	∘
5	15	176	r	×

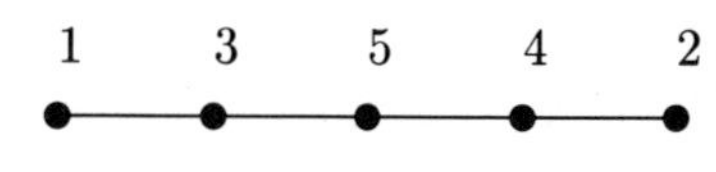

PROJECTIVES:

Nr.:	1	2	3	4	5
CC:	r	r	r	r	r
N&C:	×	×	∘	∘	×
$\chi_{17} \otimes \chi_{16}$	1		1		
$\chi_4 \otimes \chi_8$			1	1	2

Group: 2M12 Prime: 5 Block: 2

Nr.	CAS-Nr.	Degree	CC	N&C
1	2	11	r	×
2	3	11	r	×
3	4	16	5	×
4	5	16	4	×
5	7	54	r	∘

3
1 5 2
4

Group: 2M12 Prime: 5 Block: 3

Nr.	CAS-Nr.	Degree	CC	N&C
1	18	12	r	×
2	19	32	r	×
3	20	44	21	○
3	21	44	20	○

1 3 2

Group: 2M12 Prime: 11 Block: 1

Nr.	CAS-Nr.	Degree	CC	N&C
1	1	1	r	×
2	4	16	5	○
2	5	16	4	○
3	6	45	r	×
4	7	54	r	○
5	13	120	r	○
6	14	144	r	×

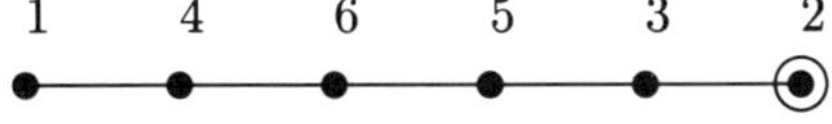

PROJECTIVES:

Nr.:	1	2	3	4	5	6
CC:	r	r	r	r	r	r
N&C:	×	○	×	○	○	×
$\chi_2 \otimes \chi_2$	1			1		
$\chi_7 \otimes \chi_2$				1		1
$\chi_6 \otimes \chi_2$					1	1

Group: 2M12 Prime: 11 Block: 2

Nr.	CAS-Nr.	Degree	CC	N&C
1	16	10	17	∘
2	17	10	16	∘
3	18	12	r	×
4	19	32	r	∘
5	24	120	r	∘
6	25	160	26	×
6	26	160	25	×

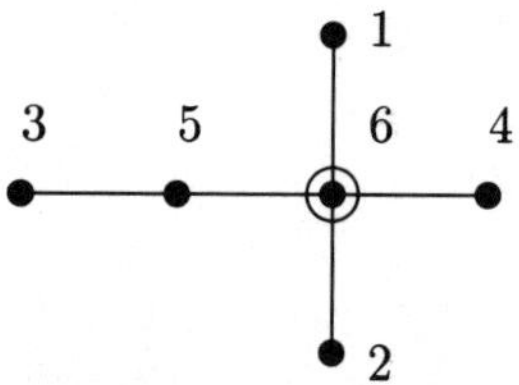

PROJECTIVES:

Nr.:	1	2	3	4	5	6
CC:	2	1	r	r	r	r
N&C:	∘	∘	×	∘	∘	×
$\chi_{16} \otimes \chi_8$		1				1
$\chi_{18} \otimes \chi_2$			1		1	
$\chi_{19} \otimes \chi_2$				1		1
$\chi_{20} \otimes \chi_2$					1	1

6.3 The Janko Group J1

Group: J1 Prime: 2 Block: 2

Nr.	CAS-Nr.	Degree	CC	N&C
1	4	76	r	×
2	5	76	r	∘

1 — 2

Group: J1 Prime: 3 Block: 1

Nr.	CAS-Nr.	Degree	CC	N&C
1	1	1	r	×
2	4	76	r	×
3	6	77	r	∘

1 — 3 — 2

Group: J1 Prime: 3 Block: 2

Nr.	CAS-Nr.	Degree	CC	N&C
1	2	56	r	×
2	8	77	r	×
3	13	133	r	∘

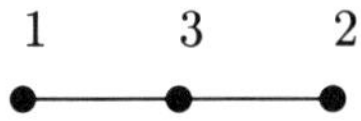

1 — 3 — 2

Group: J1 Prime: 3 Block: 3

Nr.	CAS-Nr.	Degree	CC	N&C
1	3	56	r	×
2	7	77	r	×
3	14	133	r	∘

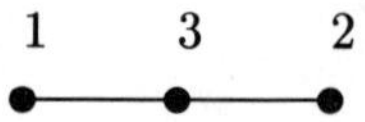

1 3 2

Group: J1 Prime: 3 Block: 4

Nr.	CAS-Nr.	Degree	CC	N&C
1	5	76	r	×
2	12	133	r	×
3	15	209	r	∘

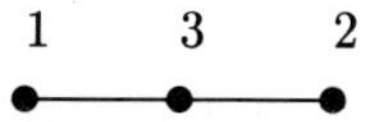

1 3 2

Group: J1 Prime: 5 Block: 1

Nr.	CAS-Nr.	Degree	CC	N&C
1	1	1	r	×
2	5	76	r	×
3	7	77	r	∘
3	8	77	r	∘

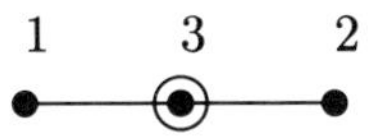

1 3 2

Group: J1 Prime: 5 Block: 2

Nr.	CAS-Nr.	Degree	CC	N&C
1	2	56	r	×
1	3	56	r	×
2	6	77	r	×
3	12	133	r	∘

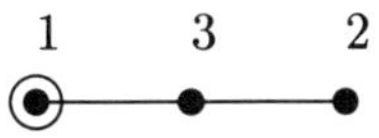

1 3 2

Group: J1 Prime: 5 Block: 3

Nr.	CAS-Nr.	Degree	CC	N&C
1	4	76	r	×
2	13	133	r	×
2	14	133	r	×
3	15	209	r	∘

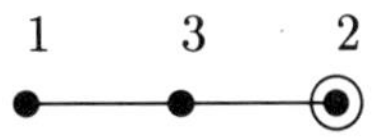

1 3 2

Group: J1 Prime: 7 Block: 1

Nr.	CAS-Nr.	Degree	CC	N&C
1	1	1	r	×
2	4	76	r	∘
3	5	76	r	∘
4	9	120	r	×
5	10	120	r	×
6	11	120	r	×
7	15	209	r	∘

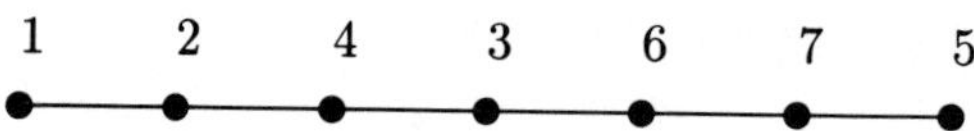

PROJECTIVES:

Nr.:	1	2	3	4	5	6	7
CC:	r	r	r	r	r	r	r
N&C:	×	∘	∘	×	×	×	∘
χ_2^{2+}	1	2		1	1	1	2
$\chi_3 \otimes \chi_2$		1	1	2	2	2	4

Without loss of generality we can assume that of the three algebraic conjugate characters χ_9, χ_{10} and χ_{11} (nodes 4, 5 and 6), χ_9 comes first on the tree. Then only the following two possibilities are consistent with the above set of projectives:

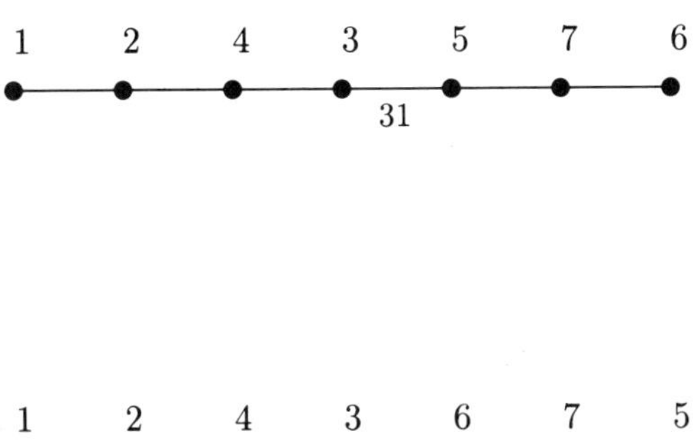

In the first of these trees, let 31 denote the irreducible Brauer character corresponding to the edge joining node 3 with node 5. Then $\Phi_{31} = \chi_5 + \chi_{10}$ is its projective cover. The contradiction $(31^{2+}, \Phi_{31}) = -1$ excludes this tree.

Group: J1 Prime: 11 Block: 1

Nr.	CAS-Nr.	Degree	CC	N&C
1	1	1	r	×
2	2	56	r	×
3	3	56	r	×
4	4	76	r	∘
5	5	76	r	∘
6	9	120	r	∘
7	10	120	r	∘
8	11	120	r	∘
9	12	133	r	×
10	13	133	r	×
11	14	133	r	×

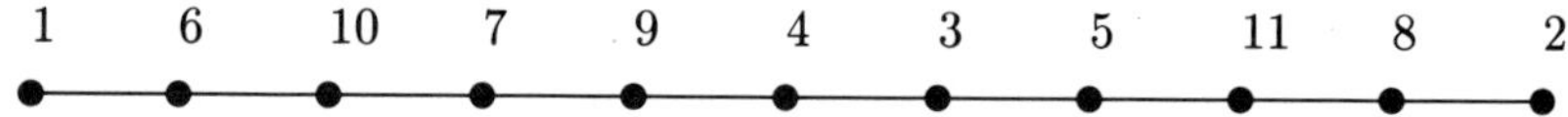

PROJECTIVES:

Nr.:	1	2	3	4	5	6	7	8	9	10	11
CC:	r	r	r	r	r	r	r	r	r	r	r
N&C:	×	×	×	∘	∘	∘	∘	∘	×	×	×
χ_6^{2+}	1	1	1	3		2	2	2	4	1	1
$\chi_{10} \otimes \chi_6$		3	3	4	4	6	7	6	7	7	7
$\chi_{11} \otimes \chi_6$		3	3	4	4	6	6	7	7	7	7
$\chi_7 \otimes \chi_6$		2	1	2	3	4	4	4	4	5	5
$\chi_3 \otimes \chi_7$		2	1	2	2	3	3	3	4	3	3
$\chi_2 \otimes \chi_6$		2	1	2	2	3	3	3	3	4	3
$\chi_2 \otimes \chi_8$		1	2	2	2	3	3	3	4	3	3
$\chi_3 \otimes \chi_6$		1	2	2	2	3	3	3	3	3	4
χ_6^{2-}		1	1	1	2	2	2	2	1	3	3
χ_7^{2-}		1	1		3	2	2	2	1	3	3

We have three orbits of algebraically conjugate characters. Two orbits of length 2 each, namely χ_2, χ_3 (nodes 2 and 3) and χ_{13}, χ_{14} (nodes 10 and 11). The characters in these orbits have their irrationalities on elements of order 5, whereas the characters in the third orbit have their irrationalities on the elements of order 19. This orbit consists of the 3 characters χ_9, χ_{10} and χ_{11} (nodes 6, 7 and 8). Without loss of generality we assume that of the characters in this last orbit χ_9 comes first on the tree. Since the irrationalities of the characters in the third orbit lie in a field disjoint from the one containing the irrationalities of the characters in the other two orbits, we can also assume that χ_{13} is located left of χ_{14} on the tree. With these assumptions only the following four labelled trees are consistent with the above set of projectives:

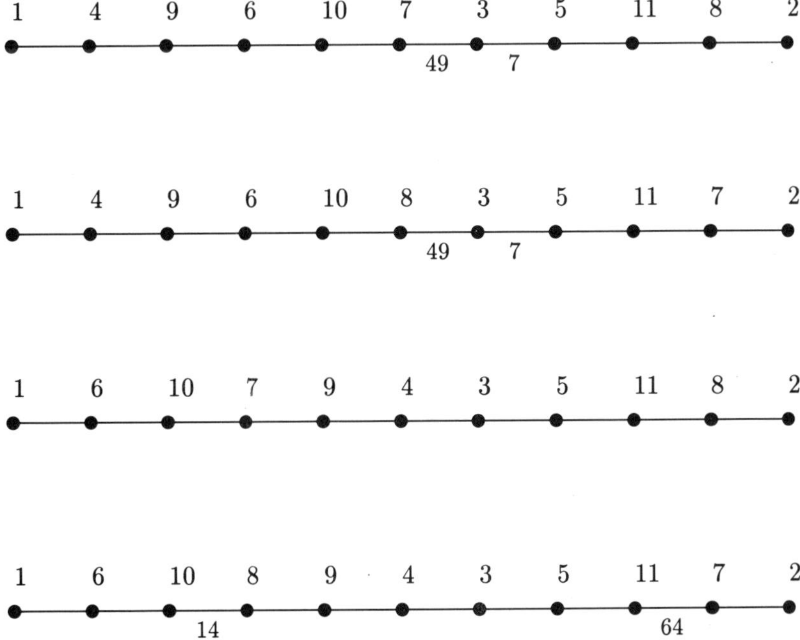

Let 49 and 7 be the irreducible Brauer characters as indicated in the first two pictures. Let Φ_{49} resp. Φ_7 denote their projective covers. Then $(7^{2+}, \Phi_{49}) = -1$, showing that the first two trees are impossible. With an analogous notation, the fourth implies $(14^{2+}, \Phi_{64}) = -1$, leaving the third possibility as the answer.

Group: J1 Prime: 19 Block: 1

Nr.	CAS-Nr.	Degree	CC	N&C
1	1	1	r	×
2	2	56	r	∘
3	3	56	r	∘
4	6	77	r	×
5	7	77	r	×
6	8	77	r	×
7	9	120	r	∘
7	10	120	r	∘
7	11	120	r	∘

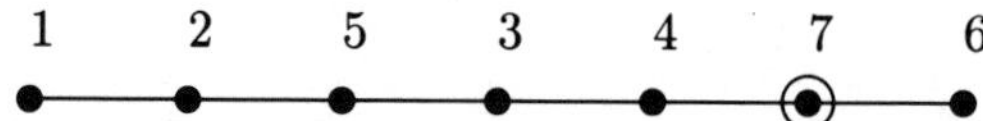

PROJECTIVES:

Nr.:	1	2	3	4	5	6	7
CC:	r	r	r	r	r	r	r
N&C:	×	∘	∘	×	×	×	∘
χ_4^{2+}	1	1	1	3			2
$\chi_6 \otimes \chi_4$		1	1	2	1	1	2
χ_4^{2-}		1	1		2	2	2

Without loss of generality we can assume that of the two algebraically conjugate characters χ_2 and χ_3 (nodes 2 and 3), χ_2 comes first on the Brauer tree. Then the following two trees are consistent with the above set of projectives:

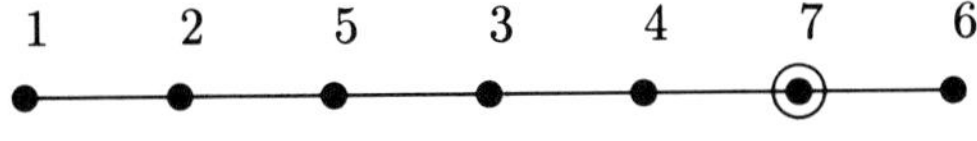

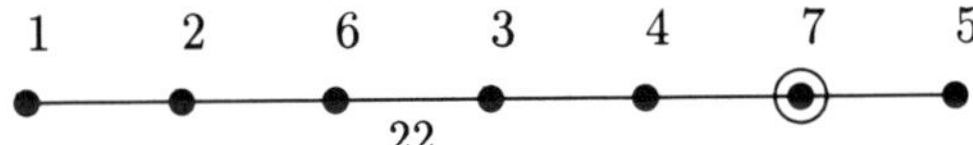

The second of these trees is impossible since it implies $(22^{2+}, \Phi_{22}) = -1$, where the notation is chosen as in the proofs above.

6.4 The Covering Group of the Mathieu Group M22

Remark: Our proof in characteristic 11 follows Benson (1985).

Group: 12M22 Prime: 2 Block: 4

Nr.	CAS-Nr.	Degree	CC	N&C
1	44	384	45	×
2	81	384	80	∘
2	106	384	109	∘
2	108	384	107	∘

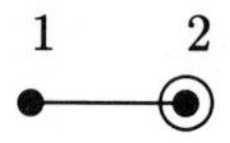

1 2

Group: 12M22 Prime: 2 Block: 5

Complex conjugate to Block 4.

Remark. These are the only blocks of cyclic defect larger than 1 occurring in covering groups of sporadic groups.

Group: 4M22 Prime: 3 Block: 2

Nr.	CAS-Nr.	Degree	CC	N&C
1	2	21	r	×
2	8	210	r	×
3	9	231	r	∘

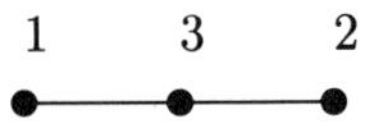

1 3 2

Group: 4M22 Prime: 3 Block: 4

Nr.	CAS-Nr.	Degree	CC	N&C
1	16	120	r	×
2	21	210	r	×
3	22	330	r	∘

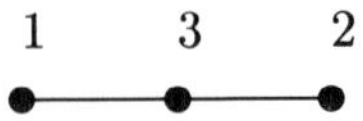

1 3 2

Remark. The numbering of these two blocks is relative to 4M22. In the full covering group 12M22 there is another 3-block with relative number 4.

Group: 12M22 Prime: 3 Block: 3

Nr.	CAS-Nr.	Degree	CC	N&C
1	3	45	4	×
2	26	45	29	∘
2	27	45	28	∘

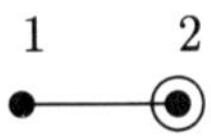

1 2

Group: 12M22 Prime: 3 Block: 4

Complex conjugate to Block 3.

Group: 12M22 Prime: 3 Block: 5

Nr.	CAS-Nr.	Degree	CC	N&C
1	6	99	r	×
2	30	99	31	∘
2	31	99	30	∘

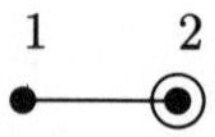

1 2

Group: 12M22 Prime: 3 Block: 8

Nr.	CAS-Nr.	Degree	CC	AC	N&C
1	17	126	18	ar	∘
2	68	126	71	ar	×
2	69	126	70	ar	×

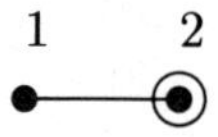

1 2

Group: 12M22 Prime: 3 Block: 9

Complex conjugate to Block 8.

Group: 12M22 Prime: 3 Block: 12

Nr.	CAS-Nr.	Degree	CC	N&C
1	50	144	53	×
2	92	144	95	∘
2	93	144	94	∘

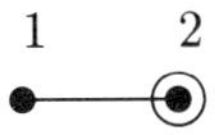

1 2

Group: 12M22 Prime: 3 Block: 13

Nr.	CAS-Nr.	Degree	CC	N&C
1	51	144	52	×
2	90	144	97	∘
2	91	144	96	∘

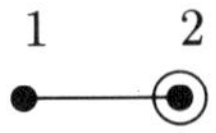

1 2

Group: 12M22 Prime: 3 Block: 14

Complex conjugate to Block 13.

Group: 12M22 Prime: 3 Block: 15

Complex conjugate to Block 12.

Group: 12M22 Prime: 5 Block: 1

Nr.	CAS-Nr.	Degree	CC	N&C
1	1	1	r	×
2	2	21	r	×
3	6	99	r	○
4	7	154	r	○
5	9	231	r	×

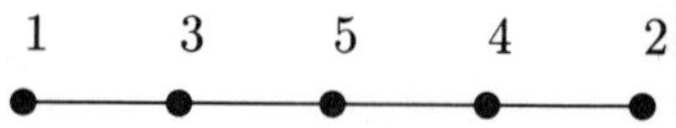

1 — 3 — 5 — 4 — 2

PROJECTIVES:

Nr.	1	2	3	4	5
CC	r	r	r	r	r
N&C	×	×	○	○	×
$\chi_{14} \otimes \chi_{13}$	1		1		
$\chi_5 \otimes \chi_3$			1		1

Group: 12M22 Prime: 5 Block: 2

Nr.	CAS-Nr.	Degree	CC	N&C
1	15	56	r	×
2	17	126	18	×
3	18	126	17	×
4	19	154	20	○
5	20	154	19	○

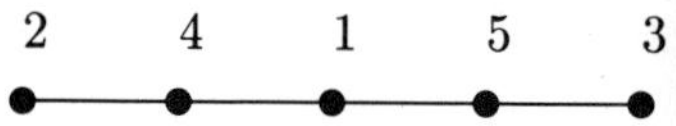

2 — 4 — 1 — 5 — 3

PROJECTIVES:

Nr.	1	2	3	4	5
CC	r	3	2	5	4
N&C	×	×	×	○	○
$\chi_{10} \otimes \chi_{13}$	1		1	1	1

The real stem of this tree consists just of node 1. Since the irrationalities of the pair 2, 3 are distinct of those of 4, 5, we can assume without loss of generality that node 2 is joined to node 4. The problem then is, however, to get the trees for the other blocks consistent with this choice.

Group: 12M22 Prime: 5 Block: 3

Nr.	CAS-Nr.	Degree	CC	AC	N&C
1	24	21	25	ar	×
2	30	99	31	ar	∘
3	38	231	39	ar	×
4	40	231	41	ar	×
5	44	384	45	ar	∘

PROJECTIVES:

Nr.	1	2	3	4	5
AC	ar	ar	ar	ar	ar
N&C	×	∘	×	×	∘
$\chi_{65} \otimes \chi_{13}$		1		1	
$\chi_{69} \otimes \chi_{13}$			1		1
$\chi_{2} \otimes \chi_{26}$				1	1

Group: 12M22 Prime: 5 Block: 4

Complex conjugate to Block 3.

Group: 12M22 Prime: 5 Block: 5

Nr.	CAS-Nr.	Degree	CC	AC	N&C
1	46	56	49	48	×
2	48	56	47	46	×
3	50	144	53	52	○
4	52	144	51	50	○
5	58	176	59	ar	×

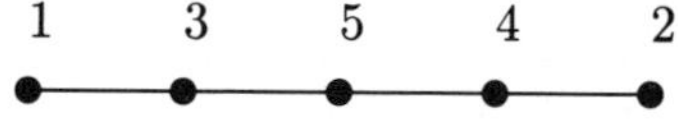

PROJECTIVES:

Nr.	1	2	3	4	5
AC	2	1	4	3	ar
N&C	×	×	○	○	×
$\chi_{46} \otimes \chi_3$		1	1	1	1
$\chi_{55} \otimes \chi_{13}$				1	1

The irrationalities of the pair 3, 4 lie in a field which is an extension of degree 2 of the fourth cyclotomic field. The intersection of this with the character field of the other relevant characters in this and the second block is contained in the field of fourth roots of unity. Hence there is a splitting 5-modular system such that the tree of the second block is as given and node 1 is joined to node 3 in the present block.

Group: 12M22 Prime: 5 Block: 6

Complex conjugate to Block 5.

Group: 12M22 Prime: 5 Block: 7

Nr.	CAS-Nr.	Degree	CC	AC	N&C
1	62	66	65	64	×
2	64	66	63	62	×
3	68	126	71	ar	×
4	70	126	69	ar	×
5	80	384	81	ar	∘

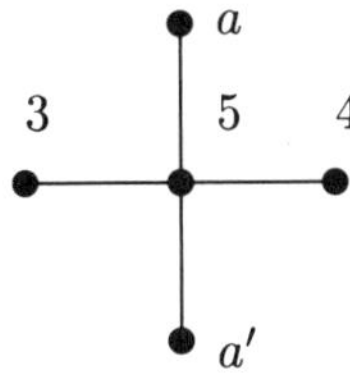

Here, $\{a, a'\} = \{1, 2\}$. The problem is to find the correct order of 1 and 2 after choosing the modular system in the preceding blocks.

Group: 12M22 Prime: 5 Block: 8

Complex conjugate to Block 7.

Group: 12M22 Prime: 5 Block: 9

Nr.	CAS-Nr.	Degree	CC	AC	N&C
1	90	144	97	94	∘
2	94	144	93	90	∘
3	98	336	105	ar	×
4	102	336	101	ar	×
5	106	384	109	ar	∘

Here, $\{a, a'\} = \{3, 4\}$ and $\{b, b'\} = \{1, 2\}$. Note that this is a problem of consistency with the previous blocks. For every choice of a, a' and b, b' there is of course a 5-modular system such that the tree is as given above for these parameters.

Group: 12M22 Prime: 5 Block: 10

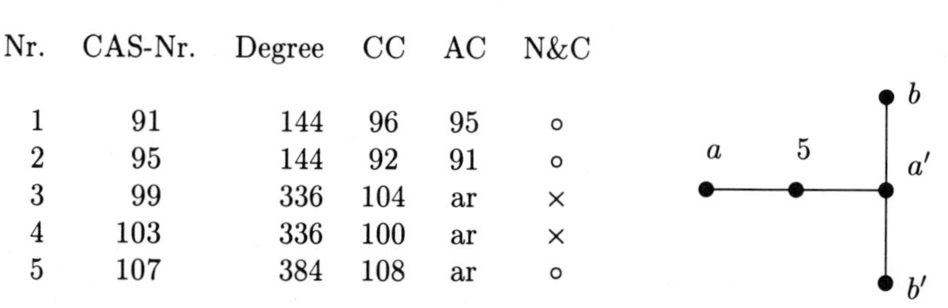

Nr.	CAS-Nr.	Degree	CC	AC	N&C
1	91	144	96	95	∘
2	95	144	92	91	∘
3	99	336	104	ar	×
4	103	336	100	ar	×
5	107	384	108	ar	∘

The same remark as above applies.

Group: 12M22 Prime: 5 Block: 11

Complex conjugate to Block 10.

Group: 12M22 Prime: 5 Block: 12

Complex conjugate to Block 9.

Group: 12M22 Prime: 7 Block: 1

Nr.	CAS-Nr.	Degree	CC	N&C
1	1	1	r	×
2	3	45	4	∘
2	4	45	3	∘
3	5	55	r	∘
4	6	99	r	×

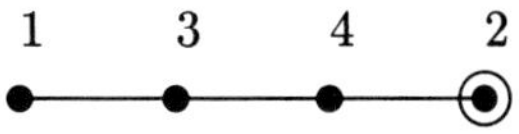

1 — 3 — 4 — 2 (exceptional)

PROJECTIVES:

Nr.	1	2	3	4
CC	r	r	r	r
N&C	×	∘	∘	×
$\chi_2 \otimes \chi_2$	1		1	

Group: 12M22 Prime: 7 Block: 2

Nr.	CAS-Nr.	Degree	CC	N&C
1	13	10	14	∘
1	14	10	13	∘
2	16	120	r	×
3	22	330	r	×
4	23	440	r	∘

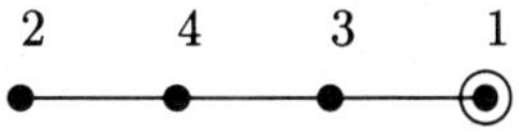

2 — 4 — 3 — 1 (exceptional)

PROJECTIVES:

Nr.	1	2	3	4
CC	r	r	r	r
N&C	∘	×	×	∘
$\chi_{13} \otimes \Phi_2$	1		1	

Φ_2: $\chi_2 \otimes \chi_2$

Group: 12M22 Prime: 7 Block: 3

Nr.	CAS-Nr.	Degree	CC	AC	N&C
1	26	45	29	28	o
1	28	45	27	26	o
2	30	99	31	ar	×
3	42	330	43	ar	×
4	44	384	45	ar	o

We have the following two possible Brauer trees:

In Humphreys (1982b), a proof is given that the first of these is correct. This proof uses a certain induced character. We sketch two other ways of proving the same result. First observe that $10_B \otimes 54_H \approx_C \hat{\chi}_{42}$. Here, the Brauer characters are denoted by there degrees, and $\hat{\chi}_{42}$ is the restriction of χ_{42} to the 7-regular classes. The letters B and H are used to indicate Blocks 2 and 8 respectively. At this stage we could use Green correspondence to show that $10_B \otimes 54_H$ must have more than one indecomposable direct summand. Since we have not proved the general formula which is necessary to do so, we proceed as follows.

We have $10_B \otimes 45_C \approx_H 66_H + 54_H + 330_H$. From the knowledge of the indecomposable modules in a block with cyclic defect group, we see that 54_H must be in the head or in the socle of $10_B \otimes 45_C$. Hence, by a general argument, 45_C is in the socle or head of $10_B \otimes 54_H$. It follows that $\hat{\chi}_{42}$ contains 45_C as a constituent. So the first tree must be correct.

Group: 12M22 Prime: 7 Block: 4

Complex conjugate to Block 3.

Group: 12M22 Prime: 7 Block: 5

Nr.	CAS-Nr.	Degree	CC	AC	N&C
1	50	144	53	52	×
1	52	144	51	50	×
2	54	160	57	ar	∘
3	56	160	55	ar	∘
4	58	176	59	ar	×

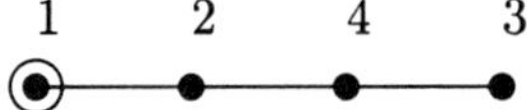

There is a field automorphism interchanging nodes 2 and 3 and fixing all other characters of defect larger than 0. Hence we can assume without loss of generality or consistency that the tree is as given above.

Group: 12M22 Prime: 7 Block: 6

Complex conjugate to Block 5.

Group: 12M22 Prime: 7 Block: 7

Nr.	CAS-Nr.	Degree	CC	AC	N&C
1	62	66	65	64	∘
1	64	66	63	62	∘
2	66	120	67	ar	×
3	78	330	79	ar	×
4	80	384	81	ar	∘

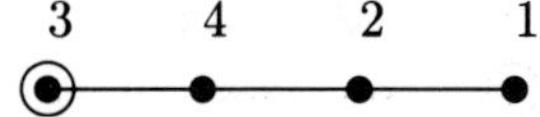

PROJECTIVES:

Nr.	1	2	3	4
CC	r	r	r	r
N&C	∘	×	×	∘
$\chi_{25} \otimes \chi_{15}$	1	2		1

Group: 12M22 Prime: 7 Block: 8

Complex conjugate to Block 7.

Group: 12M22 Prime: 7 Block: 9

Nr.	CAS-Nr.	Degree	CC	AC	N&C
1	82	120	89	86	×
2	86	120	85	82	×
3	92	144	95	96	×
3	96	144	91	92	×
4	108	384	107	ar	∘

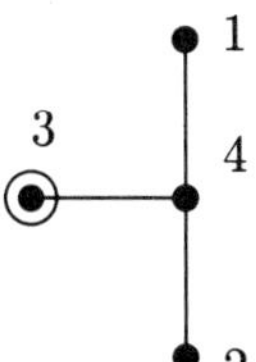

Group: 12M22 Prime: 7 Block: 10

Nr.	CAS-Nr.	Degree	CC	AC	N&C
1	83	120	88	87	×
2	87	120	84	83	×
3	93	144	94	97	×
3	97	144	90	93	×
4	109	384	106	ar	∘

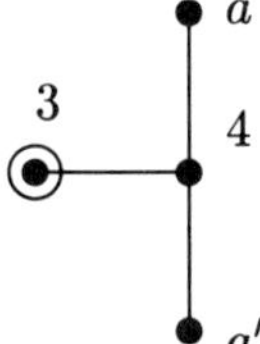

Here, $\{a, a'\} = \{1, 2\}$.

Group: 12M22 Prime: 7 Block: 11

Complex conjugate to Block 10.

Group: 12M22 Prime: 7 Block: 12

Complex conjugate to Block 9.

Group: 12M22 Prime: 11 Block: 1

Nr.	CAS-Nr.	Degree	CC	N&C
1	1	1	r	×
2	2	21	r	∘
3	3	45	4	×
4	4	45	3	×
5	8	210	r	×
6	10	280	11	∘
6	11	280	10	∘

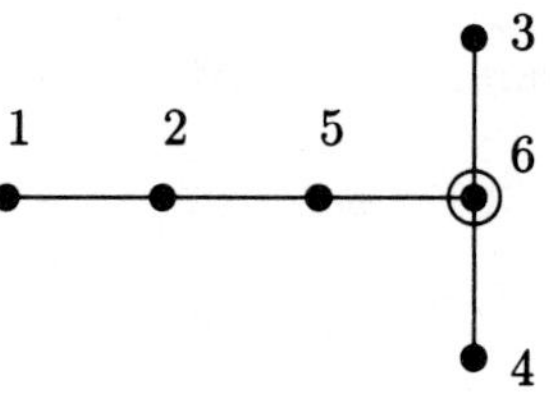

PROJECTIVES:

Nr.	1	2	3	4	5	6
CC	r	r	4	3	r	r
N&C	×	∘	×	×	×	∘
χ_5^{2+}	1	1				
$\chi_3 \otimes \chi_5$				1	1	2

Group: 12M22 Prime: 11 Block: 2

Nr.	CAS-Nr.	Degree	CC	N&C
1	13	10	14	∘
2	14	10	13	∘
3	15	56	r	×
4	16	120	r	∘
5	17	126	18	∘
5	18	126	17	∘
6	21	210	r	×

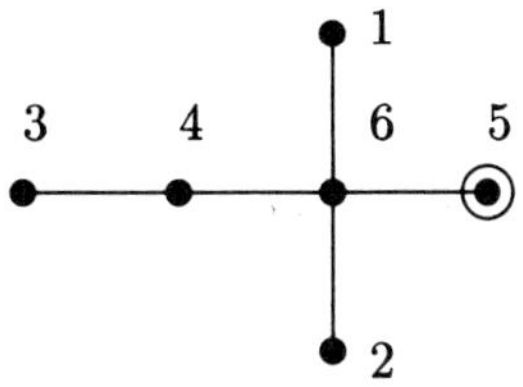

PROJECTIVES:

Nr.	1	2	3	4	5	6
CC	2	1	r	r	r	r
N&C	∘	∘	×	∘	∘	×
$\chi_{13} \otimes \chi_5$		1				1
$\chi_{15} \otimes \chi_5$			1	1		
$\chi_2 \otimes \chi_{19}$				1	1	2

Since the complex conjugate characters in this block have the same irrationalities as the ones in the principal block, we have to prove the consistency of the planar embeddings. For this purpose consider the tensor product

$$\chi_3 \otimes \chi_{13} \approx_B \chi_{14},$$

which in terms of nodes reads

$$3_A \otimes 1_B \approx_B 2_B.$$

Here the subscripts A and B indicate the principal and the second block respectively and $\approx_B$ denotes the tensor product restricted to Block 2. Let $1B0$ be the Green correspondent of the irreducible module belonging to node 3. Let x be the label at the edge joining node 1_B with the real stem, i.e. the Green correspondent of this edge is $1Bx$. We have $x \in \{2,4\}$. Now $1_B \otimes 3_A \approx_B 2_B$. It follows from $1Bx \otimes 1A2 = 1B(x+2)$ that $1B(x+2)$ is the Green correspondent of the edge joining node 2_B with the real stem. Hence $x+2 \in \{2,4\}$. Since $x+2$ has to be taken modulo 5, we have $x = 2$ and the assertion follows.

Group: 12M22 Prime: 11 Block: 3

Nr.	CAS-Nr.	Degree	CC	AC	N&C
1	24	21	25	ar	∘
2	26	45	29	28	×
3	28	45	27	26	×
4	32	105	35	ar	×
4	34	105	33	ar	×
5	36	210	37	ar	×
6	44	384	45	ar	∘

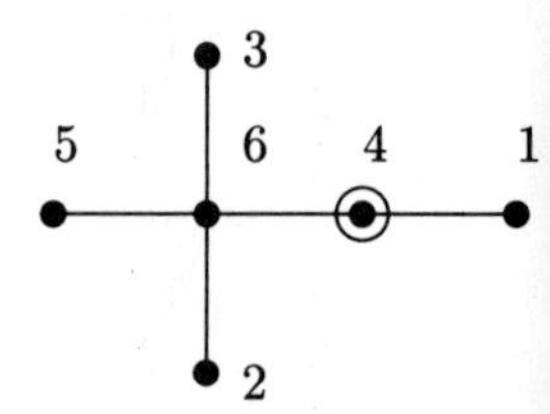

PROJECTIVES:

Nr.	1	2	3	4	5	6
AC	ar	3	2	ar	ar	ar
N&C	∘	×	×	×	×	∘
$\chi_2 \otimes \chi_{40}$		1	1		2	4
$\chi_2 \otimes \chi_{30}$				1	1	2

The planar embedding and consistency is proved with the following tensor product:

$$\chi_3 \otimes \chi_{28} \approx_C (\chi_{32} + \chi_{34}) + \chi_{36} + \chi_{44}$$

which in terms of nodes reads

$$3_A \otimes 3_C \approx_C 4_C + 5_C + 6_C.$$

Since node 2_C does not occur in this product, the edge joining node 3_C with the real stem cannot have label 4 relative to the leaf at node 5_C.

Group: 12M22 Prime: 11 Block: 4

Complex conjugate to Block 3.

Group: 12M22 Prime: 11 Block: 5

Nr.	CAS-Nr.	Degree	CC	AC	N&C
1	46	56	49	48	×
2	48	56	47	46	×
3	50	144	53	52	×
4	52	144	51	50	×
5	54	160	57	ar	×
5	56	160	55	ar	×
6	60	560	61	ar	∘

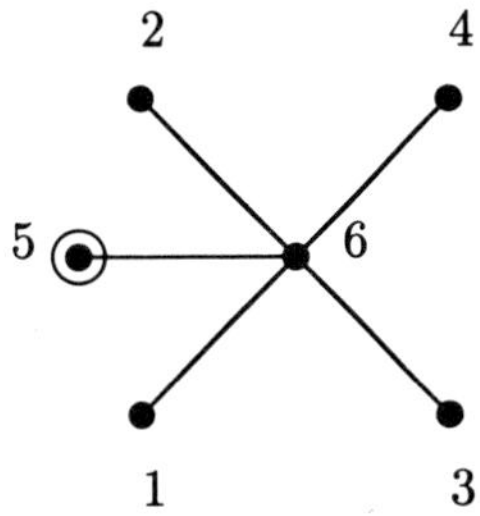

Let $1E0$ denote the Green correspondent of the leaf at the exceptional node. Since this is the only module of type cross which is invariant under the composition of complex conjugation with the outer automorphism, $1E0$ is an irreducible character of the normalizer of the defect group, which acts trivially on an element of order five. Also $1B0 = 1D0 \otimes 1D0$ is the Green correspondent of the leaf 3_B. Consider the scalar products

$$((\chi_{54} + \chi_{56}) \otimes \chi_{46}, \chi_{13}) = 0$$
$$((\chi_{54} + \chi_{56}) \otimes \chi_{48}, \chi_{13}) = 0$$
$$((\chi_{54} + \chi_{56}) \otimes \chi_{50}, \chi_{13}) = 0$$

which are in terms of nodes:

$$(5_E \otimes 1_E, 1_B) = 0$$
$$(5_E \otimes 2_E, 1_B) = 0$$
$$(5_E \otimes 3_E, 1_B) = 0$$

By the remarks above, they show that none of the nodes 1_E, 2_E and 3_E has label 2. Since the two nodes 1_E and 2_E have irrationalities independent of the characters previously considered, we can assume without loss of generality that the planar embedding is as given above.

Group: 12M22 Prime: 11 Block: 6

Complex conjugate to Block 5.

Group: 12M22 Prime: 11 Block: 7

Nr.	CAS-Nr.	Degree	CC	AC	N&C
1	66	120	67	ar	∘
2	68	126	71	ar	∘
2	70	126	69	ar	∘
3	72	210	73	ar	×
4	74	210	77	ar	×
5	76	210	75	ar	×
6	80	384	81	ar	∘

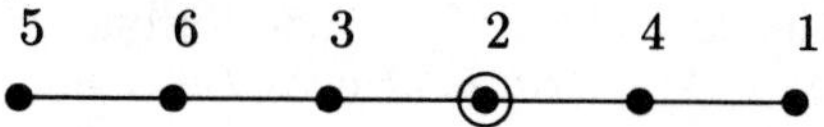

PROJECTIVES:

Nr.	1	2	3	4	5	6
AC	ar	ar	ar	ar	ar	ar
N&C	∘	∘	×	×	×	∘
$\chi_{68} \otimes \Phi_3$		2	2	1	1	2
$\chi_{13} \otimes \chi_{31}$			1			1

Φ_3: χ_5^{2+}

The following four trees are consistent with the above set of projectives. Benson (1985) shows that one of the last two of these is correct, using an induced character.

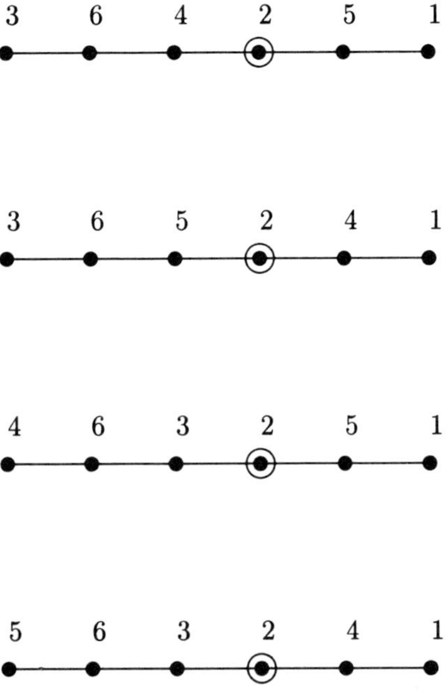

Let $1C0$ denote the Green correspondent of the leaf 5_C. Then this is an irreducible character of the normalizer having an element of order five in its kernel. Furthermore $1H0 = 1B0 \otimes 1C0$ is the Green correspondent of the leaf of type cross in the dual Block 8. From $\chi_{13} \otimes \chi_{24} \approx_H \chi_{73}$, which in terms of nodes reads $1_B \otimes 1_C \approx_H 3_H$, we now conclude that node 3_H must have an edge with label 4. Dualizing, we find that the first two of the above trees are impossible. We shall show in the proof for Block 9 that we can choose an 11-modular system such that the last tree is correct and consistent with the trees of the other blocks.

Group: 12M22 Prime: 11 Block: 8

Complex conjugate to Block 7.

Group: 12M22 Prime: 11 Block: 9

Nr.	CAS-Nr.	Degree	CC	AC	N&C
1	82	120	89	86	○
2	86	120	85	82	○
3	92	144	95	96	×
4	96	144	91	92	×
5	100	336	103	ar	×
5	104	336	99	ar	×
6	108	384	107	ar	○

PROJECTIVES:

Nr.	1	2	3	4	5	6
CC	2	1	4	3	r	r
N&C	○	○	×	×	×	○
$\chi_{86} \otimes \chi_5$	1	2	2	2	5	6

The projective above allows the following two possible trees. Benson (1985) shows that not all four trees for the Blocks 9–12 are of the first type. By mapping a primitive 12th root of unity to its 5th or 7th power if necessary, we can assume that there is an 11-modular system such that in the present block the tree is of the second type. This does not affect the planar embedding and consistency of the trees for Blocks 1, 2 and 3. In Block 5 of the fourfold cover, we may have to interchange nodes 1 and 2. We prefer to leave them as they are and adjust nodes 1 and 2 in the present block:

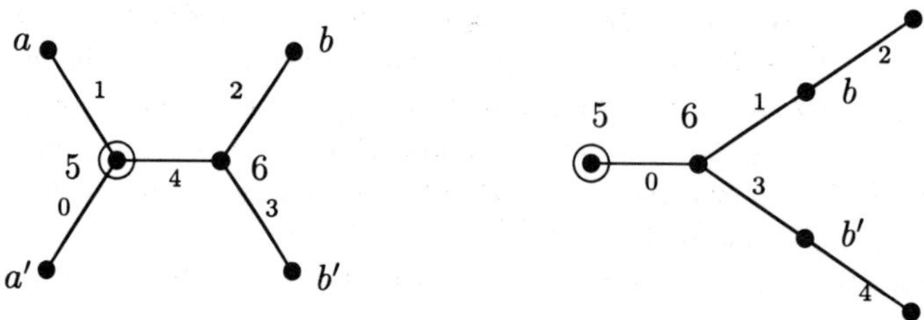

$\{a, a'\} = \{1, 2\}$ and $\{b, b'\} = \{3, 4\}$.

We first show that the tree for Block 11 is of the first type. This follows from

$$(\chi_{92} - \chi_{82}) \otimes \chi_{13} = (\chi_{92} - \chi_{86}) \otimes \chi_{13} = \chi_{106} - \chi_{90}.$$

Since this must be a proper Brauer character, the tree cannot have the second type.

Let $1I0$ denote the Green correspondent of the exceptional node. By the usual argument we know that this is an irreducible character which has an element of order five in its kernel. Since the tensor product of a character from Block 3 with one from Block 5 has nontrivial components in the present block, we have $1C0 \otimes 1E0 = 1I0$.

We find

$$\chi_{24} \otimes \chi_{46} \approx_I \chi_{82} + (\chi_{100} + \chi_{104}) + \chi_{108},$$

which in terms of nodes is

$$1_C \otimes 1_E \approx_I 1_I + 5_I + 6_I.$$

The planar embedded trees for Blocks 3 and 5 now tell us

$$10C3 \otimes 1E4 = 10I2,$$

so that a equals 1_I. You see that swapping nodes 1_E and 2_E will result in interchanging 1_I with 2_I.

In blocks of the first type the Green correspondent of the middle edge has length 8. Since it is also invariant under complex conjugation followed by an outer automorphism, it must have label 4. Hence the edges can be labelled as above. The tensor product of characters of Block 2 with characters of Block 9 lies in Block 11 (which receives label K). We have

$$\chi_{13} \otimes \chi_{92} \approx_K \chi_{98} + \chi_{102} + 2\chi_{106},$$

which is

$$1_B \otimes 3_I \approx_K 5_K + 6_K^2$$

in terms of nodes. From

$$10B2 \otimes 1I3 = 10K0$$

we see that b must equal 3_I.

This completes the proof of the planar embedding and consistency for the present block.

It remains to prove the embedding and consistency for the Blocks 7 and 10. Consider the tensor product

$$\chi_{13} \otimes \chi_{82} \approx_K \chi_{90} + (\chi_{98} + \chi_{102}) + \chi_{106},$$

which is

$$1_B \otimes 1_I \approx_K 3_K + 5_K + 6_K$$

in terms of nodes. The identity

$$10B2 \otimes 10I2 = 1K3$$

shows that node 3_K must be incident to an edge with label 3. By duality, the same is true for node 3_J in Block 10 (note that the distribution of the nodes in Block 10 is exactly the same as in its dual block, since the dual of 1_J is 2_J and the dual of 3_J is 4_J. Finally, let 24_I^1 denote the irreducible module of dimension 24 ($\hat{\chi}_{92} - \hat{\chi}_{82}$). We have $24_I^1 \otimes 21_C \approx_J \hat{\chi}_{83} + \hat{\chi}_{109}$. Here, 21_C is $\hat{\chi}_{24}$. Now the Green correspondent of 21_C is $10C3$ and the Green correspondent of 24_I^1 is $2I1$. Thus $\hat{\chi}_{83} + \hat{\chi}_{109}$ is the Brauer character of the indecomposable module with Green correspondent $10C3 \otimes 2I1 = 9J0$. By the general theory (we omit the details) this module has the edge with label 0 in its head. Hence a' equals 1_J, since 2_J is not a constituent in the above tensor product.

Finally we have to decide which of the two yet possible trees for Block 7 is consistent with the choices made so far. For this purpose, let 56_E^2 denote $\hat{\chi}_{48}$, i.e. node 2_E. We have $24_I^1 \otimes 56_E^2 \approx_H \hat{\chi}_{73} + 2\hat{\chi}_{77} + \hat{\chi}_{81}$. This is $3_H + 5_H^2 + 6$ in terms of nodes. The Green correspondent of the tensor product is $2I1 \otimes 1E1 = 2H2$. This shows that, in Block 8, the dual of Block 7, node 5 is incident to the edge with label 2. Again we omit the details. It follows that node 4_H, and so its dual 5_G, are leaves as claimed.

Group: 12M22 Prime: 11 Block: 10

Nr.	CAS-Nr.	Degree	CC	AC	N&C
1	83	120	88	87	∘
2	87	120	84	83	∘
3	93	144	94	97	×
4	97	144	90	93	×
5	101	336	102	ar	×
5	105	336	98	ar	×
6	109	384	106	ar	∘

Group: 12M22 Prime: 11 Block: 11

Complex conjugate to Block 10.

Group: 12M22 Prime: 11 Block: 12

Complex conjugate to Block 10.

6.5 The Double Cover of the Hall–Janko Group J2

Group: 2J2 Prime: 3 Block: 2

Nr.	CAS-Nr.	Degree	CC	N&C
1	6	36	r	×
2	10	90	r	×
3	11	126	r	∘

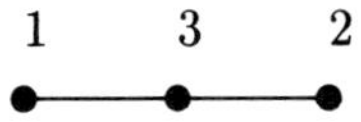

1 — 3 — 2

Group: 2J2 Prime: 3 Block: 3

Nr.	CAS-Nr.	Degree	CC	N&C
1	7	63	r	×
2	18	225	r	×
3	19	288	r	∘

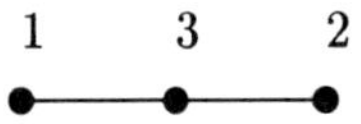

1 — 3 — 2

Group: 2J2 Prime: 3 Block: 5

Nr.	CAS-Nr.	Degree	CC	N&C
1	32	126	r	∘
2	33	126	r	∘
3	35	252	r	×

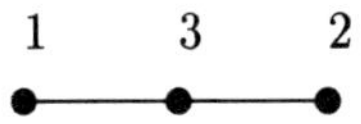

1 — 3 — 2

Group: 2J2 Prime: 5 Block: 2

Nr.	CAS-Nr.	Degree	CC	N&C
1	8	70	r	×
1	9	70	r	×
2	10	90	r	×
3	12	160	r	∘

1 3 2

Group: 2J2 Prime: 7 Block: 1

Nr.	CAS-Nr.	Degree	CC	N&C
1	1	1	r	×
2	6	36	r	×
3	10	90	r	∘
4	12	160	r	∘
5	18	225	r	×
6	19	288	r	×
7	20	300	r	∘

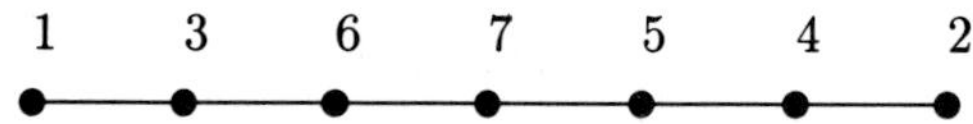

PROJECTIVES:

Nr.:	1	2	3	4	5	6	7
CC:	r	r	r	r	r	r	r
N&C:	×	×	∘	∘	×	×	∘
$\chi_2 \otimes \chi_2$	1		1				
$\chi_8 \otimes \chi_2$			1			1	
$\chi_{27} \otimes \chi_{24}$					1		1
$\chi_7 \otimes \chi_2$						1	1

Group: 2J2 Prime: 7 Block: 2

Nr.	CAS-Nr.	Degree	CC	N&C
1	22	6	r	∘
2	23	6	r	∘
3	25	50	26	×
4	26	50	25	×
5	29	64	r	×
6	30	64	r	×
7	34	216	r	∘

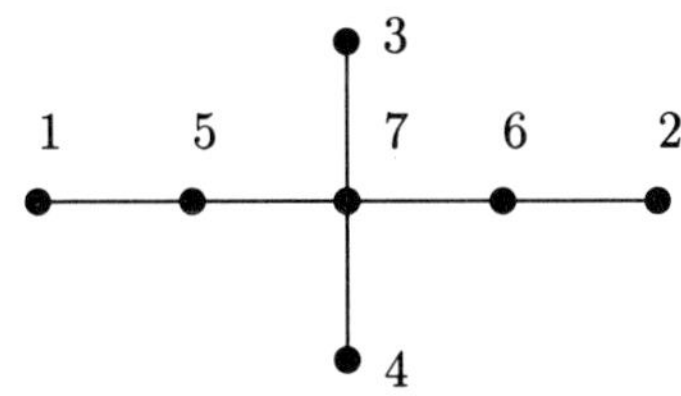

PROJECTIVES:

Nr.:	1	2	3	4	5	6	7
CC:	r	r	4	3	r	r	r
N&C:	∘	∘	×	×	×	×	∘
$\chi_{22} \otimes \chi_2$	1				1		
$\chi_{31} \otimes \chi_2$		1				1	
$\chi_{25} \otimes \chi_4$				1			1
$\chi_{27} \otimes \chi_2$					1		1
$\chi_{36} \otimes \chi_2$						1	1

6.6 The Mathieu Group M23

Group: M23 Prime: 3 Block: 2

Nr.	CAS-Nr.	Degree	CC	N&C
1	6	231	r	×
2	7	231	r	○
2	8	231	r	○

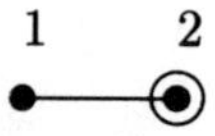

1 2

Group: M23 Prime: 5 Block: 1

Nr.	CAS-Nr.	Degree	CC	N&C
1	1	1	r	×
2	6	231	r	×
3	12	896	13	×
4	13	896	12	×
5	17	2024	r	○

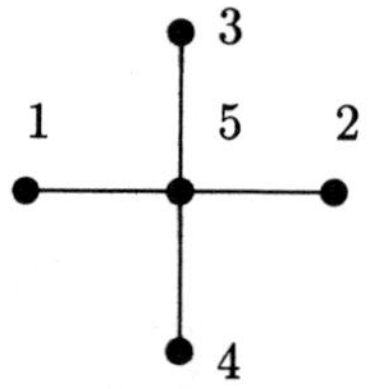

3
1 5 2
4

Group: M23 Prime: 5 Block: 2

Nr.	CAS-Nr.	Degree	CC	N&C
1	2	22	r	×
2	7	231	r	×
2	8	231	r	×
3	9	253	r	○

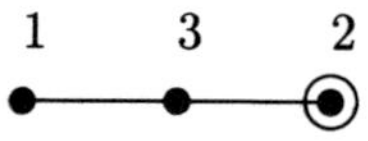

1 3 2

Group: M23 Prime: 7 Block: 1

Nr.	CAS-Nr.	Degree	CC	N&C
1	1	1	r	×
2	14	990	r	∘
2	15	990	r	∘
3	16	1035	r	∘
4	17	2024	r	×

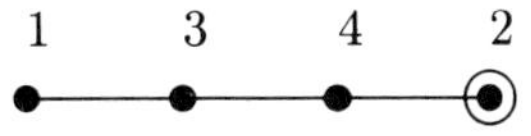

1 — 3 — 4 — 2 (2 circled)

PROJECTIVES:

Nr.:	1	2	3	4
CC:	r	r	r	r
N&C:	×	∘	∘	×
$\chi_2 \otimes \Phi_2$	1		2	1

Φ_2: $\chi_2 \otimes \chi_6$ (restricted to Block 2)

Group: M23 Prime: 7 Block: 2

Nr.	CAS-Nr.	Degree	CC	N&C
1	2	22	r	×
2	3	45	r	∘
2	4	45	r	∘
3	5	230	r	∘
4	9	253	r	×

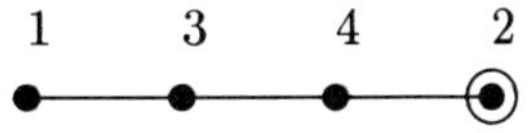

1 — 3 — 4 — 2 (2 circled)

PROJECTIVES:

Nr.:	1	2	3	4
CC:	r	r	r	r
N&C:	×	∘	∘	×
$\chi_2 \otimes \chi_6$	1		1	

Group: M23 Prime: 11 Block: 1

Nr.	CAS-Nr.	Degree	CC	N&C
1	1	1	r	×
2	3	45	4	×
3	4	45	3	×
4	5	230	r	∘
5	12	896	r	∘
5	13	896	r	∘
6	16	1035	r	×

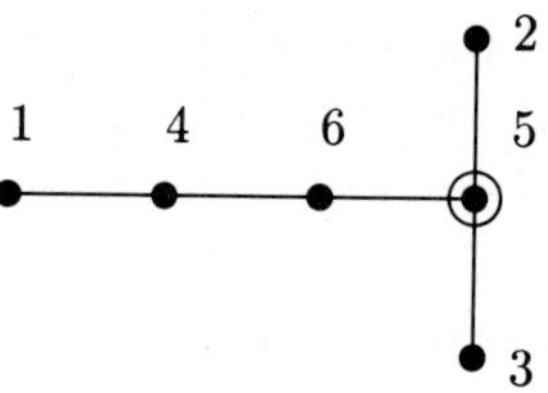

PROJECTIVES:

Nr.:	1	2	3	4	5	6
CC:	r	3	2	r	r	r
N&C:	×	×	×	∘	∘	×
$\chi_2 \otimes \chi_2$	1			1		
$\chi_{15} \otimes \chi_2$			1		2	1

Group: M23 Prime: 23 Block: 1

Nr.	CAS-Nr.	Degree	CC	N&C
1	1	1	r	×
2	2	22	r	∘
3	3	45	4	∘
4	4	45	3	∘
5	6	231	r	×
6	7	231	7	×
7	8	231	6	×
8	10	770	r	∘
8	11	770	r	∘
9	12	896	10	∘
10	13	896	9	∘
11	14	990	12	×
12	15	990	11	×

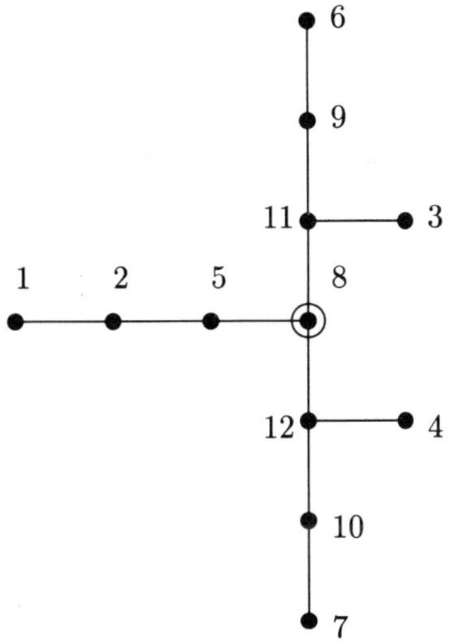

PROJECTIVES:

Nr.:	1	2	3	4	5	6	7	8	9	10	11	12
CC:	r	r	4	3	r	7	6	r	10	9	12	11
N&C:	×	∘	∘	∘	×	×	×	∘	∘	∘	×	×
χ_5^{2+}	1	2			1	1	1		2	2	1	1
$\chi_3 \otimes (\chi_1 + \chi_2)$			1								1	
$\chi_2 \otimes \chi_{16}$					1	1	1	1	2	2	1	1
$\chi_3 \otimes \chi_5$								1	1	1	1	2

Projective χ_5^{2+} shows in a first run that node 1 is joined to node 2, i.e. that $\chi_1 + \chi_2$ is projective. A second run then produces projective Φ_2 which proves that node 3 is joined to node 11 (and hence node 4 to node 12), solving the problem of algebraic conjugacy. The two pairs of nodes $\{9, 10\}$ and $\{6, 7\}$ have irrationalities independent of each other, and independent of the pairs $\{3, 4\}$ and $\{11, 12\}$.

The planar embedding of the tree has been calculated in Kawata (1987). It easily follows from the symmetrization $\chi_3^{2-} \approx \chi_{15}$, using the methods outlined in the chapter on Green correspondence. We omit the details, since there will be many more interesting examples of how this is done.

6.7 The Double Cover of the Higman–Sims Group

Group: 2HS Prime: 3 Block: 3

Nr.	CAS-Nr.	Degree	CC	N&C
1	8	231	r	×
2	13	825	r	×
3	16	1056	r	∘

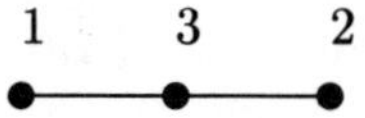

1 — 3 — 2

Group: 2HS Prime: 3 Block: 5

Nr.	CAS-Nr.	Degree	CC	N&C
1	30	924	31	×
2	31	924	30	×
3	36	1848	r	∘

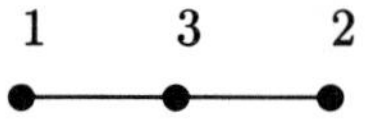

1 — 3 — 2

Group: 2HS Prime: 5 Block: 2

Nr.	CAS-Nr.	Degree	CC	N&C
1	7	175	r	×
2	13	825	r	∘
3	20	1925	r	×
4	21	1925	r	×
5	24	3200	r	∘

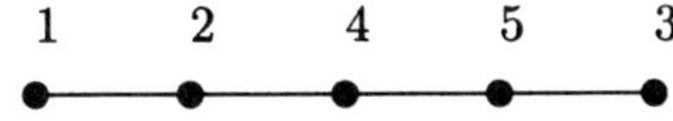

PROJECTIVES:

Nr.:	1	2	3	4	5
CC:	r	r	r	r	r
N&C:	×	∘	×	×	∘
$\chi_{25} \otimes \chi_{32}$			1	1	2

The following two trees are consistent with the projective above:

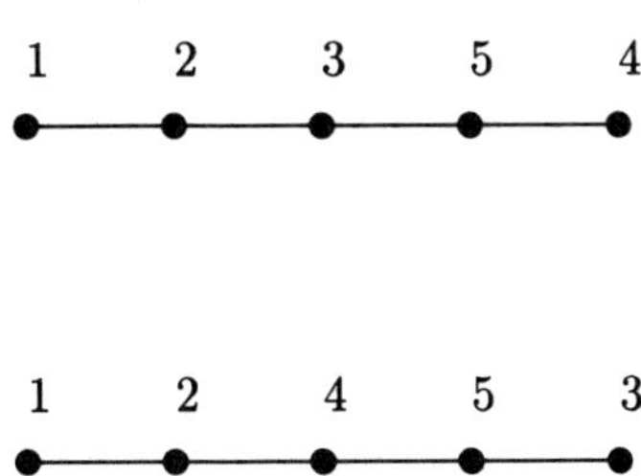

Let 2M22 denote the inverse image in 2HS of the maximal subgroup M22 of HS. If χ_5 is the irreducible character of 2M22 of degree 55, which has the centre in its kernel, then we have $\mathrm{Ind}_{2M22}(\chi_5) = \chi_3 + \chi_4 + \chi_9 + \chi_{10} + \chi_{13} + \chi_{16} + \chi_{21}$. This shows that $\chi_{13} + \chi_{21}$ is projective, so the second tree is correct.

Group: 2HS Prime: 7 Block: 1

Nr.	CAS-Nr.	Degree	CC	N&C
1	1	1	r	×
2	2	22	r	×
3	13	825	r	∘
4	16	1056	r	∘
5	18	1408	r	×
6	23	2750	r	∘
7	24	3200	r	×

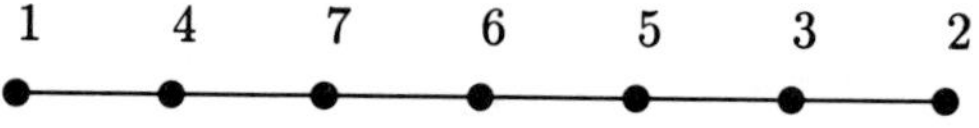

PROJECTIVES:

Nr.:	1	2	3	4	5	6	7
CC:	r	r	r	r	r	r	r
N&C:	×	×	∘	∘	×	∘	×
$\chi_{25} \otimes \chi_{25}$	1			1			
$\chi_5 \otimes \chi_4$				1		1	2
$\chi_2 \otimes \chi_{11}$					1	2	1

Group: 2HS Prime: 7 Block: 2

Nr.	CAS-Nr.	Degree	CC	N&C
1	26	176	27	×
2	27	176	26	×
3	32	1000	r	∘
4	37	1980	38	∘
5	38	1980	37	∘
6	39	2304	40	×
7	40	2304	39	×

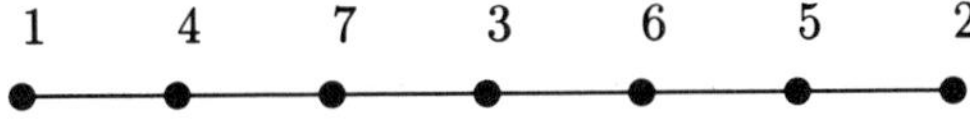

PROJECTIVES:

Nr.:	1	2	3	4	5	6	7
CC:	2	1	r	5	4	7	6
N&C:	×	×	∘	∘	∘	×	×
$\chi_{26} \otimes \chi_5$		2		1	3	1	1
$\chi_{27} \otimes \chi_3$		1		1	2	1	1
$\chi_2 \otimes \chi_{33}$			1		1	1	1
$\chi_{25} \otimes \chi_4$				1	1	1	1

It follows from the projectives above that χ_{38} is joined to χ_{39} or to χ_{40}. Since the character values of χ_{38} lie in a field disjoint from the one containing the values of χ_{39} and χ_{40}, we can assume without loss of generality that χ_{38} is joined to χ_{39}. Then only one tree is consistent with all the given projectives. Observe that in this example the real stem of the tree consists just of the character χ_{32}.

Group: 2HS Prime: 11 Block: 1

Nr.	CAS-Nr.	Degree	CC	N&C
1	1	1	r	×
2	7	175	r	∘
3	14	896	15	∘
3	15	896	14	∘
4	19	1750	r	×
5	22	2520	r	×
6	24	3200	r	∘

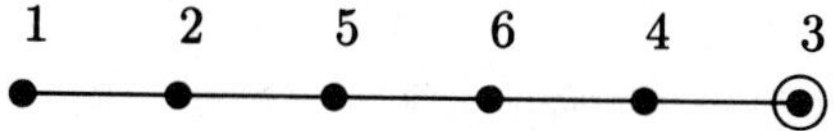

PROJECTIVES:

Nr.:	1	2	3	4	5	6
CC:	r	r	r	r	r	r
N&C:	×	∘	∘	×	×	∘
$\chi_2 \otimes \chi_2$	1	1				
$\chi_5 \otimes \chi_2$		1			1	
$\chi_{11} \otimes \chi_2$					1	1

Group: 2HS Prime: 11 Block: 2

Nr.	CAS-Nr.	Degree	CC	N&C
1	25	56	r	×
2	32	1000	r	∘
3	35	1792	r	∘
4	39	2304	40	∘
4	40	2304	39	∘
5	41	2520	r	×
6	42	2520	r	×

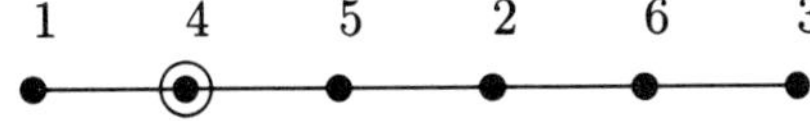

PROJECTIVES:

Nr.:	1	2	3	4	5	6
CC:	r	r	r	r	r	r
N&C:	×	∘	∘	∘	×	×
$\chi_{28} \otimes \chi_2$	1			1		
$\chi_{32} \otimes \chi_2$		3		1	2	2

Without loss of generality we can assume that of the two algebraically conjugate characters χ_{41} and χ_{42}, χ_{41} comes first on the tree. Then there is just one tree consistent with the projectives above.

6.8 The Triple Cover of the Janko Group J3

Group: J3 Prime: 3 Block: 2

Nr.	CAS-Nr.	Degree	CC	N&C
1	6	324	r	×
2	20	2754	r	×
3	21	3078	r	∘

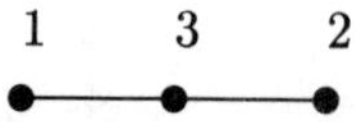

1 3 2

Remark: The inverse image of this block in 3J3 has a noncyclic defect group of order 9.

Group: 3J3 Prime: 3 Block: 3

Nr.	CAS-Nr.	Degree	CC	N&C
1	11	1215	r	×
2	38	1215	39	∘
2	39	1215	38	∘

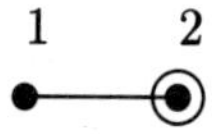

1 2

Group: 3J3 Prime: 3 Block: 4

Nr.	CAS-Nr.	Degree	CC	N&C
1	12	1215	r	×
2	40	1215	41	∘
2	41	1215	40	∘

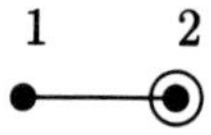

1 2

Group: 3J3 Prime: 5 Block: 1

Nr.	CAS-Nr.	Degree	CC	N&C
1	1	1	r	×
2	4	323	r	×
2	5	323	r	×
3	6	324	r	∘

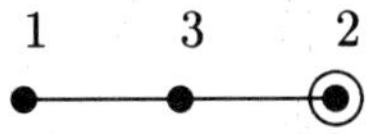

1 3 2

Group: 3J3 Prime: 5 Block: 2

Nr.	CAS-Nr.	Degree	CC	N&C
1	7	646	r	×
1	8	646	r	×
2	19	2432	r	×
3	21	3078	r	∘

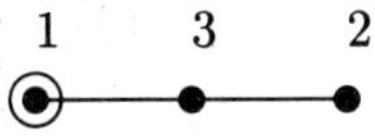

1 3 2

Group: 3J3 Prime: 5 Block: 3

Nr.	CAS-Nr.	Degree	CC	N&C
1	9	816	r	×
2	17	1938	r	×
2	18	1938	r	×
3	20	2754	r	∘

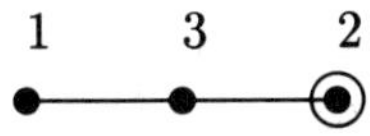

1 3 2

Group: 3J3 Prime: 5 Block: 4

Nr.	CAS-Nr.	Degree	CC	AC	N&C
1	22	18	23	24	×
1	24	18	25	22	×
2	46	2736	47	ar	×
3	48	2754	49	ar	∘

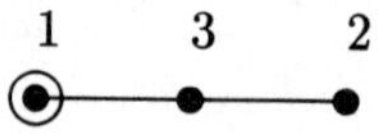

1 3 2

Group: 3J3 Prime: 5 Block: 5

Complex conjugate to Block 4.

Group: 3J3 Prime: 5 Block: 6

Nr.	CAS-Nr.	Degree	CC	AC	N&C
1	26	153	27	28	×
1	28	153	29	26	×
2	30	171	31	ar	×
3	36	324	37	ar	∘

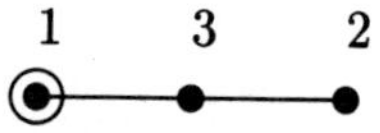

1 3 2

Group: 3J3 Prime: 5 Block: 7

Complex conjugate to Block 6.

Group: 3J3 Prime: 5 Block: 8

Nr.	CAS-Nr.	Degree	CC	AC	N&C
1	32	171	33	34	×
1	34	171	35	32	×
2	50	2907	51	ar	×
3	54	3078	55	ar	∘

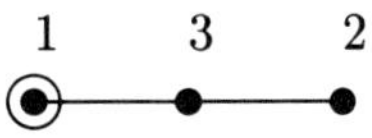

1 3 2

Group: 3J3 Prime: 5 Block: 9

Complex conjugate to Block 8.

Group: 3J3 Prime: 17 Block: 1

Nr.	CAS-Nr.	Degree	CC	N&C
1	1	1	r	×
2	6	324	r	×
3	10	1140	r	×
4	11	1215	r	∘
4	12	1215	r	∘
5	14	1920	r	∘
6	15	1920	r	∘
7	16	1920	r	∘
8	19	2432	r	×
9	21	3078	r	×

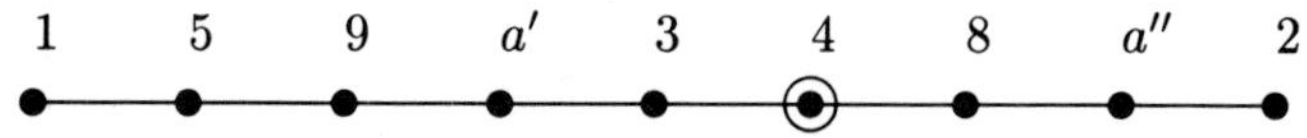

$\{a', a''\} = \{6, 7\}$.

PROJECTIVES:

Nr.:	1	2	3	4	5	6	7	8	9
CC:	r	r	r	r	r	r	r	r	r
N&C:	×	×	×	∘	∘	∘	∘	×	×
$\chi_3 \otimes \chi_2$	1	1	1		1	1	1		
$\chi_{33} \otimes \chi_{26}$			1	1	1	1	1	2	1
$\chi_2 \otimes \chi_2$			1	1					
$\chi_4 \otimes \chi_2$				1	1	1	1	2	2
$\chi_{31} \otimes \chi_{26}$					1	1	1	1	2

Without loss of generality we can assume that of the three algebraically conjugate characters χ_{14}, χ_{15} and χ_{16} (nodes 5, 6 and 7), χ_{14} is located left of the others on the tree. Then the following four trees are consistent with the above projectives:

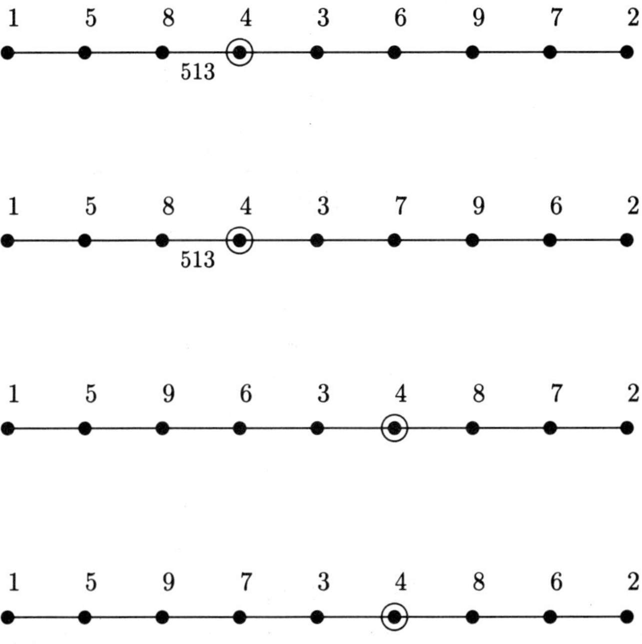

Let 513 denote the irreducible Brauer character corresponding to the third edge of the first two trees. The contradiction $(513^{2-}, \chi_2) = -1$ allows only one of the other two possibilities. However, we are not able to decide which of them is correct.

Group: 3J3 Prime: 17 Block: 2

Nr.	CAS-Nr.	Degree	CC	AC	N&C
1	22	18	23	24	×
2	24	18	25	22	×
3	30	171	31	ar	×
4	32	171	33	34	×
5	34	171	35	32	×
6	36	324	37	ar	×
7	38	1215	39	ar	∘
7	40	1215	41	ar	∘
8	46	2736	47	ar	∘
9	54	3078	55	ar	×

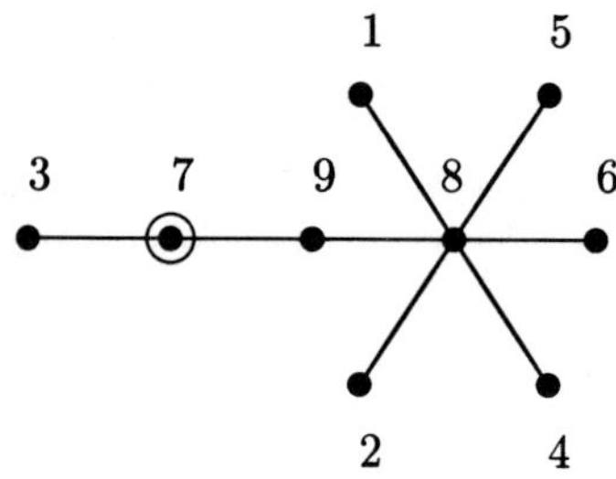

PROJECTIVES:

Nr.:	1	2	3	4	5	6	7	8	9
AC:	2	1	ar	5	4	ar	ar	ar	ar
N&C:	×	×	×	×	×	×	∘	∘	×
$\chi_{24} \otimes \chi_5$		1						1	
χ_{27}^{2+}			1				1		
$\chi_{24} \otimes \chi_8$					1			1	
$\chi_{30} \otimes \chi_2$							1		1
$\chi_{26} \otimes \chi_2$								1	1

The planar embedding has already been obtained by Feit in (1984b). It follows from $\chi_6 \otimes \chi_{22} \approx_B \chi_{24}$ and $\chi_{19} \otimes \chi_{22} \approx_B \chi_{32} + (\chi_{38} + \chi_{40}) + 2\chi_{46} + 3\chi_{54}$. We leave the proof as an exercise.

Group: 3J3 Prime: 17 Block: 3

Complex conjugate to Block 2.

Group: 3J3 Prime: 19 Block: 1

Nr.	CAS-Nr.	Degree	CC	N&C
1	1	1	r	×
2	2	85	3	∘
2	3	85	2	∘
3	6	324	r	×
4	9	816	r	∘
5	11	1215	r	∘
6	12	1215	r	∘
7	14	1920	r	×
8	15	1920	r	×
9	16	1920	r	×
10	20	2754	r	∘

PROJECTIVES:

Nr.:	1	2	3	4	5	6	7	8	9	10
CC:	r	r	r	r	r	r	r	r	r	r
N&C:	×	∘	×	∘	∘	∘	×	×	×	∘
$\chi_4 \otimes \chi_4$	1		1	2	3	3	4	4	4	6
$\chi_{31} \otimes \chi_{30}$	1		1		2	2	1	1	1	1
$\chi_2 \otimes \chi_{10}$		1		1	3	3	4	4	4	4
$\chi_5 \otimes \chi_4$			2	2	3	3	4	4	4	6
$\chi_{27} \otimes \chi_{30}$				1			1	1	1	2

We have two orbits of algebraically conjugate characters in the principal block. One has length 2 and consists of the characters χ_{11} and χ_{12} (nodes 5 and 6) of degree 1215. In the two blocks of 3J3 containing faithful characters, we also have 2 orbits of length 2 containing characters of the same degree. The irrationalities of these 6 characters all lie in the same field. We are going to fix an ordering of these characters in one of the non principal blocks. Then this ordering determines the ordering in the other blocks.

The second orbit of algebraically conjugate characters has length 3 and contains the characters χ_{14}, χ_{15} and χ_{16} (nodes 7, 8 and 9) of degree 1920. Their irrationalities lie in a field which is disjoint from the field containing the values of the characters of degree 1215. Hence we can assume without loss of generality that of the characters in this orbit χ_{14} comes first on the tree. With this assumption we have four possible trees consistent with the projectives:

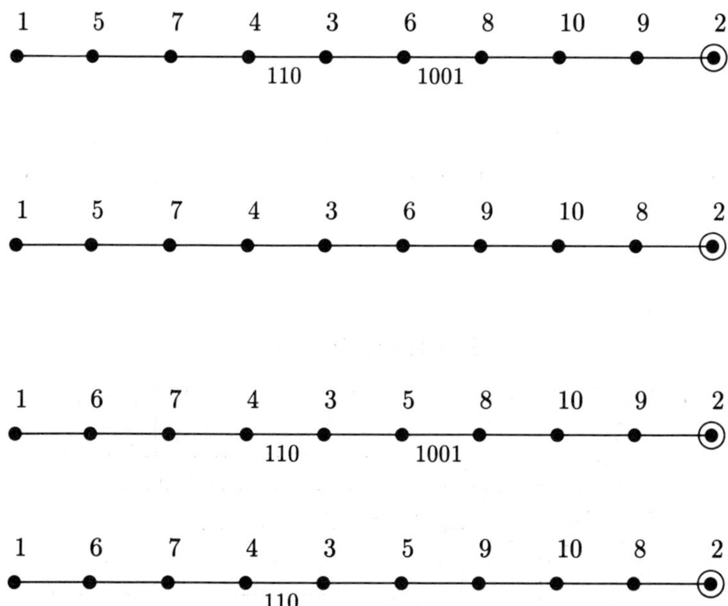

The first and third trees are impossible since they imply $(110^{2-}, \Phi_{1001}) = -1$, where the notation is chosen as in the proof for the principal 17-block. The last implies $(110 \otimes \hat{\chi}_{22}, \chi_{36} + \chi_{38}) = -1$, where $\hat{\chi}_{22}$ is an irreducible Brauer character in Block 2 and $\chi_{36} + \chi_{38}$ is the projective cover of $\hat{\chi}_{36}$ in the same block. This leaves the second tree as the only possibility.

Group: 3J3 Prime: 19 Block: 2

Nr.	CAS-Nr.	Degree	CC	AC	N&C
1	22	18	23	24	∘
2	24	18	25	22	∘
3	26	153	27	28	×
4	28	153	29	26	×
5	36	324	37	ar	×
6	38	1215	39	ar	∘
7	40	1215	41	ar	∘
8	42	1530	45	ar	×
8	44	1530	43	ar	×
9	48	2754	49	ar	∘
10	52	3060	53	ar	×

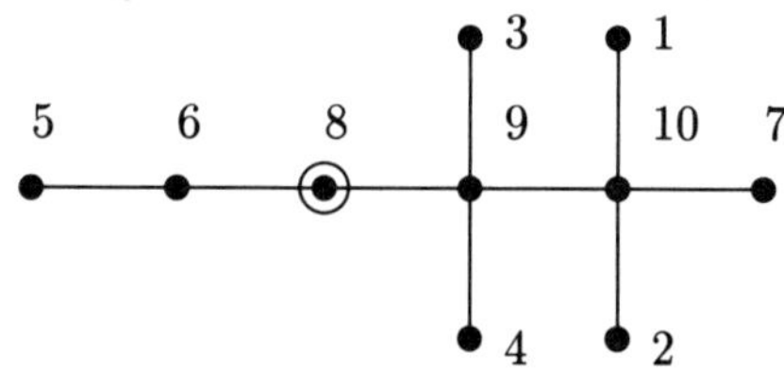

PROJECTIVES:

Nr.:	1	2	3	4	5	6	7	8	9	10
AC:	2	1	4	3	ar	ar	ar	ar	ar	ar
N&C:	∘	∘	×	×	×	∘	∘	×	∘	×
$\chi_{22} \otimes \chi_4$	1									1
$\chi_{30} \otimes \chi_4$			1		1	2	2	1	3	4
$\chi_{26} \otimes \chi_4$			1			1	1	1	3	3
$\chi_{27} \otimes \chi_{31}$			1						2	1
χ_{31}^{2+}					1	2	2	1		2
$\chi_{32} \otimes \chi_4$						1	1	2	3	3
$\chi_2 \otimes \chi_{30}$						1	1	1		1
$\chi_{33} \otimes \chi_{31}$								1	2	1

Without loss of generality we assume that of the two algebraically conjugate characters χ_{38} and χ_{40} (nodes 6 and 7) of degree 1215, χ_{38} comes first on the tree. Remember that we have carefully determined the position of the characters of degree 1215 in the principal block, using the present assumption. The tree is then uniquely determined by the above projectives.

To find its planar embedding, we use the scalar product $(\chi_{26} \otimes \chi_9, \chi_{22}) = 0$. In terms of nodes, this means that 2_B does not occur in $3_B \otimes 4_A$. Since there is an indecomposable co-trivial source module with Brauer character 4_A and Green correspondent $18A2$, this shows that the 'distance' of 3_B and 2_B on the tree cannot be 2, thus proving the embedding. It has also been obtained by Feit in (1984b).

Group: 3J3 Prime: 19 Block: 3

Complex conjugate to Block 2. Having fixed an ordering of the two algebraically conjugate characters of degree 1215 in that block, the ordering of the corresponding characters is determined in this block.

6.9 The Mathieu Group M24

Group: M24 Prime: 3 Block: 2

Nr.	CAS-Nr.	Degree	CC	N&C
1	3	45	4	×
2	12	990	13	×
3	15	1035	16	∘

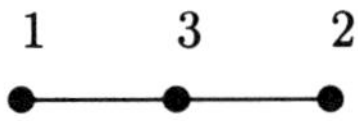

1 3 2

Group: M24 Prime: 3 Block: 3

Complex conjugate to Block 2.

Group: M24 Prime: 3 Block: 4

Nr.	CAS-Nr.	Degree	CC	N&C
1	7	252	r	×
2	24	5544	r	×
3	25	5796	r	∘

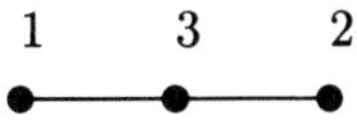

1 3 2

Group: M24 Prime: 3 Block: 5

Nr.	CAS-Nr.	Degree	CC	N&C
1	14	1035	r	×
2	20	2277	r	×
3	21	3312	r	∘

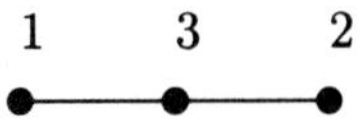

1 3 2

Group: M24 Prime: 5 Block: 1

Nr.	CAS-Nr.	Degree	CC	N&C
1	1	1	r	×
2	18	1771	r	×
3	19	2024	r	∘
4	24	5544	r	∘
5	25	5796	r	×

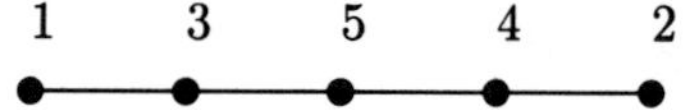

PROJECTIVES:

Nr.:	1	2	3	4	5
CC:	r	r	r	r	r
N&C:	×	×	∘	∘	×
$\chi_4 \otimes \chi_3$	1		1		
$\chi_{24} \otimes \chi_3$		2	3	5	6

Group: M24 Prime: 5 Block: 3

Nr.	CAS-Nr.	Degree	CC	N&C
1	5	231	6	×
1	6	231	5	×
2	7	252	r	×
3	9	483	r	∘

1 3 2

Group: M24 Prime: 5 Block: 2

Nr.	CAS-Nr.	Degree	CC	N&C
1	2	23	r	×
2	8	253	r	×
3	20	2277	r	∘
4	21	3312	r	∘
5	23	5313	r	×

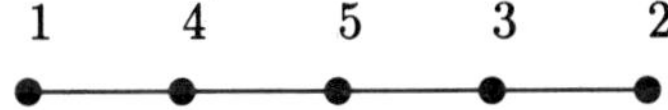

PROJECTIVES:

Nr.:	1	2	3	4	5
CC:	r	r	r	r	r
N&C:	×	×	∘	∘	×
$\mathrm{Ind}_{2^6:3S_6}(\chi_3)$	1			1	
$\mathrm{Ind}_{2^6:3S_6}(\chi_{10})$		1	1		
$\chi_{12} \otimes \chi_3$			1		1
$\chi_{10} \otimes \chi_3$				1	1

The ordinary characters χ_3 and χ_{10} of $2^6 : 3S_6$ are inflations of ordinary characters of the symmetric group S_6. They have degrees 5 and 10 resp., hence are projectives. We did not find any argument using only the ordinary character table of M24 to determine this tree.

Group: M24 Prime: 7 Block: 1

Nr.	CAS-Nr.	Degree	CC	N&C
1	1	1	r	×
2	12	990	13	∘
2	13	990	12	∘
3	14	1035	r	∘
4	19	2024	r	×

1 3 4 2

PROJECTIVES:

Nr.:	1	2	3	4
CC:	r	r	r	r
N&C:	×	∘	∘	×
$\chi_6 \otimes \chi_5$	1		1	

Group: M24 Prime: 7 Block: 2

Nr.	CAS-Nr.	Degree	CC	N&C
1	2	23	r	×
2	15	1035	16	∘
2	16	1035	15	∘
3	17	1265	r	∘
4	20	2277	r	×

1 3 4 2

PROJECTIVES:

Nr.:	1	2	3	4
CC:	r	r	r	r
N&C:	×	∘	∘	×
$\chi_2 \otimes \chi_7$	1		1	

Group: M24 Prime: 7 Block: 3

Nr.	CAS-Nr.	Degree	CC	N&C
1	3	45	4	∘
1	4	45	3	∘
2	8	253	r	×
3	21	3312	r	×
4	22	3520	r	∘

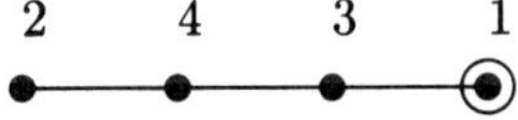

PROJECTIVES:

Nr.:	1	2	3	4
CC:	r	r	r	r
N&C:	∘	×	×	∘
χ_{10}^{2+}	1		6	5

Group: M24 Prime: 11 Block: 1

Nr.	CAS-Nr.	Degree	CC	N&C
1	1	1	r	×
2	2	23	r	×
3	3	45	4	×
4	4	45	3	×
5	7	252	r	∘
6	9	483	r	∘
7	14	1035	r	×
8	15	1035	16	×
9	16	1035	15	×
10	21	3312	r	×
11	25	5796	r	∘

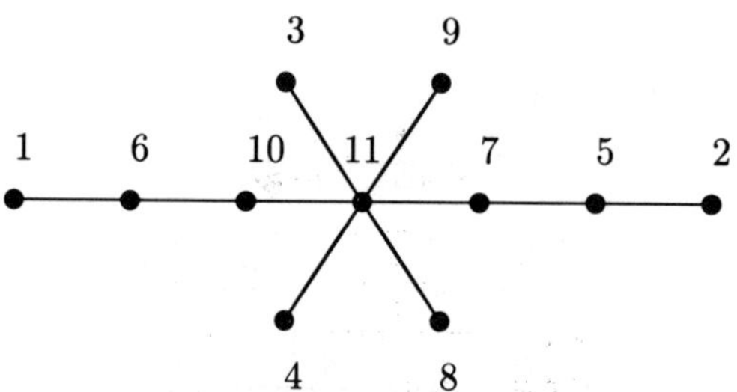

PROJECTIVES:

Nr.:	1	2	3	4	5	6	7	8	9	10	11
CC:	r	r	4	3	r	r	r	9	8	r	r
N&C:	×	×	×	×	∘	∘	×	×	×	×	∘
$\chi_6 \otimes \chi_5$	1					1	1			1	2
$\chi_3 \otimes \chi_{12}$				1							1
$\chi_2 \otimes \chi_{20}$						1				2	1
$\chi_8 \otimes \chi_5$							1			1	2
$\chi_2 \otimes \chi_{13}$									1		1

The planar embedding, which has also been determined in Kawata (1987), follows easily from $\chi_2 \otimes \chi_3 \approx \chi_{15}$.

Group: M24 Prime: 23 Block: 1

Nr.	CAS-Nr.	Degree	CC	N&C
1	1	1	r	×
2	3	45	4	∘
3	4	45	3	∘
4	5	231	6	×
5	6	231	5	×
6	7	252	r	∘
7	10	770	11	∘
7	11	770	10	∘
8	12	990	13	×
9	13	990	12	×
10	22	3520	r	×
11	24	5544	r	×
12	26	10395	r	∘

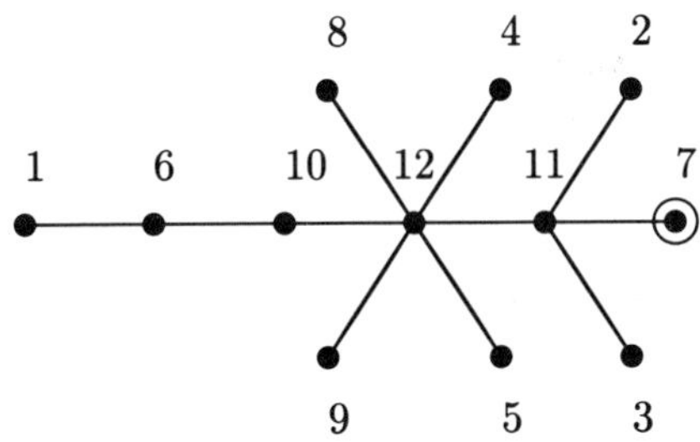

PROJECTIVES:

Nr.:	1	2	3	4	5	6	7	8	9	10	11	12
CC:	r	3	2	5	4	r	r	9	8	r	r	r
N&C:	×	∘	∘	×	×	∘	∘	×	×	×	×	∘
$\chi_2 \otimes \chi_2$	1					1						
$\chi_3 \otimes \chi_{14}$			1								2	1
$\chi_6 \otimes \chi_9$					1					2	2	5
$\chi_7 \otimes \chi_2$						1				1		
$\chi_4 \otimes \chi_8$									1			1

The planar embedding, which has also been determined in Kawata (1987), follows easily from $\chi_3 \otimes \chi_3 \approx \chi_{13}$.

6.10 The Triple Cover of the McLaughlin Group

Group: 3MCL Prime: 2 Block: 2

Nr.	CAS-Nr.	Degree	CC	N&C
1	10	3520	r	×
2	11	3520	r	∘

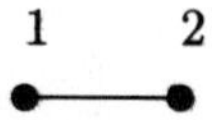

1 ●——● 2

Group: 3MCL Prime: 2 Block: 5

Nr.	CAS-Nr.	Degree	CC	AC	N&C
1	47	6336	48	ar	×
2	49	6336	50	ar	∘

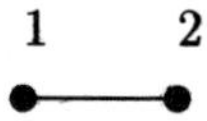

1 ●——● 2

Group: 3MCL Prime: 2 Block: 6

Complex conjugate to Block 5.

Group: 3MCL Prime: 3 Block: 2

Nr.	CAS-Nr.	Degree	CC	N&C
1	14	5103	r	×
2	45	5103	46	∘
2	46	5103	45	∘

1 2

Group: 3MCL Prime: 3 Block: 3

Nr.	CAS-Nr.	Degree	CC	AC	N&C
1	16	8019	17	ar	×
2	55	8019	54	ar	∘
2	56	8019	53	ar	∘

1 2

Group: 3MCL Prime: 3 Block: 4

Complex conjugate to Block 3.

Group: 3MCL Prime: 7 Block: 1

Nr.	CAS-Nr.	Degree	CC	N&C
1	1	1	r	×
2	11	3520	r	∘
3	12	4500	r	∘
4	16	8019	17	×
4	17	8019	16	×

1 3 4 2

PROJECTIVES:

Nr.:	1	2	3	4
CC:	r	r	r	r
N&C:	×	∘	∘	×
$\chi_3 \otimes \chi_3$	1		1	

Group: 3MCL Prime: 7 Block: 2

Nr.	CAS-Nr.	Degree	CC	N&C
1	2	22	r	×
2	10	3520	r	∘
3	13	4752	r	∘
4	18	8250	19	×
4	19	8250	18	×

1 2 4 3

PROJECTIVES:

Nr.:	1	2	3	4
CC:	r	r	r	r
N&C:	×	∘	∘	×
$\chi_2 \otimes \chi_3$	1	1		

Group: 3MCL Prime: 7 Block: 3

Nr.	CAS-Nr.	Degree	CC	AC	N&C
1	29	792	30	ar	×
2	33	2376	36	ar	∘
2	35	2376	34	ar	∘
3	43	4752	44	ar	∘
4	49	6336	50	ar	×

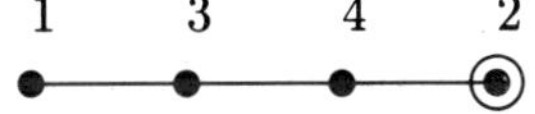

PROJECTIVES:

Nr.:	1	2	3	4
AC:	ar	ar	ar	ar
N&C:	×	∘	∘	×
$\chi_{29} \otimes \chi_3$	1		1	

Group: 3MCL Prime: 7 Block: 4

Complex conjugate to Block 3.

Group: 3MCL Prime: 7 Block: 5

Nr.	CAS-Nr.	Degree	CC	AC	N&C
1	31	1980	32	ar	∘
2	47	6336	48	ar	×
3	53	8019	56	ar	×
3	55	8019	54	ar	×
4	65	12375	66	ar	∘

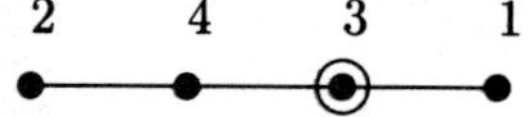

PROJECTIVES:

Nr.:	1	2	3	4
AC:	ar	ar	ar	ar
N&C:	∘	×	×	∘
$\chi_{29} \otimes \chi_3$	1		2	1

Group: 3MCL Prime: 7 Block: 6

Complex conjugate to Block 5.

Group: 3MCL Prime: 11 Block: 1

Nr.	CAS-Nr.	Degree	CC	N&C
1	1	1	r	×
2	4	252	r	∘
3	7	896	8	∘
3	8	896	7	∘
4	9	1750	r	×
5	12	4500	r	×
6	14	5103	r	∘

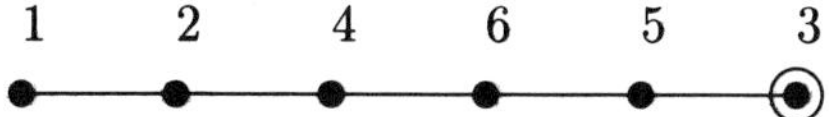

PROJECTIVES:

Nr.:	1	2	3	4	5	6
CC:	r	r	r	r	r	r
N&C:	×	∘	∘	×	×	∘
$\chi_2 \otimes \chi_2$	1	1				
$\chi_4 \otimes \chi_2$		1		1		
$\chi_2 \otimes \chi_{18}$				1		1

Group: 3MCL Prime: 11 Block: 2

Nr.	CAS-Nr.	Degree	CC	AC	N&C
1	25	126	28	27	∘
1	27	126	26	25	∘
2	37	2520	40	ar	×
3	39	2520	38	ar	×
4	45	5103	46	ar	∘
5	51	7875	52	ar	∘
6	57	8064	58	ar	×

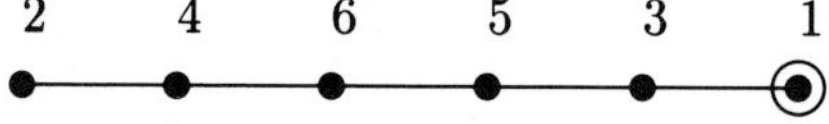

PROJECTIVES:

Nr.:	1	2	3	4	5	6
AC:	ar	ar	ar	ar	ar	ar
N&C:	∘	×	×	∘	∘	×
$\chi_2 \otimes \chi_{41}$	1	1	1	1		
$\chi_2 \otimes \chi_{53}$				1	1	2

Without loss of generality we can assume that of the 2 characters of degree 2520 χ_{37} comes first on the Brauer tree. Then there is only one tree consistent with the above projectives.

Group: 3MCL Prime: 11 Block: 3

Complex conjugate to Block 2.

6.11 The Held Group

Group: HE Prime: 3 Block: 3

Nr.	CAS-Nr.	Degree	CC	AC	N&C
1	4	153	5	ar	×
2	17	7497	18	ar	×
3	20	7650	21	ar	∘

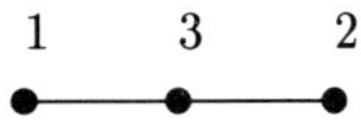

1 3 2

Group: HE Prime: 3 Block: 4

Complex conjugate to Block 3.

Group: HE Prime: 3 Block: 5

Nr.	CAS-Nr.	Degree	CC	N&C
1	19	7650	r	×
2	27	14400	r	×
3	32	22050	r	∘

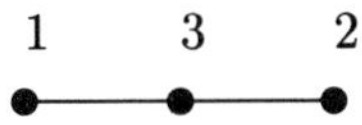

1 3 2

Group: HE Prime: 5 Block: 2

Nr.	CAS-Nr.	Degree	CC	N&C
1	6	680	r	×
2	12	1920	r	∘
3	13	4080	r	×
4	22	10880	r	×
5	26	13720	r	∘

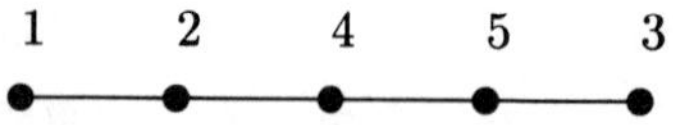

1 — 2 — 4 — 5 — 3

PROJECTIVES:

Nr.:	1	2	3	4	5
CC:	r	r	r	r	r
N&C:	×	∘	×	×	∘
$\chi_2 \otimes \chi_9$	1	1			
$\chi_7 \otimes \chi_{10}$		1		4	3

Group: HE Prime: 7 Block: 2

Nr.	CAS-Nr.	Degree	CC	N&C
1	15	6272	r	×
2	17	7497	18	∘
2	18	7497	17	∘
3	29	20825	r	∘
4	32	22050	r	×

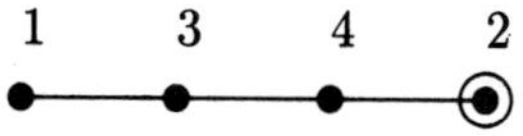

1 — 3 — 4 — 2

PROJECTIVES:

Nr.:	1	2	3	4
CC:	r	r	r	r
N&C:	×	∘	∘	×
$\chi_2 \otimes \chi_7$	1		1	

Group: HE Prime: 17 Block: 1

Nr.	CAS-Nr.	Degree	CC	N&C
1	1	1	r	×
2	7	1029	r	×
2	8	1029	r	×
3	12	1920	r	○
4	15	6272	r	○
5	26	13720	r	×
6	27	14400	r	×
7	30	21504	r	○
8	31	21504	r	○
9	32	22050	r	×

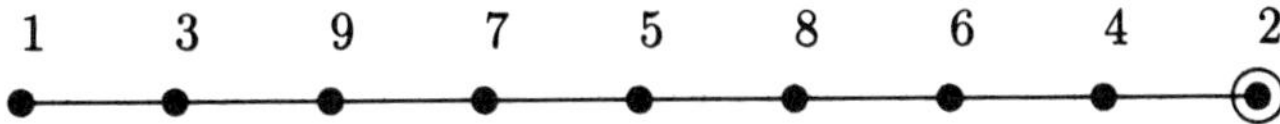

PROJECTIVES:

Nr.:	1	2	3	4	5	6	7	8	9
CC:	r	r	r	r	r	r	r	r	r
N&C:	×	×	○	○	×	×	○	○	×
$\chi_3 \otimes \chi_2$	1		1						
$\chi_{11} \otimes \chi_2$			1	1		1			1
$\chi_9 \otimes \chi_2$			1						1
$\chi_{26} \otimes \chi_2$				1	4	2	4	4	3
$\chi_{13} \otimes \chi_2$					1		1	1	1
$\chi_6 \otimes \chi_4$						1	1	1	1

Of the two algebraically conjugate characters χ_{30} and χ_{31} we assume without loss of generality that χ_{30} comes first on the tree. Then only one tree is consistent with all the given projectives.

6.12 The Double Cover of the Rudvalis Group

Group: 2RU Prime: 3 Block: 2

Nr.	CAS-Nr.	Degree	CC	N&C
1	6	3276	r	×
2	8	20475	r	×
3	10	23751	r	∘

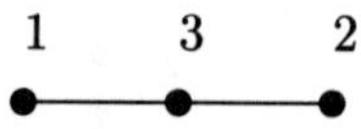

1 — 3 — 2

Group: 2RU Prime: 3 Block: 3

Nr.	CAS-Nr.	Degree	CC	N&C
1	7	3654	r	×
2	28	91350	r	×
3	29	95004	r	∘

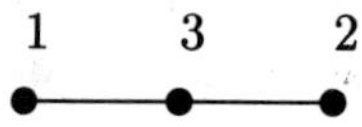

1 — 3 — 2

Group: 2RU Prime: 3 Block: 5

Nr.	CAS-Nr.	Degree	CC	N&C
1	41	3276	42	×
2	43	4032	44	×
3	47	7308	48	∘

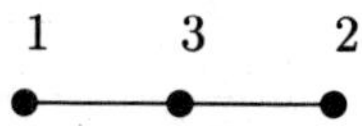

1 — 3 — 2

Group: 2RU Prime: 3 Block: 6

Complex conjugate to Block 5.

Group: 2RU Prime: 5 Block: 2

Nr.	CAS-Nr.	Degree	CC	N&C
1	8	20475	r	×
2	23	65975	r	×
3	25	75400	r	∘
4	28	91350	r	×
5	32	102400	r	∘

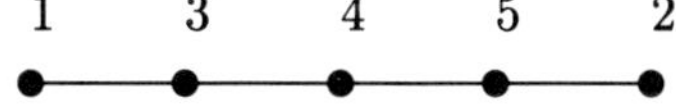

PROJECTIVES:

Nr.:	1	2	3	4	5
CC:	r	r	r	r	r
N&C:	×	×	∘	×	∘
$\chi_4 \otimes \Phi_4$	3	15	19	27	26
$\chi_4 \otimes \chi_{11}$	2	6	5	5	8
Φ_4	1		1		

Remark: The projective $\chi_4 \otimes \chi_{11}$ shows that node 1 is linked to node 3.

Group: 2RU Prime: 7 Block: 1

Nr.	CAS-Nr.	Degree	CC	N&C
1	1	1	r	×
2	5	783	r	∘
3	9	21750	r	×
4	27	81432	r	×
5	34	110592	r	∘
6	35	110592	r	∘
7	36	118784	r	×

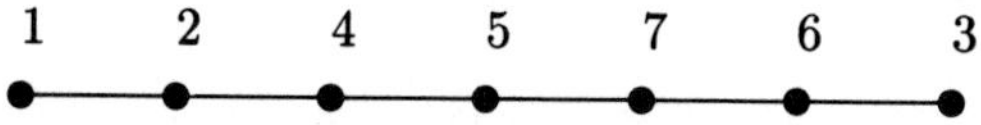

1 — 2 — 4 — 5 — 7 — 6 — 3

PROJECTIVES:

Nr.:	1	2	3	4	5	6	7
CC:	r	r	r	r	r	r	r
N&C:	×	∘	×	×	∘	∘	×
$\chi_3 \otimes \chi_2$	1	1					
$\chi_4 \otimes \chi_2$		1		1			
$\chi_{50} \otimes \chi_{37}$					1		1
$\chi_{49} \otimes \chi_{37}$						1	1

Without loss of generality we can assume that of the two algebraic conjugate characters χ_{34} and χ_{35} (nodes 5 and 6), χ_{34} comes first on the tree. Then only the one tree given above is consistent with all the projectives.

Group: 2RU Prime: 7 Block: 2

Nr.	CAS-Nr.	Degree	CC	N&C
1	11	27000	r	×
1	12	27000	r	×
1	13	27000	r	×
2	25	75400	r	×
3	32	102400	r	∘

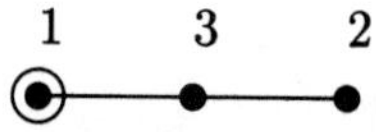

1 (exceptional) — 3 — 2

Group: 2RU Prime: 7 Block: 3

Nr.	CAS-Nr.	Degree	CC	N&C
1	39	1248	40	×
2	40	1248	39	×
3	49	8192	r	×
4	50	8192	r	×
5	56	48256	r	∘
6	60	221184	r	∘
7	61	250560	r	×

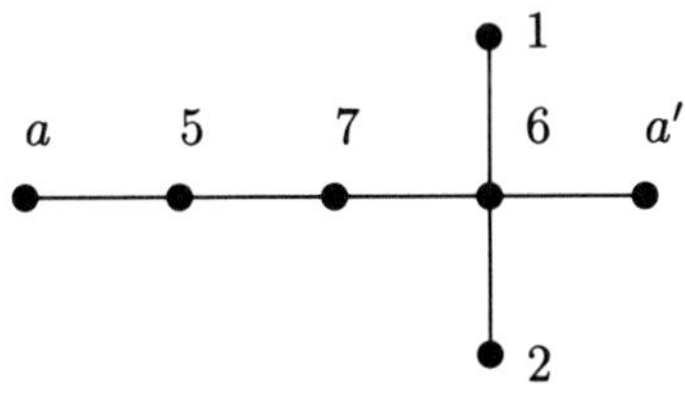

$\{a, a'\} = \{3, 4\}$

PROJECTIVES:

Nr.:	1	2	3	4	5	6	7
CC:	2	1	r	r	r	r	r
N&C:	×	×	×	×	∘	∘	×
$\chi_{39} \otimes \chi_4$		1				1	
$\chi_{43} \otimes \chi_2$					1	2	3

Since we have made a choice in the principal block, we have to allow for the two possibilities above.

Group: 2RU Prime: 13 Block: 1

Nr.	CAS-Nr.	Degree	CC	N&C
1	1	1	r	×
2	2	378	3	×
3	3	378	2	×
4	7	3654	r	×
5	9	21750	r	×
6	11	27000	r	∘
7	12	27000	r	∘
8	13	27000	r	∘
9	14	27405	r	×
10	28	91350	r	∘
11	32	102400	r	∘
12	34	110592	r	×
13	35	110592	r	×

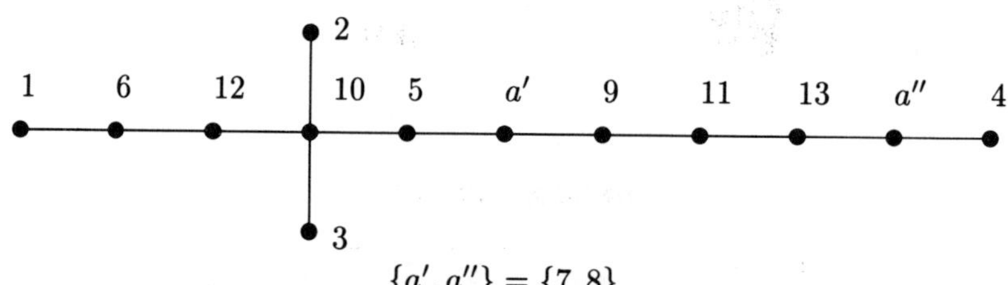

$\{a', a''\} = \{7, 8\}$

PROJECTIVES:

Nr.:	1	2	3	4	5	6	7	8	9	10	11	12	13
CC:	r	3	2	r	r	r	r	r	r	r	r	r	r
N&C:	×	×	×	×	×	∘	∘	∘	×	∘	∘	×	×
$\chi_{37} \otimes \chi_{41}$			1							1			
$\chi_{37} \otimes \chi_{58}$				1	1	1	1	1		1	2	2	2
χ_{39}^{2+}					1	1	1	1				1	1
$\chi_{41} \otimes \chi_{39}$					1					4	3	3	3
$\chi_5 \otimes \chi_6$						1	1	1	2	1	2	2	2
χ_{39}^{2-}									1		1		
$\chi_2 \otimes \chi_6$										1	1	1	1

We have two orbits of algebraically conjugate characters, which have independent irrationalities. The first orbit, of length three, consists of the characters χ_{11}, χ_{12} and χ_{13} of degree 27000 (nodes 6, 7 and 8). We assume without loss of generality that node 6 comes first on the tree. The second orbit contains the characters χ_{34} and χ_{35} (nodes 12 and 13). If we assume that node 12 comes first, then only the two trees above are consistent with all the projectives.

Group: 2RU Prime: 13 Block: 2

Nr.	CAS-Nr.	Degree	CC	N&C
1	4	406	r	×
2	5	783	r	×
3	17	43848	r	∘
3	18	43848	r	∘
3	19	43848	r	∘
4	26	76125	r	∘
5	36	118784	r	×

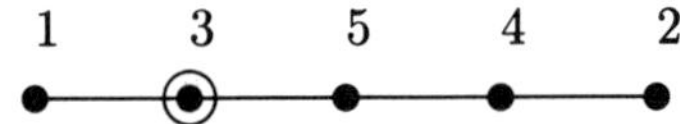

PROJECTIVES:

Nr.:	1	2	3	4	5
CC:	r	r	r	r	r
N&C:	×	×	∘	∘	×
χ_{39}^{2+}	1		1		
$\chi_{38} \otimes \chi_{51}$		1		1	
$\chi_{40} \otimes \chi_{39}$			1		1
$\chi_2 \otimes \chi_6$				1	1

Group: 2RU Prime: 13 Block: 3

Nr.	CAS-Nr.	Degree	CC	N&C
1	37	28	38	×
2	38	28	37	×
3	43	4032	44	×
4	44	4032	43	×
5	47	7308	48	×
6	48	7308	47	×
7	49	8192	r	×
8	50	8192	r	×
9	54	38976	55	×
10	55	38976	54	×
11	57	87696	r	∘
12	60	221184	r	×
13	61	250560	r	∘

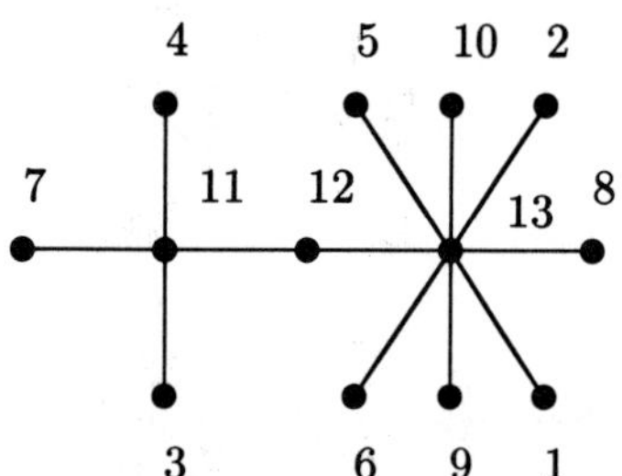

PROJECTIVES:

Nr.:	1	2	3	4	5	6	7	8	9	10	11	12	13
CC:	2	1	4	3	6	5	r	r	10	9	r	r	r
N&C:	×	×	×	×	×	×	×	×	×	×	∘	×	∘
$\chi_{38} \otimes \chi_6$				1							1		
$\mathrm{Ind}_{2\mathrm{Tits2}}(\chi_{37})$					1								1
$\chi_{38} \otimes \chi_{21}$										1		2	3
$\chi_4 \otimes \chi_{39}$											1	1	

Remark: χ_{37} is a faithful character of degree 78, i.e. of 13-defect 0, of the subgroup of shape 2Tits2.

The following trees are consistent with the above set of projectives:

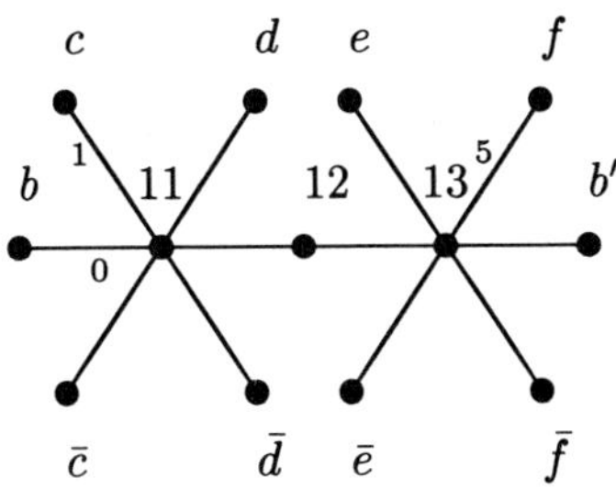

(here, $\{b, b'\} = \{7, 8\}$, $\{c, \bar{c}, d, \bar{d}\} = \{1, 2, 3, 4\}$ and $\{e, \bar{e}, f, \bar{f}\} = \{5, 6, 9, 10\}$)

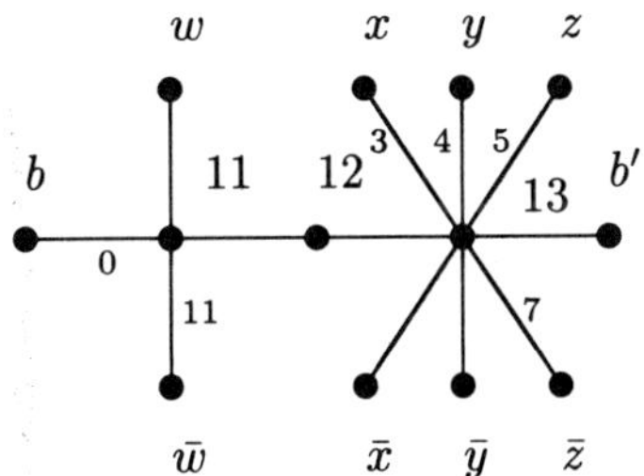

(here, $\{b, b'\} = \{7, 8\}$, $\{w, \bar{w}\} = \{3, 4\}$ and $\{x, \bar{x}, y, \bar{y}, z, \bar{z}\} = \{1, 2, 5, 6, 9, 10\}$).

To determine which tree is correct and to find its planar embedding, we shall use the following tensor products:

$$
\begin{aligned}
\chi_{37} \otimes \chi_{37} &\approx_A \chi_2 \\
\chi_{37} \otimes \chi_{50} &\approx_A \chi_{34} \\
\chi_{37} \otimes \chi_{43} &\approx_A \chi_7 \\
\chi_{37} \otimes \chi_{47} &\approx_A \chi_3 \\
\chi_{37} \otimes \chi_{55} &\approx_A \chi_{32} + \chi_{34} + \chi_{35}
\end{aligned}
$$

The symbol $\approx_A$ denotes the restriction of the characters to the principal block. For the convenience of the reader, we list these tensor products in terms of nodes:

$$
\begin{aligned}
1_B \otimes 1_B &\approx_A 2_A \\
1_B \otimes 8_B &\approx_A 12_A \\
1_B \otimes 3_B &\approx_A 4_A \\
1_B \otimes 5_B &\approx_A 3_A \\
1_B \otimes 10_B &\approx_A 11_A + 12_A + 13_A
\end{aligned}
$$

We shall need the following labelled edges in the tree for the principal block:

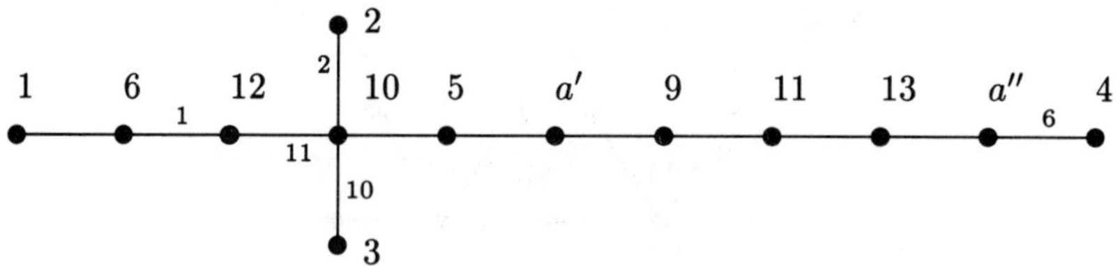

Let $1B0$ denote the Green correspondent of the edge incident to node b. Since the corresponding module is self-dual, we have $1B0 \otimes 1B0 = 1A0$. We label the edges of the Brauer tree according to this convention. Assume that the edge incident to 1_B has label n. From $1Bn \otimes 1Bn = 1A2n$ and the first tensor product we have that $2n$ equals 2. Since we have to calculate modulo 12, either $n = 1$ or $n = 7$. Suppose the first type of the above possibilities is correct. Then n equals 1. The third tensor product and $1B1 \otimes 1Bm = 1A(m+1)$ now imply that the edge meeting 3_B must have label 5. This is impossible, showing that the second type of tree must be correct, that $n = 7$ and that $z = 2_B$. Furthermore it shows that the edge meeting 3_B must have label 11. Hence $w = 4_B$.

It is now clear from the second tensor product that node 8_B cannot be incident to an edge with label 0. Thus we have $b = 7$ and $b' = 8$. The other tensor products determine the embedding of the remaining nodes. The easy proof is left as an exercise.

Group: 2RU   Prime: 29   Block: 1

Nr.	CAS-Nr.	Degree	CC	N&C
1	1	1	r	×
2	2	378	3	×
3	3	378	2	×
4	6	3276	r	∘
5	8	20475	r	×
6	11	27000	r	×
7	12	27000	r	×
8	13	27000	r	×
9	15	34944	r	∘
10	16	34944	r	∘
11	20	45500	r	∘
12	30	98280	r	∘
13	31	98280	r	∘
14	32	102400	r	×
15	34	110592	r	×
15	35	110592	r	×

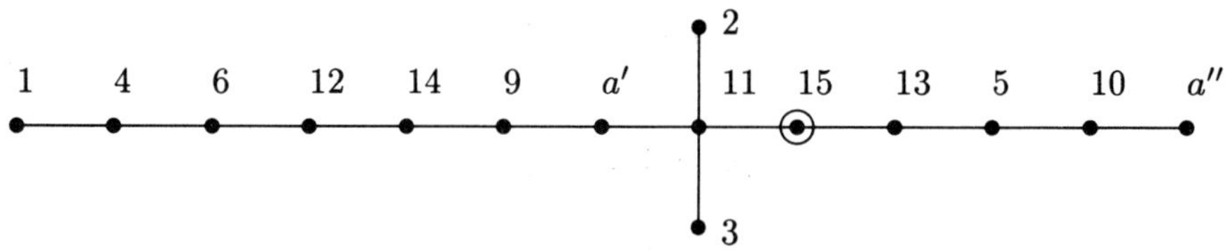

2

1 4 6 12 14 9 a' 11 15 13 5 10 a''

3

$\{a', a''\} = \{7, 8\}$

PROJECTIVES:

Nr.:	1	2	3	4	5	6	7	8	9	10	11	12	13	14	15
CC:	r	3	2	r	r	r	r	r	r	r	r	r	r	r	r
N&C:	×	×	×	∘	×	×	×	×	∘	∘	∘	∘	∘	×	×
Φ_{11}^{2+}	2			5	1	3	3	3	4	4	3	2	2	5	3
Φ_{11}	1			1											
χ_{47}^{2-}		2	1		7	2	2	2	5	5	9	18	18	21	18
$\chi_2 \otimes \chi_{14}$			1		3	1	1	1	2	2	4	7	7	8	7
$\chi_3 \otimes \chi_5$			1								1				
$\chi_{15} \otimes \chi_4$				1	2	4	4	4	5	6	5	9	9	10	11
$\chi_6 \otimes \chi_4$				1		1	1	1	1	1	2			1	1
$\chi_{42} \otimes (\chi_{37} + \chi_{43})$					3	1	1	1	2	2	4	9	9	10	10
χ_7^{2-}						2	2	2	2	2	2	5	5	4	6
$\chi_2 \otimes \chi_7$												1	1	1	1

Φ_{11}: $\chi_4 \otimes \chi_4$

We have three orbits of algebraic conjugate characters in this block. The members of any two distinct orbits generate disjoint fields. The first orbit contains the characters χ_{11}, χ_{12} and χ_{13} (nodes 6, 7 and 8). We assume without loss of generality that node 6 comes first on the tree. The second and third orbit contain the characters χ_{15} and χ_{16} (nodes 9 and 10) resp. χ_{30} and χ_{31} (nodes 12 and 13). Again we assume that node 9 resp. node 12 appear on the tree left of their conjugates. Then the following four trees are consistent with the above projectives:

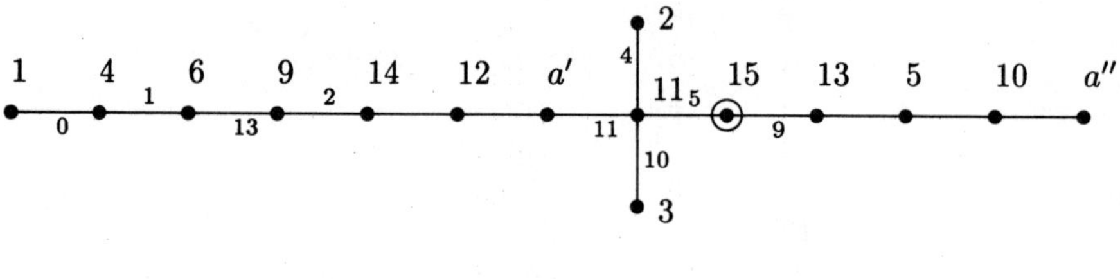

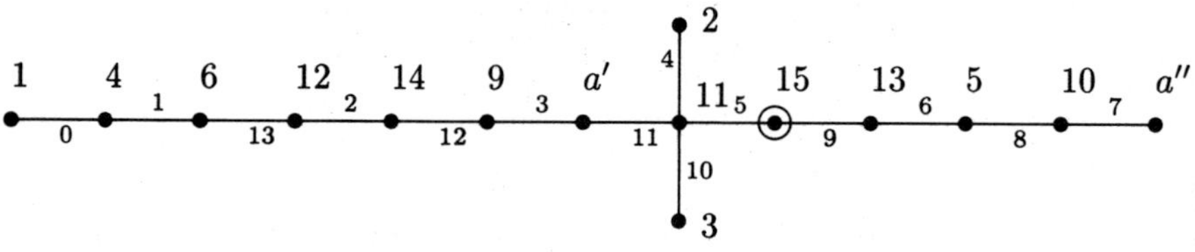

$$\{a', a''\} = \{7, 8\}.$$

We shall show in the proof for Block 2 that node 9 must be incident to an edge with label 12. This rules out the first two of these possibilities. We shall make use of the Green correspondence labelling as indicated above in the proof for the faithful block.

Group: 2RU Prime: 29 Block: 2

Nr.	CAS-Nr.	Degree	CC	N&C
1	37	28	38	∘
2	38	28	37	∘
3	39	1248	40	×
4	40	1248	39	×
5	41	3276	42	∘
6	42	3276	41	∘
7	43	4032	44	×
8	44	4032	43	×
9	45	7280	46	×
10	46	7280	45	×
11	49	8192	r	∘
11	50	8192	r	∘
12	53	34944	r	∘
13	58	98280	59	∘
14	59	98280	58	∘
15	60	221184	r	×

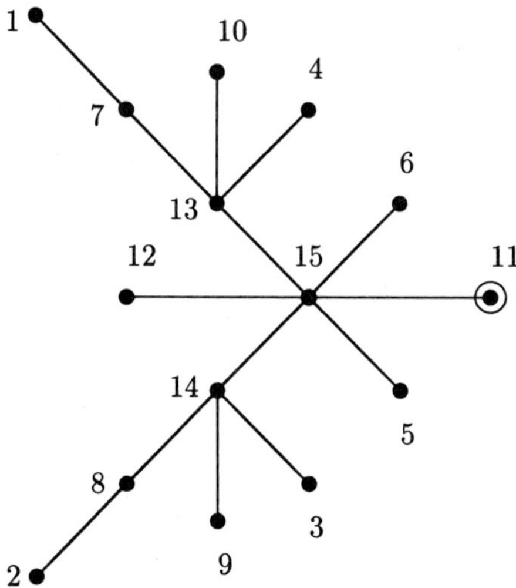

This planar embedded tree is consistent with the choices made in the principal block.

PROJECTIVES:

Nr.:	1	2	3	4	5	6	7	8	9	10	11	12	13	14	15
CC:	2	1	4	3	6	5	8	7	10	9	r	r	14	13	r
N&C:	∘	∘	×	×	∘	∘	×	×	×	×	∘	∘	∘	∘	×
$\chi_{37} \otimes \chi_4$		1						1							
$\chi_{39} \otimes \chi_4$				1									1	1	1
$\chi_{41} \otimes \chi_4$						1						1			2
$\chi_{37} \otimes \chi_7$								1						1	
$\chi_{45} \otimes \chi_4$									1	1			3	3	4
$\chi_{37} \otimes \chi_9$														1	1

These projectives show that all nodes in $\{1,2,3,4,5,6,9,10,11,12\}$ are leaves. The leaves on the real stem are 11 and 12, the exceptional node is 11. The leaves 1, 2, 5, 6, 11, and 12 are of type nought, the others of type cross.

We shall make use of the following tensor products:

$$\begin{array}{lll}
\chi 37 \otimes \chi_{37} & \approx_A & \chi_2 \\
\chi 37 \otimes \chi_{39} & \approx_A & \chi_{15} \\
\chi 37 \otimes \chi_{40} & \approx_A & \chi_{16} \\
\chi 37 \otimes \chi_{41} & \approx_A & \chi_3 \\
\chi 37 \otimes \chi_{43} & \approx_A & \chi_{20} \\
\chi 37 \otimes \chi_{44} & \approx_A & \chi_6 \\
\chi 37 \otimes \chi_{45} & \approx_A & \chi_{30} \\
\chi 37 \otimes \chi_{53} & \approx_A & \chi_{20} + \chi_{32} + \chi_{34} + \chi_{35}
\end{array}$$

Recall that $\approx_A$ denotes the characters restricted to the principal block. For the convenience of the reader we give these tensor products in terms of nodes:

$$\begin{array}{lll}
1_B \otimes 1_B & \approx_A & 2_A \\
1_B \otimes 3_B & \approx_A & 9_A \\
1_B \otimes 4_B & \approx_A & 10_A \\
1_B \otimes 5_B & \approx_A & 3_A \\
1_B \otimes 7_B & \approx_A & 11_A \\
1_B \otimes 8_B & \approx_A & 4_A \\
1_B \otimes 9_B & \approx_A & 12_A \\
1_B \otimes 12_B & \approx_A & 11_A + 14_A + 15_A
\end{array}$$

Let $28B0$ denote the Green correspondent of the leaf 1_B. Then from the first tensor product and the labels in the principal block we have $28B0 \otimes 28B0 = 1A4$. Let $28Bn$ be the Green correspondent of the dual module. Then $1A0 = 28B0 \otimes 28Bn = (28B0 \otimes 28B0) \otimes 1An$. This implies $n = 10$. Since only two edges are incident to node 7, we now have the following two subtrees with labelling according to our assumption.

It is now clear that the two leaves of the real stem must have the labels 12 and 5 resp. and that the remaining leaves have labels between 1 and 8.

Suppose the leaf of the real stem with label 5 is node 12_B. From $28B0 \otimes 28B5 = 1A9$ and the last tensor product, we conclude that $11_A + 14_A$ is projective. This contradicts all four possibilities for the tree in the principal block. Hence the leaf 12_B has label 12. The fourth of the above tensor products shows that node 5_B has label 6.

We are going to determine $28B0 \otimes 1B0 = 28An$. From $1_B \otimes 7_B \approx_A 11_A$ we conclude that $n \in \{4, 5, 10, 11\}$. From $1_B \otimes 8_B \approx_A 4_A$ and $28B0 \otimes 1B9 = (28B0 \otimes 1B0) \otimes 1A9$ we conclude that $n \in \{5, 6\}$. It follows $28B0 \otimes 1B0 = 28A5$.

The third tensor product now readily shows that leaf 4_B has label 2. But then its complex conjugate has label 7 and the second product implies that node 9_A must be incident to an edge with label 12. This completes the proof for the principal block.

Finally, the remaining tensor product shows that node 9_B must have label 8. We have now determined the labels of all edges, so the planar embedded tree is determined.

6.13 The Covering Group of the Suzuki Group

Group: SUZ Prime: 3 Block: 3

Nr.	CAS-Nr.	Degree	CC	N&C
1	16	18954	r	×
2	38	189540	r	×
3	41	208494	r	o

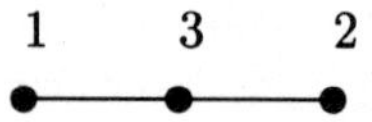

1 3 2

Remark: The inverse image of this block in the covering group has a non-cyclic defect group.

Group: 6SUZ Prime: 3 Block: 5

Nr.	CAS-Nr.	Degree	CC	N&C
1	60	61236	r	×
2	187	61236	189	o
2	189	61236	187	o

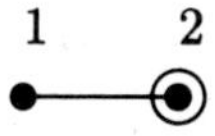

1 2

Group: 6SUZ Prime: 3 Block: 6

Nr.	CAS-Nr.	Degree	CC	N&C
1	61	61236	r	×
2	188	61236	190	o
2	190	61236	188	o

1 2

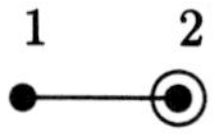

Group: 6SUZ Prime: 5 Block: 2

Nr.	CAS-Nr.	Degree	CC	N&C
1	4	780	r	×
2	13	15015	r	×
2	14	15015	r	×
3	15	15795	r	∘

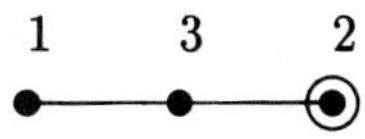

1 3 2

Group: 6SUZ Prime: 5 Block: 3

Nr.	CAS-Nr.	Degree	CC	N&C
1	7	5005	8	×
2	8	5005	7	×
3	31	93555	r	×
4	32	93555	r	×
5	40	197120	r	∘

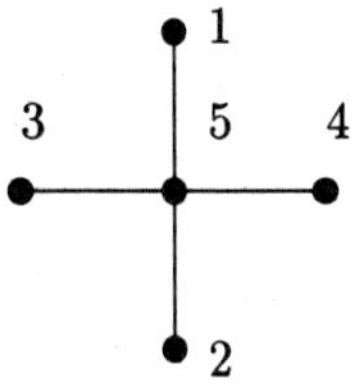

1
3 5 4
2

Group: 6SUZ Prime: 5 Block: 4

Nr.	CAS-Nr.	Degree	CC	N&C
1	9	5940	r	×
2	20	40040	r	×
3	27	66560	r	∘
4	37	168960	r	∘
5	38	189540	r	×

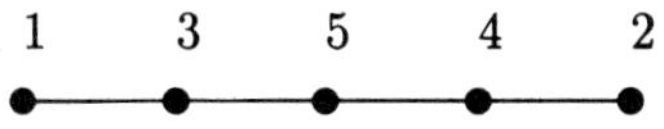

1 3 5 4 2

PROJECTIVES:

Nr.	1	2	3	4	5
CC	r	r	r	r	r
N&C	×	×	∘	∘	×
$\chi_2 \otimes \chi_{10}$	1		1		
$\chi_2 \otimes \chi_{28}$		3		5	2
$\chi_2 \otimes \chi_{17}$			2	1	3

Group: 6SUZ Prime: 5 Block: 5

Nr.	CAS-Nr.	Degree	CC	N&C
1	44	220	r	∘
2	53	20020	r	∘
3	64	80080	r	×
4	69	137280	r	×
5	73	197120	r	∘

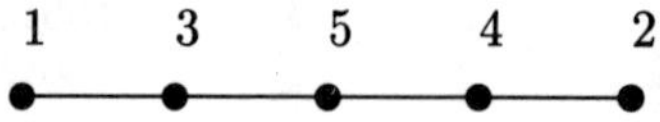

1 — 3 — 5 — 4 — 2

PROJECTIVES:

Nr.	1	2	3	4	5
CC	r	r	r	r	r
N&C	∘	∘	×	×	∘
$\chi_2 \otimes \chi_{56}$		2		3	1
$\chi_2 \otimes \chi_{54}$			1	1	2

Group: 6SUZ Prime: 5 Block: 7

Nr.	CAS-Nr.	Degree	CC	N&C
1	52	20020	r	∘
2	58	60060	r	∘
2	59	60060	r	∘
3	65	80080	r	×

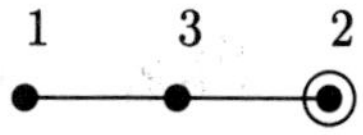

1 — 3 — 2 (circled)

Group: 6SUZ Prime: 5 Block: 10

Nr.	CAS-Nr.	Degree	CC	AC	N&C
1	83	1365	84	ar	×
2	95	4290	96	ar	×
3	119	60060	120	ar	∘
4	135	135135	136	ar	∘
5	143	189540	144	ar	×

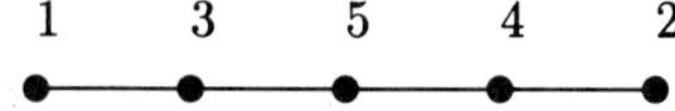

PROJECTIVES:

Nr.	1	2	3	4	5
AC	ar	ar	ar	ar	ar
N&C	×	×	∘	∘	×
$\chi_2 \otimes \chi_{117}$		1		5	4
$\chi_2 \otimes \chi_{121}$			1	6	7

Group: 6SUZ Prime: 5 Block: 11

Complex conjugate to Block 10.

Group: 6SUZ Prime: 5 Block: 12

Nr.	CAS-Nr.	Degree	CC	AC	N&C
1	87	2145	88	ar	×
2	101	6720	102	ar	×
3	139	160380	140	ar	∘
4	141	180180	142	ar	∘
5	151	331695	152	ar	×

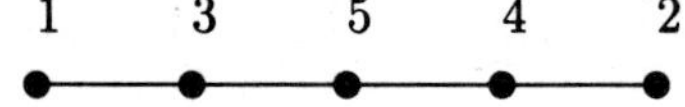

PROJECTIVES:

Nr.	1	2	3	4	5
AC	ar	ar	ar	ar	ar
N&C	×	×	∘	∘	×
$\chi_2 \otimes \chi_{115}$		1	1	5	5
$\chi_{153} \otimes \chi_{56}$		1		1	
$\chi_2 \otimes \chi_{107}$			2		2

Group: 6SUZ Prime: 5 Block: 13

Complex conjugate to Block 12.

Group: 6SUZ Prime: 5 Block: 16

Nr.	CAS-Nr.	Degree	CC	AC	N&C
1	155	780	156	ar	∘
2	163	8580	164	ar	∘
3	191	77220	192	ar	×
4	195	112320	196	ar	×
5	205	180180	206	ar	∘

PROJECTIVES:

Nr.	1	2	3	4	5
CC	r	r	r	r	r
N&C	∘	∘	×	×	∘
$\chi_2 \otimes \chi_{185}$	1	3	4	1	1
$\chi_2 \otimes \chi_{181}$			1	2	3

Group: 6SUZ Prime: 5 Block: 17

Complex conjugate to Block 16.

Group: 6SUZ Prime: 7 Block: 1

Nr.	CAS-Nr.	Degree	CC	N&C
1	1	1	r	×
2	10	10725	r	×
3	12	14300	r	∘
4	35	146432	r	∘
5	37	168960	r	×
6	38	189540	r	×
7	41	208494	r	∘

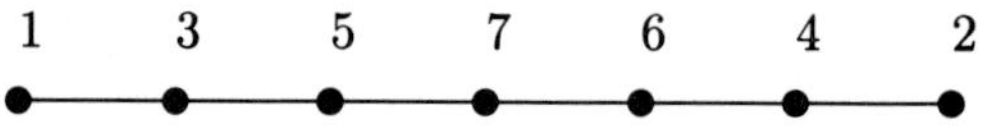

1 — 3 — 5 — 7 — 6 — 4 — 2

PROJECTIVES:

Nr.	1	2	3	4	5	6	7
CC	r	r	r	r	r	r	r
N&C	×	×	∘	∘	×	×	∘
$\chi_2 \otimes \chi_7$		1		1			
$\chi_2 \otimes \chi_{11}$			1		1		
$\chi_2 \otimes \chi_{21}$				2	3	3	4
$\chi_2 \otimes \chi_{13}$					1		1

Group: 6SUZ Prime: 7 Block: 2

Nr.	CAS-Nr.	Degree	CC	N&C
1	2	143	r	×
2	4	780	r	×
3	9	5940	r	∘
4	15	15795	r	×
5	27	66560	r	∘
6	39	193050	r	∘
7	43	248832	r	×

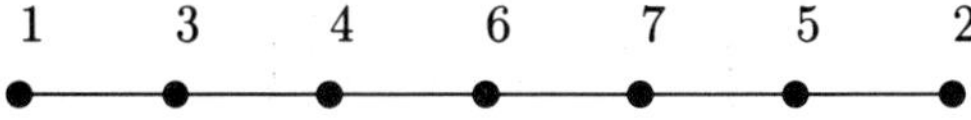

PROJECTIVES:

Nr.	1	2	3	4	5	6	7
CC	r	r	r	r	r	r	r
N&C	×	×	∘	×	∘	∘	×
$\chi_2 \otimes \chi_7$	1	1	1		1		
$\chi_{153} \otimes \chi_{158}$	1		1				
$\chi_2 \otimes \chi_{11}$			1	1		1	1
$\chi_2 \otimes \chi_{20}$				1		4	3
$\chi_2 \otimes \chi_{17}$					2		2

Group: 6SUZ Prime: 7 Block: 3

Nr.	CAS-Nr.	Degree	CC	N&C
1	6	3432	r	×
2	16	18954	r	∘
3	25	64350	r	∘
3	26	64350	r	∘
4	29	79872	r	×

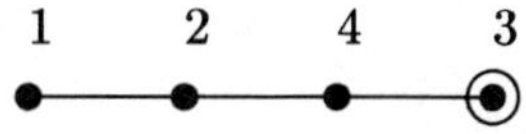

PROJECTIVES:

Nr.	1	2	3	4
CC	r	r	r	r
N&C	×	∘	∘	×
$\chi_2 \otimes \chi_7$	2	2		

Group: 6SUZ Prime: 7 Block: 4

Nr.	CAS-Nr.	Degree	CC	N&C
1	44	220	r	∘
2	54	35100	r	×
2	55	35100	r	×
3	67	102400	r	×
4	69	137280	r	∘

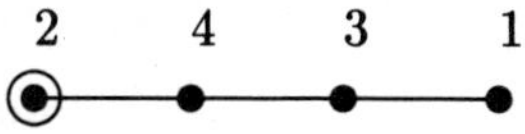

PROJECTIVES:

Nr.	1	2	3	4
CC	r	r	r	r
N&C	∘	×	×	∘
$\chi_2 \otimes \chi_{49}$	2		2	

Group: 6SUZ Prime: 7 Block: 5

Nr.	CAS-Nr.	Degree	CC	N&C
1	47	572	48	∘
2	48	572	47	∘
3	56	35100	57	×
4	57	35100	56	×
5	62	79872	r	×
6	63	79872	r	×
7	74	228800	r	∘

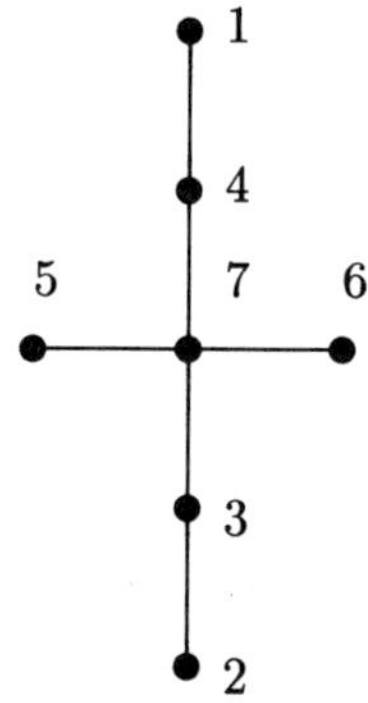

PROJECTIVES:

Nr.	1	2	3	4	5	6	7
CC	2	1	4	3	r	r	r
N&C	∘	∘	×	×	×	×	∘
$\chi_2 \otimes \chi_{45}$	1			1			
$\chi_2 \otimes \chi_{50}$			1				1

Group: 6SUZ Prime: 7 Block: 6

Nr.	CAS-Nr.	Degree	CC	AC	N&C
1	77	66	78	ar	×
2	87	2145	88	ar	×
3	113	30888	114	ar	∘
4	115	42900	116	ar	∘
5	129	104247	130	ar	×
6	139	160380	140	ar	×
7	145	193050	146	ar	∘

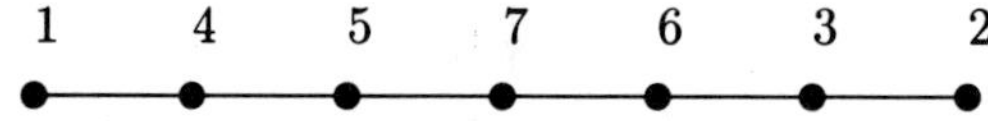

PROJECTIVES:

Nr.	1	2	3	4	5	6	7
AC	ar	ar	ar	ar	ar	ar	ar
N&C	×	×	∘	∘	×	×	∘
$\chi_2 \otimes \chi_{101}$	1	2	2	1			
$\chi_2 \otimes \chi_{131}$		1	5	2	3	7	4
$\chi_2 \otimes \chi_{117}$		1	3	2	2	3	1
$\chi_2 \otimes \chi_{93}$		1	1				
$\chi_{77} \otimes \chi_{17}$				1	1		
$\chi_2 \otimes \chi_{109}$					1	2	3

Group: 6SUZ Prime: 7 Block: 7

Complex conjugate to Block 6.

Group: 6SUZ Prime: 7 Block: 8

Nr.	CAS-Nr.	Degree	CC	AC	N&C
1	79	78	80	ar	×
2	85	1716	86	ar	×
3	95	4290	96	ar	∘
4	121	64350	122	ar	∘
5	127	85800	128	ar	×
6	143	189540	144	ar	×
7	147	208494	148	ar	∘

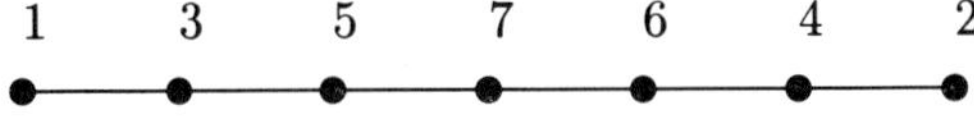

PROJECTIVES:

Nr.	1	2	3	4	5	6	7
AC	ar	ar	ar	ar	ar	ar	ar
N&C	×	×	∘	∘	×	×	∘
$\chi_2 \otimes \chi_{101}$	1	1	1	1			
$\chi_{77} \otimes \chi_8$	1		1				
$\chi_2 \otimes \chi_{117}$		1	1	4	1	4	1
$\chi_2 \otimes \chi_{93}$		1		1			
$\chi_2 \otimes \chi_{111}$			1		2		1
$\chi_2 \otimes \chi_{97}$						1	1

Group: 6SUZ Prime: 7 Block: 9

Complex conjugate to Block 8.

Group: 6SUZ Prime: 7 Block: 10

Nr.	CAS-Nr.	Degree	CC	AC	N&C
1	81	429	82	ar	×
2	89	2925	91	90	∘
2	90	2925	92	89	∘
3	103	18954	104	ar	∘
4	105	21450	106	ar	×

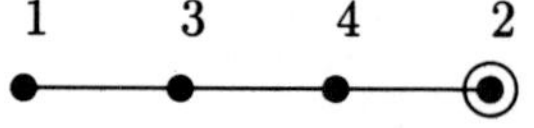

PROJECTIVES:

Nr.	1	2	3	4
AC	ar	ar	ar	ar
N&C	×	∘	∘	×
$\chi_2 \otimes \chi_{101}$	1		3	2

Group: 6SUZ Prime: 7 Block: 11

Complex conjugate to Block 10.

Group: 6SUZ Prime: 7 Block: 12

Nr.	CAS-Nr.	Degree	CC	AC	N&C
1	153	12	154	ar	∘
2	163	8580	164	ar	∘
3	177	27456	179	ar	×
4	178	27456	180	ar	×
5	193	105600	194	ar	∘
6	195	112320	196	ar	∘
7	203	171600	204	ar	×

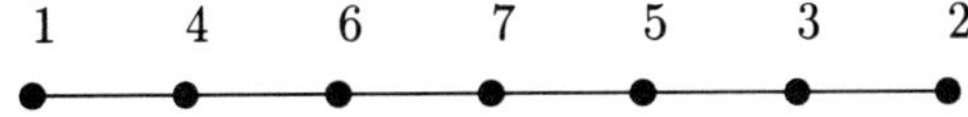

PROJECTIVES:

Nr.	1	2	3	4	5	6	7
AC	ar	ar	ar	ar	ar	ar	ar
N&C	∘	∘	×	×	∘	∘	×
$\chi_2 \otimes \chi_{157}$	1	1	1	1			
$\chi_2 \otimes \chi_{199}$		3	7	1	5	2	2
$\chi_2 \otimes \chi_{159}$		2	2				
$\chi_2 \otimes \chi_{175}$		1	2	1	2	1	1
$\chi_2 \otimes \chi_{160}$				1		2	1
$\chi_2 \otimes \chi_{165}$					1	1	2

Group: 6SUZ Prime: 7 Block: 13

Complex conjugate to Block 12.

Group: 6SUZ Prime: 7 Block: 14

Nr.	CAS-Nr.	Degree	CC	AC	N&C
1	155	780	156	ar	∘
2	181	35100	183	182	×
2	182	35100	184	181	×
3	185	42900	186	ar	×
4	191	77220	192	ar	∘

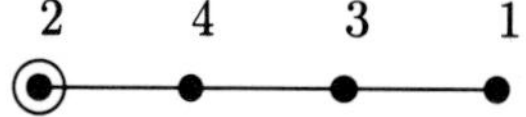

PROJECTIVES:

Nr.	1	2	3	4
CC	r	r	r	r
N&C	∘	×	×	∘
$\chi_2 \otimes \chi_{157}$	1		1	

Group: 6SUZ Prime: 7 Block: 15

Complex conjugate to Block 14.

Group: 6SUZ Prime: 11 Block: 1

Nr.	CAS-Nr.	Degree	CC	N&C
1	1	1	r	×
2	3	364	r	×
3	4	780	r	∘
4	15	15795	r	∘
5	16	18954	r	×
6	27	66560	r	∘
7	29	79872	r	×
8	30	88452	r	×
9	36	163800	r	∘
10	38	189540	r	∘
11	43	248832	r	×

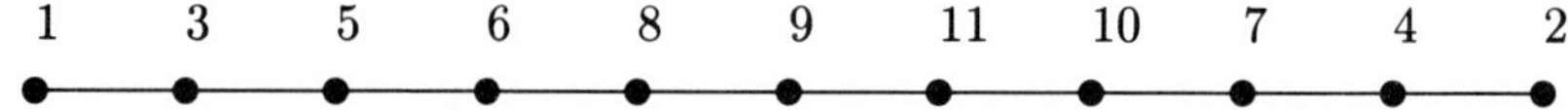

PROJECTIVES:

Nr.	1	2	3	4	5	6	7	8	9	10	11
CC	r	r	r	r	r	r	r	r	r	r	r
N&C	×	×	∘	∘	×	∘	×	×	∘	∘	×
$\chi_2 \otimes \chi_2$	1		1								
$\chi_6 \otimes \chi_2$			2		3	1					
$\chi_{16} \otimes \chi_2$			1		4	3	1			2	1
$\chi_{12} \otimes \chi_2$						1		1		1	1
$\chi_{13} \otimes \chi_2$								1	2		1

Group: 6SUZ Prime: 11 Block: 2

Nr.	CAS-Nr.	Degree	CC	N&C
1	45	364	46	×
2	46	364	45	×
3	54	35100	r	∘
4	55	35100	r	∘
5	56	35100	57	∘
6	57	35100	56	∘
7	60	61236	r	∘
8	61	61236	r	∘
9	62	79872	r	×
10	63	79872	r	×
11	67	102400	r	×

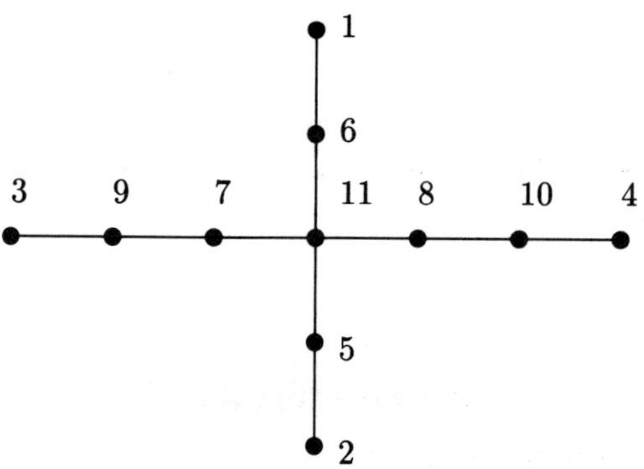

PROJECTIVES:

Nr.	1	2	3	4	5	6	7	8	9	10	11
CC	2	1	r	r	6	5	r	r	r	r	r
N&C	×	×	∘	∘	∘	∘	∘	∘	×	×	×
$\chi_{45} \otimes \chi_2$	1					1					
$\chi_{54} \otimes \chi_2$			2	2					2	2	
$\chi_{65} \otimes \chi_2$					3	3	1	1			8
$\chi_{58} \otimes \chi_2$					1	1	2	2	1	1	4
$\chi_{60} \otimes \chi_2$							3	3	2	2	2

The real valued pairs of algebraically conjugate characters have their values in pairwise disjoint fields. Hence, without loss of generality, there is just one tree which is consistent with all the projectives.

Group: 6SUZ Prime: 11 Block: 3

Nr.	CAS-Nr.	Degree	CC	AC	N&C
1	79	78	80	ar	×
2	83	1365	84	ar	×
3	89	2925	91	90	∘
4	90	2925	92	89	∘
5	97	5103	99	98	∘
6	98	5103	100	97	∘
7	101	6720	102	ar	∘
8	103	18954	104	ar	×
9	137	139776	138	ar	∘
10	143	189540	144	ar	∘
11	151	331695	152	ar	×

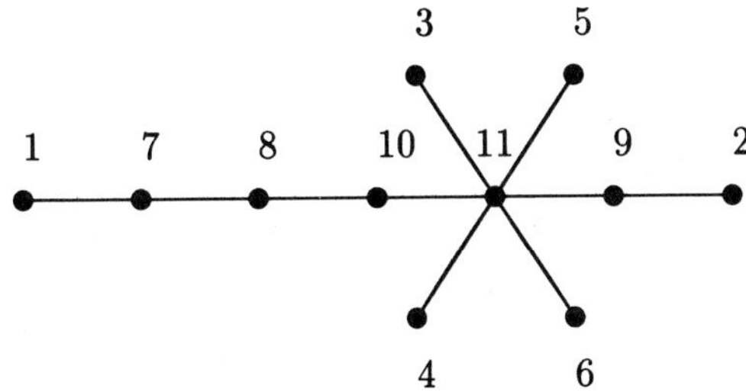

This planar embedded tree is consistent with the choice made in Block 2. The proof is given below.

PROJECTIVES:

Nr.	1	2	3	4	5	6	7	8	9	10	11
AC	ar	ar	4	3	6	5	ar	ar	ar	ar	ar
N&C	×	×	∘	∘	∘	∘	∘	×	∘	∘	×
$\chi_{77} \otimes \chi_2$	1						1				
$\chi_{83} \otimes \chi_2$		1							1		
$\chi_{145} \otimes \chi_2$			1	1					9	8	19
$\chi_{143} \otimes \chi_2$					1	1		2	6	16	22
$\chi_{147} \otimes \chi_2$					1	1			9	12	23
$\chi_{103} \otimes \chi_2$							3	4		2	1
$\chi_{81} \otimes \chi_2$							1	1			

To prove the planar embedding and consistency, we use the following tensor products:

$$\begin{array}{lll} \chi_{153} \otimes \chi_{80} & \approx_B & \chi_{45} \\ \chi_{153} \otimes \chi_{91} & \approx_B & \chi_{54} \\ \chi_{153} \otimes \chi_{99} & \approx_B & \chi_{60} \end{array}$$

As usual, $\approx_B$ denotes the restriction of the characters to Block 2. For the convenience of the reader we give these tensor products in terms of nodes:

$$\begin{array}{lll} 1_E \otimes 1_D & \approx_B & 1_B \\ 1_E \otimes 3_D & \approx_B & 3_B \\ 1_E \otimes 5_D & \approx_B & 7_B \end{array}$$

Let $1E0$ denote the Green correspondent of node 1_E, i.e. the 12-dimensional character with CAS-number 153. Similarly, $1D0$ denotes the Green correspondent of node 1_D in Block 4, dual to Block 3. The first tensor product shows that the component of $1_E \otimes 1_D$ in the second block is the irreducible trivial source character at node 1_B. If $1B0$ is its Green correspondent we have $1E0 \otimes 1D0 \approx_B 1B0$. Hence we can label the trees for Block 2 and Block 4 as follows:

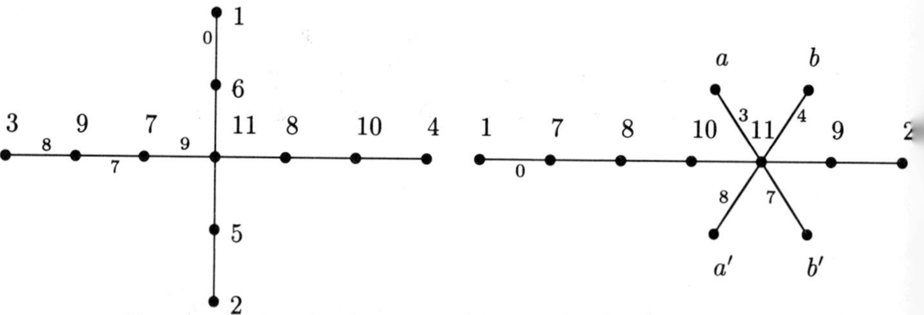

Here, $\{a, a', b, b'\} = \{3_D, 4_D, 5_D, 6_D\}$.

Let x be the label on the edge joining 3_D with 11_D. Since the only label on an edge incident to node 3_B is 8, it follows from $1E0 \otimes 1Dx \approx_B 1Bx$ that $x = 8$. Hence $a' = 3_D$. The edges incident to 7_B carry the labels 7 and 9. Hence $b' = 5_D$. Dualizing gives the result for Block 3.

Group: 6SUZ Prime: 11 Block: 4

Complex conjugate to Block 3.

Group: 6SUZ Prime: 11 Block: 5

Nr.	CAS-Nr.	Degree	CC	AC	N&C
1	153	12	154	ar	×
2	155	780	156	ar	∘
3	159	4368	161	ar	×
4	160	4368	162	ar	×
5	181	35100	183	182	∘
6	182	35100	184	181	∘
7	187	61236	189	188	∘
8	188	61236	190	187	∘
9	195	112320	196	ar	∘
10	197	139776	198	ar	∘
11	209	436800	210	ar	×

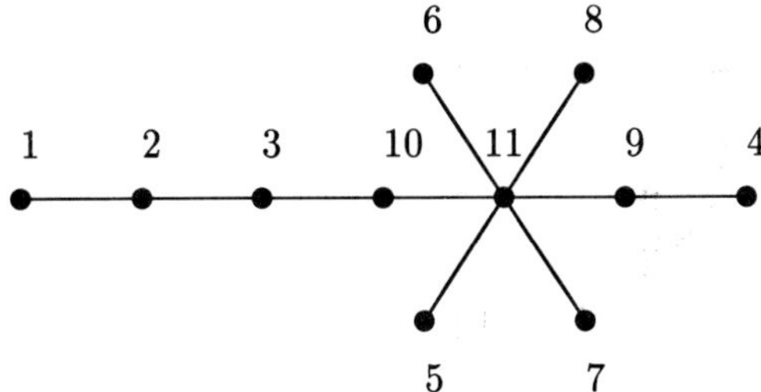

This planar embedded tree is consistent with the trees for Blocks 2, 3 and 4. A proof of this fact is given below.

PROJECTIVES:

Nr.	1	2	3	4	5	6	7	8	9	10	11
AC	ar	ar	ar	ar	6	5	8	7	ar	ar	ar
N&C	×	∘	×	×	∘	∘	∘	∘	∘	∘	×
$\chi_{155} \otimes \chi_2$	1	3	2								
$\chi_{159} \otimes \chi_2$		2	3							1	
$\chi_{160} \otimes \chi_2$				2					2		
$\chi_{181} \otimes \chi_2$					2	2			2		6
$\chi_{187} \otimes \chi_2$							3	3		3	9
$\chi_{165} \otimes \chi_2$									1	1	2

To prove the planar embedding and consistency, we use the following tensor products:

$$\begin{array}{lll} \chi_{153} \otimes \chi_{153} & \approx_D & \chi_{80} \\ \chi_{153} \otimes \chi_{181} & \approx_D & \chi_{91} \\ \chi_{153} \otimes \chi_{187} & \approx_D & \chi_{99} + \chi_{144} + \chi_{152} \end{array}$$

In terms of nodes these tensor products are as follows:

$$\begin{array}{lll} 1_E \otimes 1_E & \approx_D & 1_D \\ 1_E \otimes 5_E & \approx_D & 3_D \\ 1_E \otimes 7_E & \approx_D & 5_D + 10_D + 11_D \end{array}$$

With the notation of the proof for Block 3, the first tensor product shows $1E0 \otimes 1E0 \approx_D 1D0$. The rest of the proof is now clear. It is left to the reader as an exercise.

Group: 6SUZ Prime: 11 Block: 6

Complex conjugate to Block 5.

Group: 6SUZ Prime: 13 Block: 1

Nr.	CAS-Nr.	Degree	CC	N&C
1	1	1	r	×
2	9	5940	r	∘
3	31	93555	r	×
3	32	93555	r	×
4	34	133056	r	×
5	37	168960	r	∘
6	40	197120	r	×
7	43	248832	r	∘

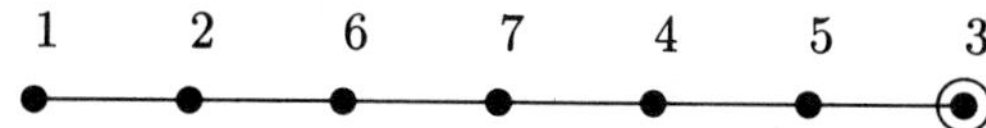

PROJECTIVES:

Nr.	1	2	3	4	5	6	7
CC	r	r	r	r	r	r	r
N&C	×	∘	×	×	∘	×	∘
$\chi_2 \otimes \chi_2$	1	1					
$\chi_{10} \otimes \chi_2$		1				2	1
$\chi_{34} \otimes \chi_2$			2	12	9	6	11

Group: 6SUZ Prime: 13 Block: 2

Nr.	CAS-Nr.	Degree	CC	N&C
1	44	220	r	∘
2	49	4928	r	×
3	60	61236	r	∘
3	61	61236	r	∘
4	67	102400	r	∘
5	73	197120	r	×
6	75	277200	r	×
7	76	315392	r	∘

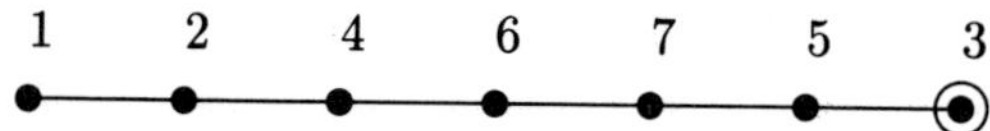

PROJECTIVES:

Nr.	1	2	3	4	5	6	7
CC	r	r	r	r	r	r	r
N&C	∘	×	∘	∘	×	×	∘
$\chi_{49} \otimes \chi_2$	2	4		2			
$\chi_{52} \otimes \chi_2$	1	4		4		1	
$\chi_{53} \otimes \chi_2$		2		4		2	
$\chi_{50} \otimes \chi_2$					1	1	2

Group: 6SUZ Prime: 13 Block: 3

Nr.	CAS-Nr.	Degree	CC	AC	N&C
1	77	66	78	ar	×
2	97	5103	99	98	×
2	98	5103	100	97	×
3	101	6720	102	ar	∘
4	107	23100	108	ar	∘
5	117	51975	118	ar	×
6	133	133056	134	ar	×
7	139	160380	140	ar	∘

PROJECTIVES:

Nr.	1	2	3	4	5	6	7
AC	ar	ar	ar	ar	ar	ar	ar
N&C	×	×	∘	∘	×	×	∘
$\chi_{77} \otimes \chi_2$	1		1				
$\chi_{143} \otimes \chi_2$		1		1	4	6	10
$\chi_{103} \otimes \chi_2$			3		4		1
$\chi_{107} \otimes \chi_2$				1		3	2

Group: 6SUZ Prime: 13 Block: 4

Complex conjugate to Block 3.

Group: 6SUZ Prime: 13 Block: 5

Nr.	CAS-Nr.	Degree	CC	AC	N&C
1	153	12	154	ar	∘
2	157	924	158	ar	×
3	165	11088	167	166	∘
4	166	11088	168	165	∘
5	175	23100	176	ar	∘
6	187	61236	189	188	∘
6	188	61236	190	187	∘
7	193	105600	194	ar	×

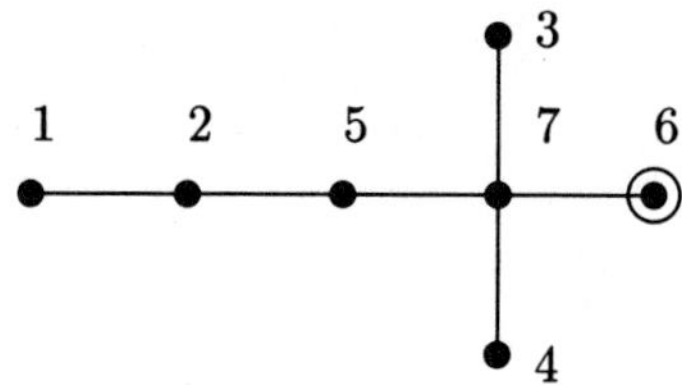

PROJECTIVES:

Nr.	1	2	3	4	5	6	7
AC	ar	ar	4	3	ar	ar	ar
N&C	∘	×	∘	∘	∘	∘	×
$\chi_{157} \otimes \chi_2$	1	2			1		
$\chi_{175} \otimes \chi_2$		1			3		2
$\chi_{165} \otimes \chi_2$				1			1
$\chi_{187} \otimes \chi_2$						3	3

Group: 6SUZ Prime: 13 Block: 6

Complex conjugate to Block 5.

6.14 The Triple Cover of the O'Nan Group

We start this section with a lemma on the normalizers of certain Sylow p-subgroups of 3ON.

Lemma 6.14.1. *Let $p \in \{11, 19, 31\}$, and let P denote a Sylow p-subgroup of 3ON. Let N denote the normalizer in 3ON of P. Then $N \cong C \times PH$, where C is the centre of 3ON and H is cyclic of order* 10, 6 *or* 15, *resp. Furthermore, there is an outer automorphism α of 3ON, which fixes N and H, centralizes H and inverts C.*

Proof. The structure of the normalizers follows from the corresponding structure in ON and the fact that there are no elements of order 18 or 45 in 3ON.

In ON2, the automorphism group of ON, the normalizer modulo the centralizer of a Sylow 31-subgroup is cylic of order 30 since the principal 31-block contains 30 irreducible Brauer characters. Hence the result follows for $p = 31$. The centralizer in ON2 of an element of type 2B is J1 $\times$ 2, a direct product of the first Janko group with a cyclic group of order 2. Since N can be chosen to be inside J1 if $p \in \{11, 19\}$, the second assertion also holds in these cases.□

Let the notation be as in Lemma 6.14.1. Let $^\#$ denote the functor on the modules of 3ON and N obtained by composing the duality operation with α. Then clearly $^\#$ commutes with Green correspondence, and fixes the three blocks of defect 1. The usual duality operation is denoted by *. It fixes the principal block and interchanges the two faithful blocks.

Let λ be the irreducible character of N of order 3 belonging (via Brauer correspondence) to Block 2. Denote by $1Bx$ the irreducible character $\lambda \times 1Ax$, and by $1Cx$ the irreducible character $\lambda^2 \times 1Ax$ of N. Then $1Bx^\# = 1B(p - 1 - x)$. We note the following lemma.

Lemma 6.14.2. *Let $p \in \{11, 19, 31\}$. Let M be an irreducible module in Block 2 with Green correspondent $1Bx$. This means that it corresponds to a leaf of the Brauer tree. Let χ be the corresponding ordinary character. Suppose $\chi \otimes \chi^\# \approx \psi$, where ψ is an ordinary irreducible character in Block 3. Then ψ^* is a leaf, i.e. ψ^* remains irreducible on reduction modulo p. Furthermore, the reduction of ψ^* has Green correspondent $1B0$.*

Proof. The Green correspondent of $M \otimes M^\#$ is $1Bx \otimes 1B(p-1-x) = 1C0$. Hence ψ^* has Green correspondent $1B0$. Obviously $\psi = \psi^\#$, so ψ is in the $^\#$-invariant stem of the Brauer tree. We shall show that the two leaves of this stem have labels $\in \{0, (p-1)/2\}$.

Let ϑ be a $^\#$-invariant irreducible ordinary character of Block 2, which occupies a leaf of the $^\#$-invariant stem. Let L be the corresponding irreducible module and suppose L has Green correspondent $1Ay$. Then $L \otimes L^*$ has Green correspondent $1By \otimes 1Cy = 1A2y$, where $2y$ has to be read modulo $p - 1$. Now 11 does not divide the dimension of L, hence the trivial module is a direct summand of $L \otimes L^*$. It follows that $1A2y = 1A0$, hence $y \in \{0, (p-1)/2\}$.

Since the #-invariant stem has two leaves which are of type cross, 0 must label a leaf. Thus ψ is at the end of the tree. The proof is complete.□

Group: 3ON Prime: 3 Block: 3

Nr.	CAS-Nr.	Degree	CC	N&C
1	21	169290	r	×
2	61	169290	62	∘
2	62	169290	61	∘

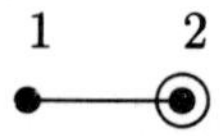

1 2

Group: 3ON Prime: 3 Block: 4

Nr.	CAS-Nr.	Degree	CC	N&C
1	22	169290	r	×
2	63	169290	64	∘
2	64	169290	63	∘

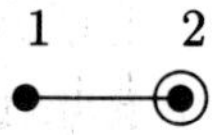

1 2

Group: 3ON Prime: 3 Block: 5

Nr.	CAS-Nr.	Degree	CC	N&C
1	25	175770	r	×
2	69	175770	70	∘
2	70	175770	69	∘

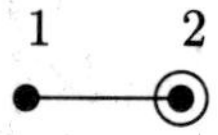

1 2

Group: 3ON Prime: 3 Block: 6

Nr.	CAS-Nr.	Degree	CC	N&C
1	26	207360	r	×
2	73	207360	74	∘
2	74	207360	73	∘

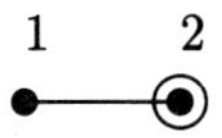

1 2

Group: 3ON Prime: 3 Block: 7

Nr.	CAS-Nr.	Degree	CC	N&C
1	27	207360	r	×
2	75	207360	76	∘
2	76	207360	75	∘

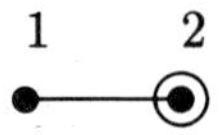

1 2

Group: 3ON Prime: 3 Block: 8

Nr.	CAS-Nr.	Degree	CC	N&C
1	28	207360	r	×
2	71	207360	72	∘
2	72	207360	71	∘

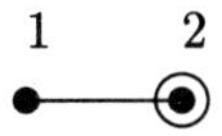

1 2

Group: 3ON Prime: 5 Block: 1

Nr.	CAS-Nr.	Degree	CC	N&C
1	1	1	r	×
2	2	10944	r	∘
3	10	37696	r	×
4	12	58311	r	×
5	18	85064	r	∘

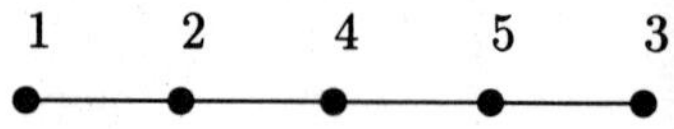

1 — 2 — 4 — 5 — 3

PROJECTIVES:

Nr.:	1	2	3	4	5
CC:	r	r	r	r	r
N&C:	×	∘	×	×	∘
$\chi_{35} \otimes \chi_{36}$	1	1			
$\chi_{35} \otimes \chi_{38}$		1		1	

Group: 3ON Prime: 5 Block: 2

Nr.	CAS-Nr.	Degree	CC	N&C
1	3	13376	4	×
2	4	13376	3	×
3	13	58311	r	×
4	14	58311	r	×
5	20	143374	r	∘

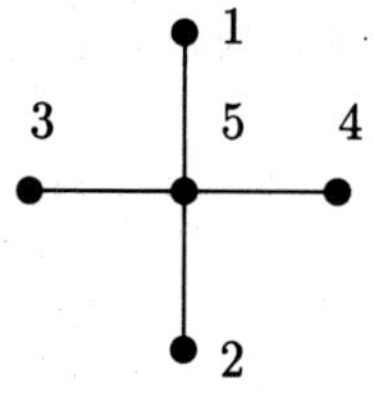

Group: 3ON Prime: 5 Block: 3

Nr.	CAS-Nr.	Degree	CC	N&C
1	5	25916	6	×
1	6	25916	5	×
2	7	26752	r	×
3	11	52668	r	∘

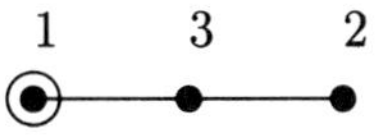

1 3 2

Group: 3ON Prime: 5 Block: 4

Nr.	CAS-Nr.	Degree	CC	N&C
1	15	58653	r	×
2	19	116963	r	×
3	23	175616	r	∘
3	24	175616	r	∘

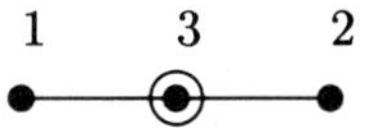

1 3 2

Group: 3ON Prime: 5 Block: 5

Nr.	CAS-Nr.	Degree	CC	AC	N&C
1	31	342	32	33	∘
2	33	342	34	31	∘
3	39	5643	40	ar	×
4	51	58653	52	ar	×
5	53	63612	54	ar	∘

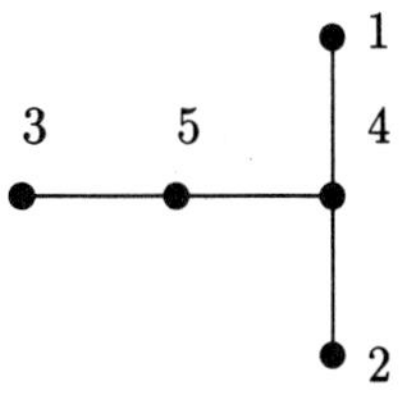

1 3 5 4 2

PROJECTIVES:

Nr.:	1	2	3	4	5
AC:	2	1	ar	ar	ar
N&C:	∘	∘	×	×	∘
$\chi_{35} \otimes \chi_{22}$		1		12	11

Group: 3ON Prime: 5 Block: 6

Complex conjugate to Block 5.

Group: 3ON Prime: 5 Block: 7

Nr.	CAS-Nr.	Degree	CC	AC	N&C
1	41	5643	42	43	×
2	43	5643	44	41	×
3	45	52668	46	ar	×
4	47	52668	48	ar	×
5	57	116622	58	ar	∘

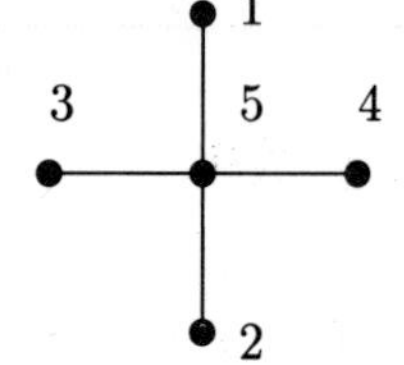

Group: 3ON Prime: 5 Block: 8

Complex conjugate to Block 7.

Group: 3ON Prime: 5 Block: 9

Nr.	CAS-Nr.	Degree	CC	AC	N&C
1	49	58311	50	ar	×
2	55	111321	56	ar	×
3	65	169632	68	ar	∘
3	67	169632	66	ar	∘

Group: 3ON Prime: 5 Block: 10

Complex conjugate to Block 9.

Group: 3ON Prime: 11 Block: 1

Nr.	CAS-Nr.	Degree	CC	N&C
1	1	1	r	×
2	2	10944	r	∘
3	10	37696	r	∘
4	15	58653	r	×
5	18	85064	r	×
6	23	175616	r	×
7	24	175616	r	×
8	25	175770	r	×
9	26	207360	r	∘
10	27	207360	r	∘
11	28	207360	r	∘

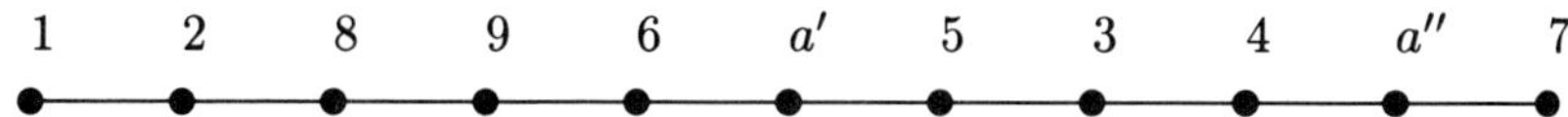

$\{a', a''\} = \{10, 11\}$

PROJECTIVES:

Nr.:	1	2	3	4	5	6	7	8	9	10	11
CC:	r	r	r	r	r	r	r	r	r	r	r
N&C:	×	∘	∘	×	×	×	×	×	∘	∘	∘
$\chi_{35} \otimes \chi_{36}$	1	1									
$\chi_2 \otimes \chi_3$		5	11	18	26	56	56	58	66	66	66
$\chi_{35} \otimes \chi_{38}$		1						1			
$\chi_{31} \otimes \chi_{40}$					1	1	1		1	1	1
$\chi_{31} \otimes \chi_{42}$						1	1	1	1	1	1

We have two orbits of algebraically conjugate characters. The first one has length two and contains the characters χ_{23} and χ_{24} (nodes 6 and 7). The second one has length three containing χ_{26}, χ_{27} and χ_{28} (nodes 9, 10 and 11). The characters in the two orbits generate disjoint fields. Hence we can assume without loss of generality that χ_{24} is to the left of χ_{25} on the tree and that χ_{26} is to the left of the other two characters in its orbit. We then have the following 8 trees consistent with all the projectives given above:

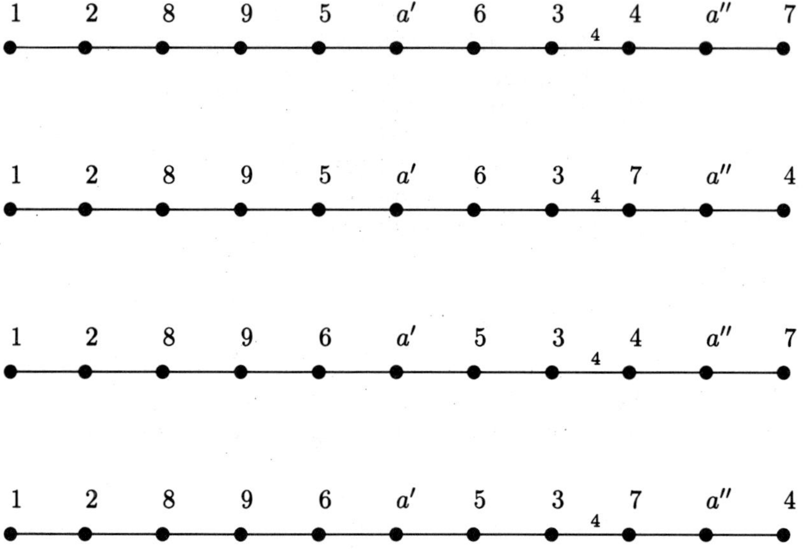

The label 4 on the third edge (counting from the right) indicates the Green correspondent of the irreducible module corresponding to that edge. We shall show later in the proof for the trees of the faithful blocks that the indecomposable trivial source module with Green correspondent $1A4$ has Brauer character χ_{15} which is node 4. It follows that node 4 must sit on the edge with label 4, hence the trees of the second and last type cannot be correct.

To exclude the two trees of the first type, consider the alternating character 1^- of the subgroup SL(3,7):2, which has index 122760. By the generalized Green correspondence due to Burry and Carlson, $(1^-)^{3\mathrm{ON}}$ has an indecomposable direct summand with trivial source in the principal 19-block. The tree for that block is known. It follows from this that χ_{18} is a constituent of $(1^-)^{3\mathrm{ON}}$. Since $(1^-)^{3\mathrm{ON}}$ is projective modulo 11, but contains no character of degree 207360, we see that node 5 cannot sit in the middle of two characters of this degree.

Group: 3ON Prime: 11 Block: 2

Nr.	CAS-Nr.	Degree	CC	AC	N&C
1	31	342	32	33	×
2	33	342	34	31	×
3	51	58653	52	ar	×
4	53	63612	54	ar	∘
5	55	111321	56	ar	×
6	65	169632	68	ar	×
7	67	169632	66	ar	×
8	69	175770	70	ar	×
9	71	207360	72	ar	∘
10	73	207360	74	ar	∘
11	75	207360	76	ar	∘

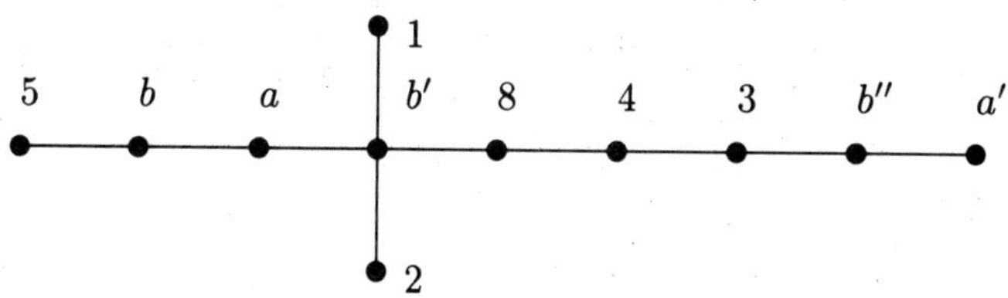

$\{a, a'\} = \{6, 7\}$, $\{b, b', b''\} = \{9, 10, 11\}$

PROJECTIVES:

Nr.:	1	2	3	4	5	6	7	8	9	10	11
AC:	2	1	ar	ar	ar	ar	ar	ar	ar	ar	ar
N&C:	×	×	×	∘	×	×	×	×	∘	∘	∘
$\chi_{35} \otimes \chi_{21}$	1		12	11	21	30	30	31	38	38	38
$\chi_{31} \otimes \chi_{12}$		1	3	3	5	7	7	7	9	9	9
$\chi_{32} \otimes \chi_{40}$		1				1	1		1	1	1
$\chi_2 \otimes \chi_{35}$			2	3	1	1	1	4	2	2	2
χ_{36}^{2+}			1	1							
$\chi_{38} \otimes \chi_{36}$				1				1			
$\chi_{31} \otimes \chi_3$					1	2	2	1	2	2	2

We have two orbits of algebraically conjugate characters. The first one has length two and contains χ_{65} and χ_{67}. The second has length three containing χ_{71}, χ_{73} and χ_{75}. The two orbits correspond to the following sets of nodes: $\{6, 7\}$ and $\{9, 10, 11\}$. We have the following four types of trees consistent with the projectives above:

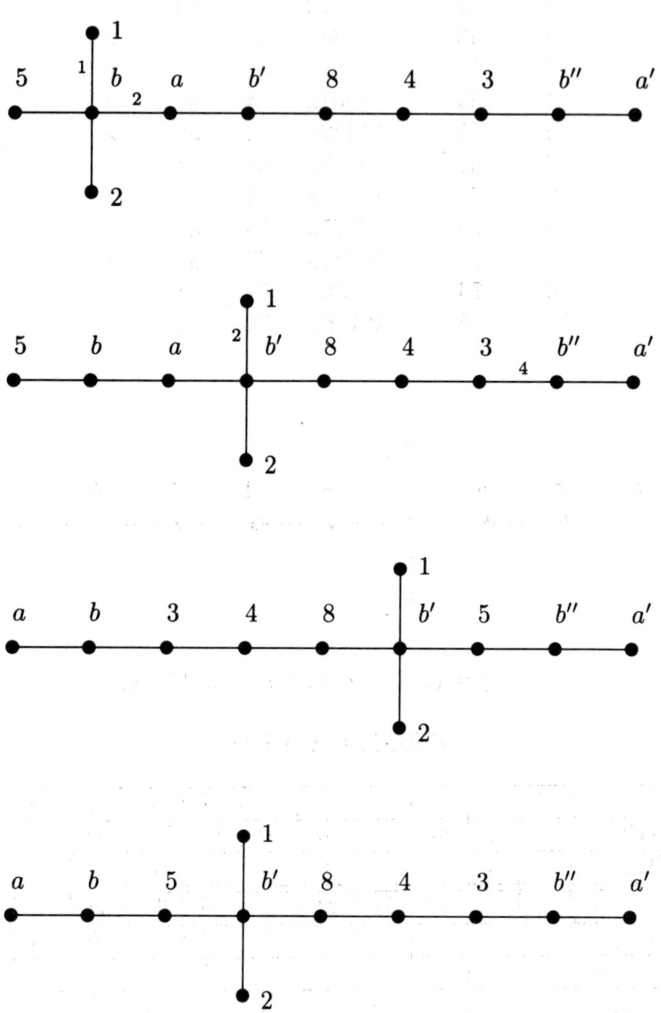

To finish the proof for the Brauer trees of 3ON modulo 11 we consider the following tensor products:

$$\begin{aligned}
\chi_{31} \otimes \chi_{31} &\approx_C \chi_{52} \\
\chi_{31} \otimes \chi_{33} &\approx_C \chi_{56} \\
\chi_{31} \otimes \chi_{34} &\approx_A \chi_{15}
\end{aligned}$$

In terms of nodes these tensor products are as follows:

$$\begin{array}{lll} 1_B \otimes 1_B & \approx_C & 3_C \\ 1_B \otimes 2_B & \approx_C & 5_C \\ 1_B \otimes 1_C & \approx_A & 4_A \end{array}$$

The last tensor product implies that node 4_A must be incident to an edge with label $2x$, if x is the label at the edge of node 1_B. We shall show in a second that x must be 2, thereby proofing the claim made in the proof for the principal block.

Since $\chi_{33} = \chi_{31}^{\#}$ it follows from the second tensor product and Lemma 6.14.2 that node 5_B must sit at the end of the $^{\#}$-invariant stem and that the edge meeting it must have label 0. Hence only trees of the first two types are possible. If the first type is correct, then, without loss of generality, the irreducible module corresponding to node 1_B has Green correspondent $1B1$. Hence $1_B \otimes 1_B$ has Green correspondent $1C2$. From the tensor product given above, node 3_C would have an edge with label 2. This contradiction shows that the second type of tree is correct and that the edge incident to node 1_B has label 2.

Group: 3ON Prime: 11 Block: 3

Complex conjugate to Block 2.

Group: 3ON Prime: 19 Block: 1

Nr.	CAS-Nr.	Degree	CC	N&C
1	1	1	r	×
2	18	85064	r	×
3	19	116963	r	○
4	23	175616	r	○
5	24	175616	r	○
6	25	175770	r	×
7	26	207360	r	×
7	27	207360	r	×
7	28	207360	r	×

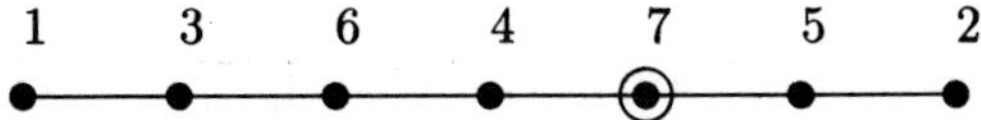

PROJECTIVES:

Nr.:	1	2	3	4	5	6	7
CC:	r	r	r	r	r	r	r
N&C:	×	×	○	○	○	×	×
$\chi_{31} \otimes \chi_{32}$	1		1				
$\chi_{31} \otimes \chi_{54}$		2	3	4	4	4	5
$\chi_{35} \otimes \chi_{40}$			1	1	1	2	1

We have two algebraically conjugate characters, namely χ_{23} and χ_{24}. If we assume without loss of generality that χ_{23} comes first on the tree, then we have only the following two trees consistent with the above set of projectives. We shall show in the proof for Block 2 that there must be an edge meeting node 6 with label 4. This excludes the first possibility.

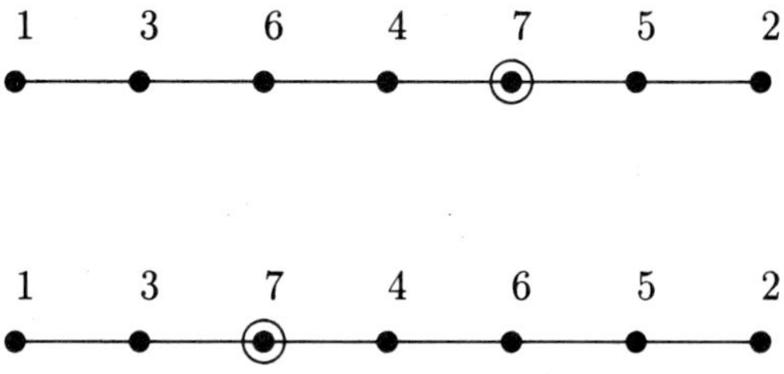

Group: 3ON Prime: 19 Block: 2

Nr.	CAS-Nr.	Degree	CC	AC	N&C
1	35	495	36	37	×
2	37	495	38	35	×
3	59	122760	60	ar	×
4	69	175770	70	ar	×
5	71	207360	72	ar	×
5	73	207360	74	ar	×
5	75	207360	76	ar	×
6	77	253440	80	ar	∘
7	79	253440	78	ar	∘

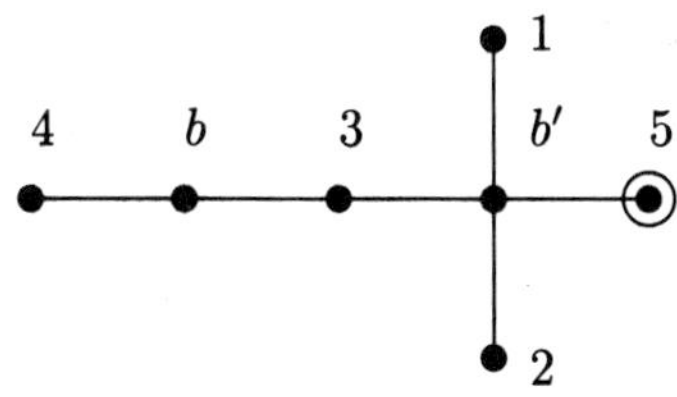

$\{b, b'\} = \{6, 7\}$

PROJECTIVES:

Nr.:	1	2	3	4	5	6	7
CC:	2	1	r	r	r	r	r
N&C:	×	×	×	×	×	∘	∘
$\chi_{35} \otimes \chi_2$	1	1		4	2	4	4
$\chi_{36} \otimes \chi_{42}$	1		2		1	2	2

We have the two algebraically conjugate characters χ_{77} and χ_{79}. There are four trees consistent with the above set of projectives:

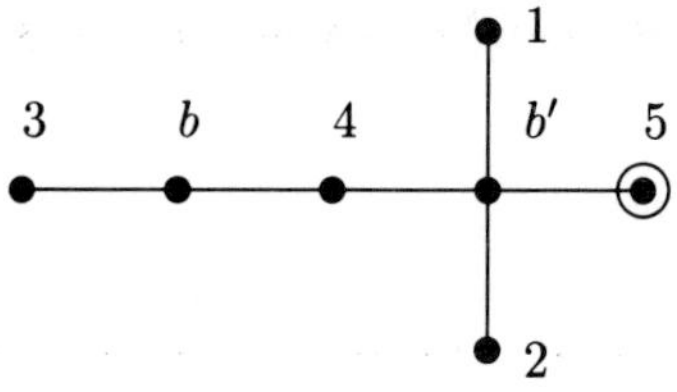

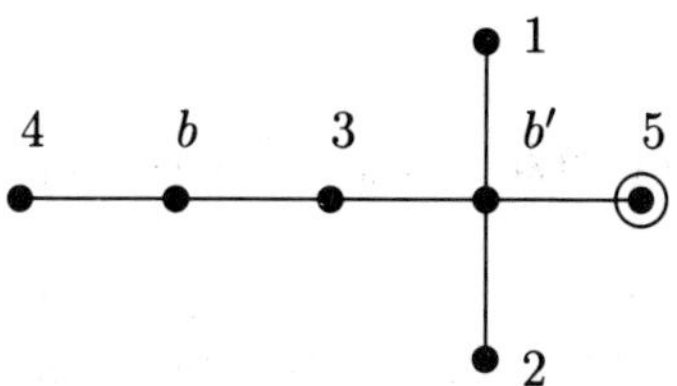

To exclude the first type, we again use Lemma 6.14.2 to show that node 4_B must be a leaf. This follows from the tensor product

$$\chi_{35} \otimes \chi_{37} \quad \approx_C \quad \chi_{70},$$

since χ_{70} is node 4_C. It also follows from Lemma 6.14.2 that the edge meeting node 4_B has label 0. Consequently, the edge at node 1_B has label 2. Therefore the Green correspondent of $\chi_{35} \otimes \chi_{38} = \chi_{35} \otimes (\chi_{35}^{\#})^*$ is $1B2 \otimes 1C2 = 1A4$. Since $\chi_{35} \otimes \chi_{38} \approx_A \chi_{25}$, and χ_{25} is node 6_A, this proves the claim made in the proof for the principal block.

Group: 3ON Prime: 19 Block: 3

Complex conjugate to Block 2.

Group: 3ON Prime: 31 Block: 1

Nr.	CAS-Nr.	Degree	CC	N&C
1	1	1	r	×
2	2	10944	r	×
3	3	13376	4	∘
3	4	13376	3	∘
4	7	26752	r	∘
5	11	52668	r	∘
6	15	58653	r	×
7	20	143374	r	∘
8	21	169290	r	∘
9	22	169290	r	∘
10	23	175616	r	×
11	24	175616	r	×
12	26	207360	r	×
13	27	207360	r	×
14	28	207360	r	×
15	29	234080	r	∘
16	30	234080	r	∘

1 — 5 — 10 — 15 — 12 — 9 — a' — 7 — 2 — 4 — a'' — 16 — 6 — 8 — 11 — 3 (exceptional)

$\{a', a''\} = \{13, 14\}$

It is a remarkable fact that the irreducible Brauer character corresponding to the edge joining node 2 with 4 has degree 1869, which is much smaller than the smallest degree of a nontrivial ordinary irreducible character.

PROJECTIVES:

Nr.:	1	2	3	4	5	6	7	8	9	10	11	12	13	14	15	16
CC:	r	r	r	r	r	r	r	r	r	r	r	r	r	r	r	r
N&C:	×	×	∘	∘	∘	×	∘	∘	∘	×	×	×	×	×	∘	∘
$\chi_{35} \otimes \chi_{54}$		3	1	4	5	3	12	11	11	12	12	14	14	14	14	14
$\chi_{35} \otimes \chi_{58}$		2	4	2	6	7	21	18	18	22	22	26	26	26	31	31
$\chi_{35} \otimes \chi_{50}$			2		2	4	9	9	9	11	11	13	13	13	17	17
$\chi_{35} \otimes \chi_{60}$			1	2	5	9	16	23	23	23	23	27	27	27	33	33
$\chi_{31} \otimes \chi_{50}$			1		2	2	6	7	7	8	8	9	9	9	11	11
$\chi_{31} \otimes \chi_{54}$				2	3	3	6	9	9	8	8	10	10	10	10	10

We have four orbits of algebraically conjugate characters, the characters of any two distinct orbits generating disjoint fields. The orbits are $\{\chi_{21}, \chi_{22}\}$, $\{\chi_{23}, \chi_{24}\}$, $\{\chi_{26}, \chi_{27}, \chi_{28}\}$ and $\{\chi_{29}, \chi_{30}\}$. The position of χ_{21} and χ_{22} is determined by a choice we made in Block 2. For the other orbits we assume as usual that the character with the smallest number comes first on the tree.

We shall show in the proof for Block 2 that node 2 is joined to node 7. We shall also show that some of the nodes must have edges with labels as follows.

node	6	8	9	2
label	6	9	13	4

We also know that the exceptional node must be a leaf. This follows from the fact that the tree for the automorphism group of ON is on one hand an unfolding around the exceptional vertex of the present tree and on the other hand a real stem. The latter observation follows easily by inspecting the Frobenius Schur indicators of the characters given in the ATLAS. Hence we have the following partially labelled tree:

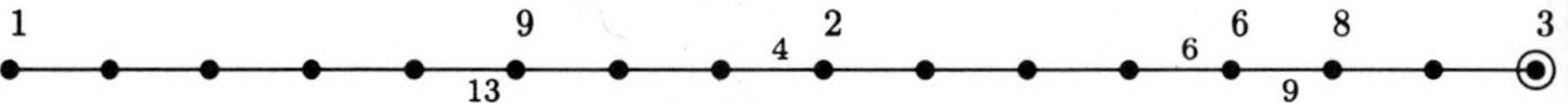

The projectives given above, together with this information, lead to the following three types of trees:

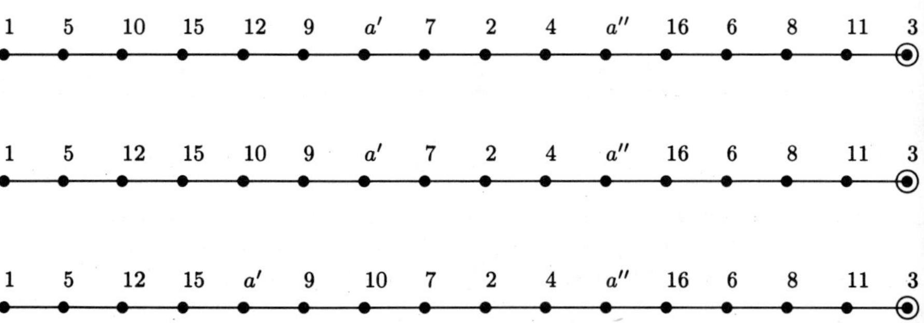

$\{a', a''\} = \{13, 14\}$

Let 1869 denote the irreducible module corresponding to the edge joining the nodes 2 and 4. We find $(\chi_1 + \chi_{11}, 1869^{2+}) = 2$ and $(\chi_{11} + \chi_{26}, 1869^{2+}) = 1$, independent of a' being 13 or 14. Now suppose that one of the last two types of trees is correct. Thus if 1 denotes the trivial module and φ the irreducible module corresponding to the second edge of the Brauer tree, the multiplicity of 1 as a constituent of 1869^{2+} is 2, whereas φ is contained just once in the symmetric square of 1869. However, the trivial module is a direct summand of 1869^{2+} and its multiplicity in the head and in the socle of $1869 \otimes 1869$ is exactly 1. Hence there must be an indecomposable direct summand of the symmetric square, which has 1 as a composition factor, but 1 sitting neither in the head nor in the socle. From the knowledge of the indecomposable modules in the block this can only be the projective cover of φ. But then φ must occur at least twice in the symmetric square. This contradiction shows that the first type of tree is correct.

Group: 3ON Prime: 31 Block: 2

Nr.	CAS-Nr.	Degree	CC	AC	N&C
1	31	342	32	33	×
2	33	342	34	31	×
3	35	495	36	37	∘
4	37	495	38	35	∘
5	39	5643	40	ar	×
6	41	5643	42	43	×
7	43	5643	44	41	×
8	45	52668	46	ar	∘
9	47	52668	48	ar	∘
10	51	58653	52	ar	×
11	61	169290	62	ar	∘
12	63	169290	64	ar	∘
13	71	207360	72	ar	×
14	73	207360	74	ar	×
15	75	207360	76	ar	×
16	77	253440	80	ar	∘
16	79	253440	78	ar	∘

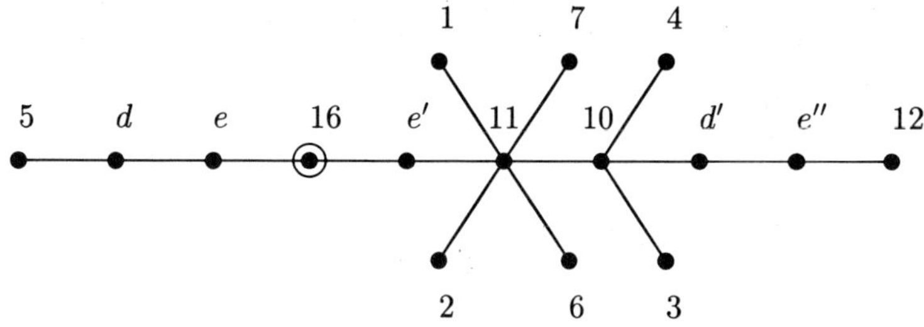

$\{d, d'\} = \{8, 9\}$ and $\{e, e', e''\} = \{13, 14, 15\}$

PROJECTIVES:

Nr.:	1	2	3	4	5	6	7	8	9	10	11	12	13	14	15	16
AC:	2	1	4	3	ar	7	6	ar	ar	ar	ar	ar	ar	ar	ar	ar
N&C:	×	×	∘	∘	×	×	×	∘	∘	×	∘	∘	×	×	×	∘
$\chi_{32} \otimes \chi_{50}$	1							2	2	1	7	7	9	9	9	11
$\chi_{31} \otimes \chi_{12}$		1						2	2	3	8	8	9	9	9	11
$\chi_{35} \otimes \chi_{8}$			1			1		2	2	4	6	6	7	7	7	9
$\chi_{35} \otimes \chi_{12}$				1	1			4	4	5	10	10	13	13	13	16
$\chi_{36} \otimes \chi_{50}$					1		1	4	4	1	9	9	13	13	13	16
$\chi_{35} \otimes \chi_{10}$						1	1	1	1	1	7	7	8	8	8	11
$\chi_{35} \otimes \chi_{14}$								3	3	5	11	11	13	13	13	16
$\chi_{31} \otimes \chi_{16}$								3	3	4	8	8	10	10	10	12
$\chi_{31} \otimes \chi_{10}$								2	2	1	4	4	6	6	6	7
$\chi_{31} \otimes \chi_{5}$								1	1	1	3	3	4	4	4	5

We use the notation $\{6,7\} = \{c, \bar{c}\}$ in the proof. The above set of projectives show that the nodes 1, 3, and c are joined to one of the nodes in the following table:

1	8, 9, 11, 12, 16
3	10, 13, 14, 15
c	11, 12, 16

From this it already follows that 1, 3 and c must sit on the stem, and so they are leaves.

We are going to use the following tensor products:

$$\begin{aligned}
\chi_{31} \otimes \chi_{31} &\approx_C \chi_{52} \\
\chi_{31} \otimes \chi_{33} &\approx_C \chi_{40} \\
\chi_{31} \otimes \chi_{35} &\approx_C \chi_{62} \\
\chi_{31} \otimes \chi_{37} &\approx_C \chi_{64} \\
\chi_{31} \otimes \chi_{41} &\approx_C \chi_{64} + \chi_{72} + \chi_{74} + \chi_{76} + \chi_{78} + \chi_{80} \\
\chi_{31} \otimes \chi_{43} &\approx_C \chi_{62} + \chi_{72} + \chi_{74} + \chi_{76} + \chi_{78} + \chi_{80} \\
\chi_{35} \otimes \chi_{35} &\approx_C \chi_{36} + \chi_{42} + \chi_{52}
\end{aligned}$$

In terms of nodes these tensor products are as follows:

$$\begin{aligned}
1_B \otimes 1_B &\approx_C 10_C \\
1_B \otimes 2_B &\approx_C 5_C \\
1_B \otimes 3_B &\approx_C 11_C \\
1_B \otimes 4_B &\approx_C 12_C \\
1_B \otimes 6_B &\approx_C 12_C + 13_C + 14_C + 15_C + 16_C \\
1_B \otimes 7_B &\approx_C 11_C + 13_C + 14_C + 15_C + 16_C \\
3_B \otimes 3_B &\approx_C 3_C + 6_C + 10_C
\end{aligned}$$

Of course, $\approx_C$ denotes restriction to Block 3. We already know that 3 is not joined to 6 by the remark above. So the Green correspondent of $3_B \otimes 3_B$ cannot be 10_C. Hence it must be 6_C and 3_C is joined to 10_C. Using this information and the above set of projectives we obtain the following subtree:

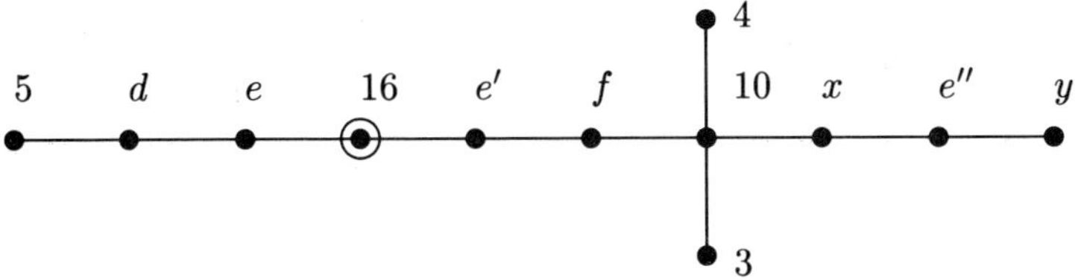

We have used the notation $d \in \{8, 9\}$, $f \in \{11, 12\}$ and $\{x, y\} = \{9, 12\}$. Also we have assumed without loss of generality that node 4 sits on the upper half of the tree.

Please note that it follows from Lemma 6.14.2 and the second tensor product given above that the edge meeting node 5 must have label 0. This is important when we are trying to find the labels of the other edges of the tree.

Let z be the label on the edge incident to node 1. We make a choice by assuming that the label on the edge incident to node 1 is smaller than the label on the edge of node 2. Although we have made a choice already, this is no loss of generality, since by mapping a square root of to onto its negative but fixing the other irrationalities involved, we can can find a 31-modular system consistent with our choice.

We now distinguish three cases.

CASE I. There is no branch of the tree coming out to the left of node 10

From our assumption we get that the labels on the edges around node 10 are as follows:

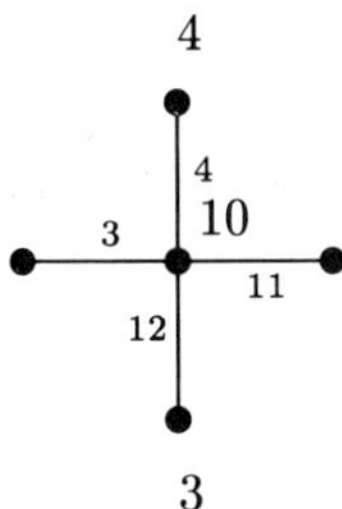

Now consider the first tensor product. In terms of Green correspondents this tells us $1Bz \otimes 1Bz = 1C2z$, where $2z$ has to be read modulo 15. Since only node 10 occurs on the right-hand side of this tensor product we must have that $2z$ labels an edge around 10. With the above restriction on z this implies $z = 6$. The last tensor product now shows that node 6_C is incident to an edge with label $(12 + 12) - 1 \equiv 8(\text{mod} 15)$. Furthermore, the third tensor product tells us that f equals 11. It is now easy to see that the tree must have the following form.

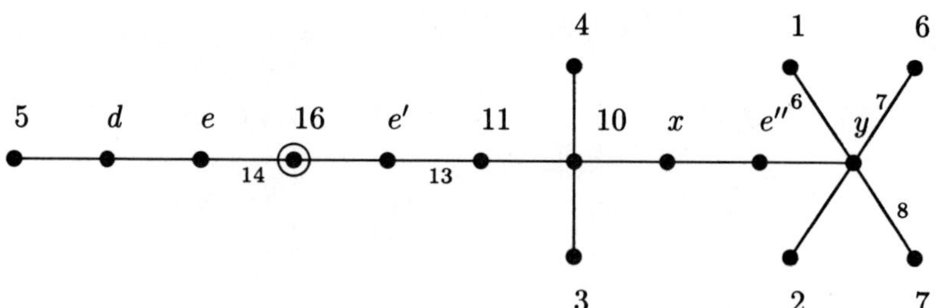

The second to the last tensor product now gives a contradiction as follows. We have $1B6 \otimes 1B8 = 1C14$. This tells us that if we subtract node e, which is one of 13, 14 or 15 from $11_C + 13_C + 14_C + 15_C + 16_C$, we obtain a projective character of Block 3. This is not consistent with the location of node 11_C already known. We conclude that CASE I is impossible. Note that our argument is independent of a permutation of the three algebraically conjugate nodes $\{13, 14, 15\}$.

CASE II There is exactly one branch coming out to the left of node 10:

Here we have the following labels on the edges around node 10.

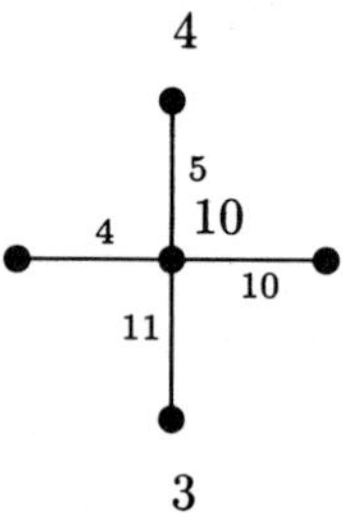

Let again z denote the label on the edge coming from node 1. The same argument as above implies that z must be equal to 2. The last tensor product implies as above that the labels on the edges corresponding to c and $\bar{c}$ must be $5 + 5 - 1 = 9$ and $11 + 11 - 1 = 6$. This leads to the following tree:

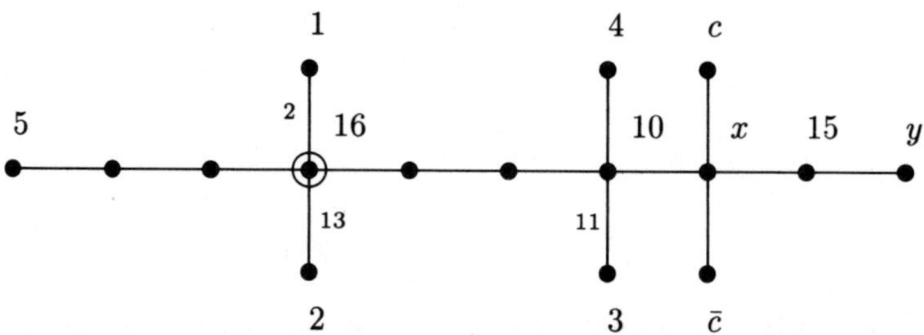

Now by the tensor products given above $1_B \otimes 3_B \approx_C 11_C$. But $1B2 \otimes 30B11 = 30c13$, a contradiction since 13 does not label any edge incident to 11_C. Hence CASE II does not occur.

It follows that the nodes which have not been located yet must sit left of node 10. Hence the four edges around 10 are labelled as follows:

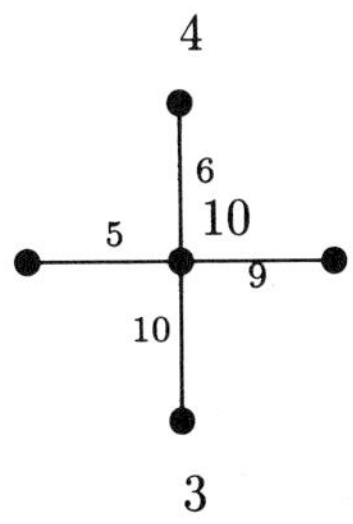

We conclude as above that the edge incident to node 1_B must have label 3 and that the edge incident to node 6_C carries the label $10+10-1=4$. Also, node 11_C is incident to an edge with label $10+3$, by the third tensor product. This implies the following tree:

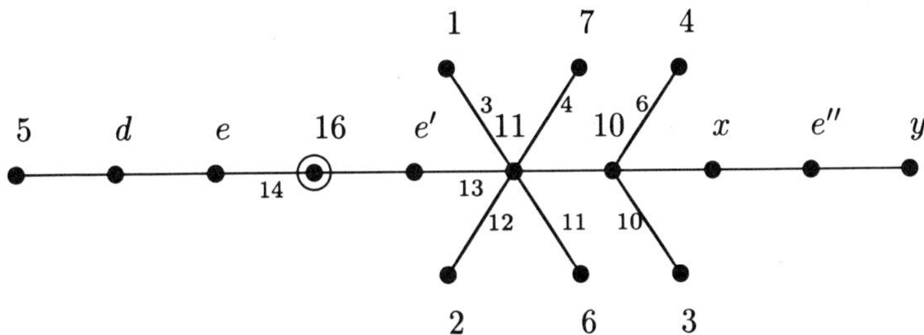

It remains to determine the characters on nodes x and y. Now $1_B \otimes 4_B$ has Green correspondent $1B3 \otimes 30B6 = 30C9$. This corresponds to node x. On the other hand, $1_B \otimes 4_B \approx_C 12_C$, which implies that x must be 12.

We now use our information on the labelling of this tree to obtain the location of some nodes in the principal block. We use the following tensor products:

$$\begin{array}{lll}
\chi_{31} \otimes \chi_{34} & \approx_A & \chi_{15} \\
\chi_{31} \otimes \chi_{36} & \approx_A & \chi_{21} \\
\chi_{31} \otimes \chi_{38} & \approx_A & \chi_{22} \\
\chi_{35} \otimes \chi_{36} & \approx_A & \chi_1 + \chi_2 + \chi_{20} \\
\chi_{35} \otimes \chi_{38} & \approx_A & \chi_2
\end{array}$$

In terms of nodes (which of course have to be interpreted in the right block) these tensor products are as follows:

$$\begin{array}{lll}
1_B \otimes 2_C & \approx_A & 6_A \\
1_B \otimes 3_C & \approx_A & 8_A \\
1_B \otimes 4_C & \approx_A & 9_A \\
3_B \otimes 3_C & \approx_A & 1_A + 2_A + 7_A \\
3_B \otimes 4_C & \approx_A & 2_A
\end{array}$$

Finally we interpret these tensor products in terms of Green correspondence which gives the claim made in the proof for the principal block:

$$\begin{aligned}
1B3 \otimes 1C3 &= 1A6\\
1B3 \otimes 30C6 &= 30A9\\
1B3 \otimes 30C10 &= 30A13\\
30B10 \otimes 30C6 &= 1A0\\
30B10 \otimes 30C10 &= 1A4
\end{aligned}$$

The second to the last tensor product proves that node 2_A is joined to node 7_A, since the Green correspondent is the trivial module. Please note that interchanging nodes 11 and 12 in Block 2 would result in swapping nodes 8 and 9 in the principal block. This completes the proof for the triple cover of the O'Nan group.

Group: 3ON Prime: 31 Block: 3

Complex conjugate to Block 2.

6.15 The Conway Group C3

Group: C3 Prime: 2 Block: 3

Nr.	CAS-Nr.	Degree	CC	N&C
1	33	129536	r	×
2	34	129536	r	∘

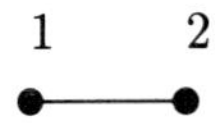

1 — 2

Group: C3 Prime: 3 Block: 2

Nr.	CAS-Nr.	Degree	CC	N&C
1	31	91125	r	×
2	32	93312	r	×
3	36	184437	r	∘

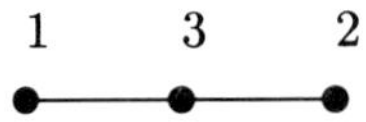

1 — 3 — 2

Group: C3 Prime: 5 Block: 2

Nr.	CAS-Nr.	Degree	CC	N&C
1	5	275	r	×
2	12	4025	r	×
3	29	73600	r	∘
4	35	177100	r	∘
5	39	246400	r	×

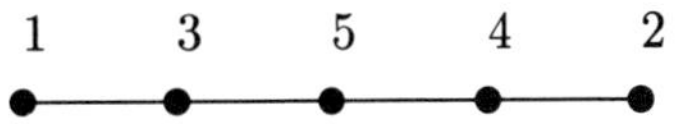

1 — 3 — 5 — 4 — 2

PROJECTIVES:

Nr.	1	2	3	4	5
CC	r	r	r	r	r
N&C	×	×	∘	∘	×
$\chi_2 \otimes \chi_{20}$		1		1	
$\chi_3 \otimes \chi_{16}$			1	1	2

Group: C3 Prime: 7 Block: 1

Nr.	CAS-Nr.	Degree	CC	N&C
1	1	1	r	×
2	3	253	r	×
3	10	3520	11	∘
4	11	3520	10	∘
5	23	31625	r	∘
6	31	91125	r	∘
7	33	129536	r	×

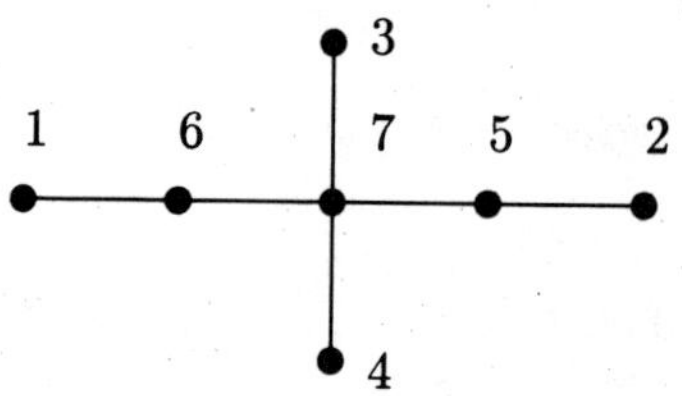

PROJECTIVES:

Nr.	1	2	3	4	5	6	7
CC	r	r	4	3	r	r	r
N&C	×	×	∘	∘	∘	∘	×
$\chi_7 \otimes \chi_6$	1					1	
$\chi_{10} \otimes \chi_{15}$				1	1	4	6

Group: C3 Prime: 7 Block: 2

Nr.	CAS-Nr.	Degree	CC	N&C
1	2	23	r	×
2	5	275	r	×
3	20	23000	r	∘
4	28	63250	r	∘
5	29	73600	r	×
6	30	80960	r	∘
7	32	93312	r	×

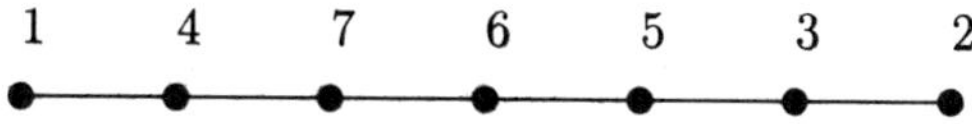

PROJECTIVES:

Nr.	1	2	3	4	5	6	7
CC	r	r	r	r	r	r	r
N&C	×	×	∘	∘	×	∘	×
$\chi_4 \otimes \chi_8$	1			1			
$\chi_2 \otimes \chi_{12}$		1	1				
$\chi_3 \otimes \chi_{16}$			1		1		
$\chi_{10} \otimes \chi_6$				1			1
$\chi_7 \otimes \chi_6$					1	1	
$\chi_3 \otimes \chi_{14}$						1	1

Group: C3 Prime: 7 Block: 3

Nr.	CAS-Nr.	Degree	CC	N&C
1	4	253	r	×
2	9	2024	r	×
3	22	31625	r	∘
4	24	31625	r	∘
5	34	129536	r	×
6	36	184437	r	×
7	41	253000	r	∘

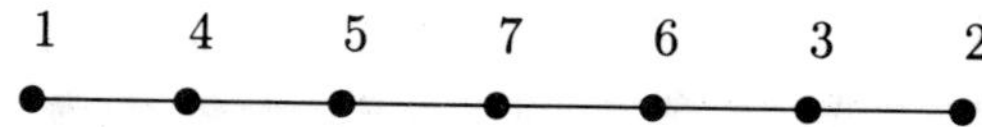

PROJECTIVES:

Nr.	1	2	3	4	5	6	7
CC	r	r	r	r	r	r	r
N&C	×	×	∘	∘	×	×	∘
$\chi_2 \otimes \chi_8$	1			1			
$\chi_5 \otimes \chi_{12}$		1	1				
$\chi_3 \otimes \chi_{16}$			1			1	
$\chi_3 \otimes \chi_8$				1	1		
$\chi_2 \otimes \chi_{21}$					1		1
$\chi_8 \otimes \chi_6$						1	1

Group: C3 Prime: 11 Block: 1

Nr.	CAS-Nr.	Degree	CC	N&C
1	1	1	r	×
2	6	896	7	∘
2	7	896	6	∘
3	20	23000	r	∘
4	21	26082	r	×
5	31	91125	r	×
6	32	93312	r	∘

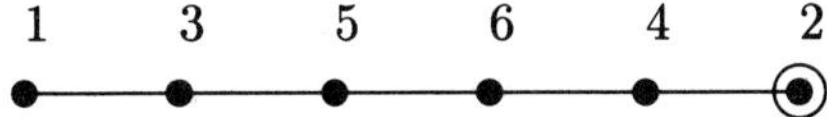

PROJECTIVES:

Nr.	1	2	3	4	5	6
CC	r	r	r	r	r	r
N&C	×	∘	∘	×	×	∘
$\chi_3 \otimes \chi_3$	1		1			
$\chi_9 \otimes \chi_3$			1		1	
$\chi_{15} \otimes \chi_3$					1	1

Group: C3 Prime: 11 Block: 2

Nr.	CAS-Nr.	Degree	CC	N&C
1	2	23	r	×
2	12	4025	r	∘
3	18	20608	19	∘
3	19	20608	18	∘
4	26	40250	r	×
5	27	57960	r	×
6	29	73600	r	∘

PROJECTIVES:

Nr.	1	2	3	4	5	6
CC	r	r	r	r	r	r
N&C	×	∘	∘	×	×	∘
$\chi_5 \otimes \chi_3$	1	1				
$\chi_4 \otimes \chi_3$		1			1	
$\chi_{18} \otimes \chi_3$			1	1		
$\chi_{16} \otimes \chi_3$					1	1

Group: C3 Prime: 23 Block: 1

Nr.	CAS-Nr.	Degree	CC	N&C
1	1	1	r	×
2	5	275	r	○
3	6	896	7	○
4	7	896	6	○
5	10	3520	11	×
6	11	3520	10	×
7	13	5544	r	×
8	16	9625	17	○
8	17	9625	16	○
9	31	91125	r	○
10	32	93312	r	×
11	39	246400	r	×
12	40	249480	r	○

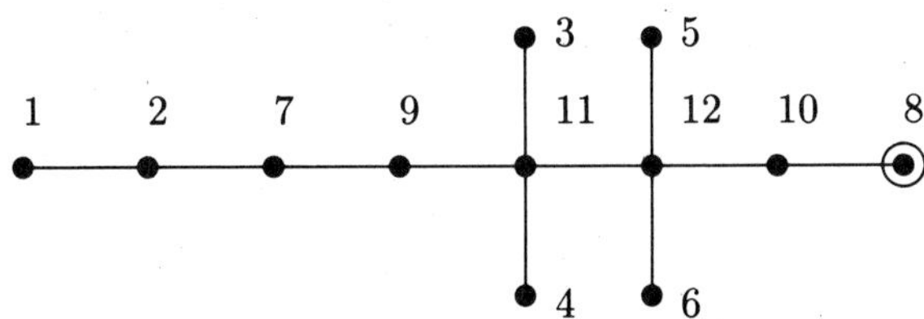

PROJECTIVES:

Nr.	1	2	3	4	5	6	7	8	9	10	11	12
CC	r	r	4	3	6	5	r	r	r	r	r	r
N&C	×	○	○	○	×	×	×	○	○	×	×	○
$\chi_2 \otimes \chi_2$	1	1										
$\chi_3 \otimes \chi_2$		1					1					
$\chi_{19} \otimes \chi_2$				1							1	
$\chi_{10} \otimes \chi_3$						1						1
$\chi_{22} \otimes \chi_2$							1		1			
$\chi_{28} \otimes \chi_2$									1		1	
$\chi_{26} \otimes \chi_2$											1	1

Since the complex conjugate pairs of characters have independent irrationalities, we do not have to worry about the planar emdedding.

6.16 The Conway Group C2

Group: C2 Prime: 3 Block: 2

Nr.	CAS-Nr.	Degree	CC	N&C
1	19	37422	r	×
2	40	430353	r	×
3	43	467775	r	∘

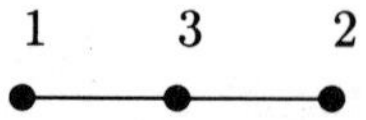

1 — 3 — 2

Group: C2 Prime: 3 Block: 3

Nr.	CAS-Nr.	Degree	CC	N&C
1	33	245916	r	×
2	36	312984	r	×
3	44	558900	r	∘

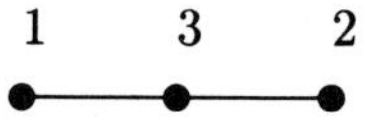

1 — 3 — 2

Group: C2 Prime: 5 Block: 2

Nr.	CAS-Nr.	Degree	CC	N&C
1	4	275	r	×
2	20	44275	r	×
3	24	113850	r	○
4	39	398475	r	○
5	43	467775	r	×

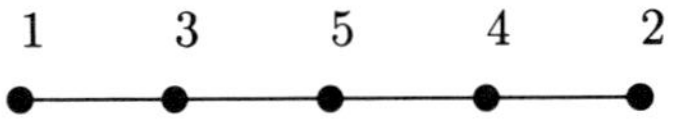

1 — 3 — 5 — 4 — 2

PROJECTIVES:

Nr.	1	2	3	4	5
CC	r	r	r	r	r
N&C	×	×	○	○	×
$\chi_{10} \otimes \chi_{10}$	1		1		
$\chi_2 \otimes \chi_{16}$		1		1	
$\chi_{11} \otimes \chi_{10}$			1		1
$\chi_4 \otimes \chi_{10}$				1	1

Group: C2 Prime: 5 Block: 3

Nr.	CAS-Nr.	Degree	CC	N&C
1	8	4025	r	×
2	14	12650	r	×
3	26	177100	r	○
4	38	398475	r	○
5	44	558900	r	×

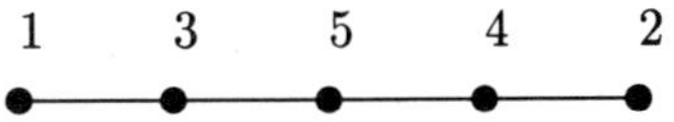

1 — 3 — 5 — 4 — 2

PROJECTIVES:

Nr.	1	2	3	4	5
CC	r	r	r	r	r
N&C	×	×	○	○	×
$\chi_2 \otimes \chi_{15}$	1		1		
$\chi_4 \otimes \chi_{16}$		1		2	1
$\chi_{12} \otimes \chi_{10}$			1		1

Group: C2 Prime: 7 Block: 1

Nr.	CAS-Nr.	Degree	CC	N&C
1	1	1	r	×
2	14	12650	r	×
3	33	245916	r	∘
4	44	558900	r	∘
5	49	853875	r	×
6	55	1943040	r	×
7	57	2004750	r	∘

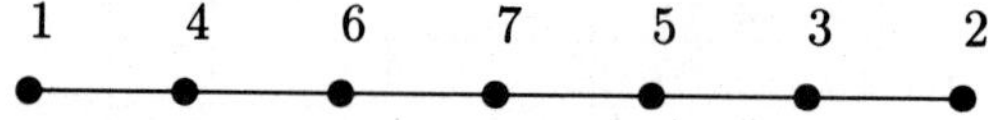

PROJECTIVES:

Nr.	1	2	3	4	5	6	7
CC	r	r	r	r	r	r	r
N&C	×	×	∘	∘	×	×	∘
$\chi_5 \otimes \chi_5$	1			1			
$\chi_8 \otimes \chi_5$				1		1	
$\chi_{12} \otimes \chi_5$					1		1
$\chi_{10} \otimes \chi_5$						1	1

Group: C2 Prime: 7 Block: 2

Nr.	CAS-Nr.	Degree	CC	N&C
1	2	23	r	×
2	7	2277	r	×
3	21	63250	r	∘
4	35	284625	r	∘
5	37	368874	r	×
6	59	2072576	r	×
7	60	2095875	r	∘

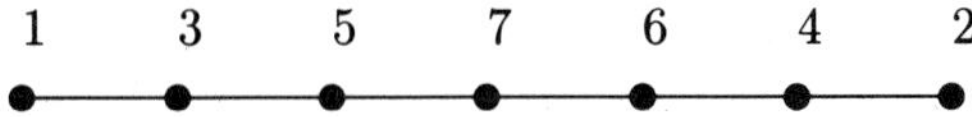

PROJECTIVES:

Nr.	1	2	3	4	5	6	7
CC	r	r	r	r	r	r	r
N&C	×	×	∘	∘	×	×	∘
$\chi_3 \otimes \chi_5$	1		1				
$\chi_4 \otimes \chi_{18}$		1		1			
$\chi_8 \otimes \chi_5$			1		1		
$\chi_2 \otimes \chi_{30}$				1		1	
$\chi_7 \otimes \chi_9$					1		1
$\chi_{10} \otimes \chi_5$						1	1

Group: C2 Prime: 7 Block: 3

Nr.	CAS-Nr.	Degree	CC	N&C
1	3	253	r	×
2	6	2024	r	×
3	16	31625	r	∘
4	17	31625	r	∘
5	25	129536	r	×
6	27	184437	r	×
7	34	253000	r	∘

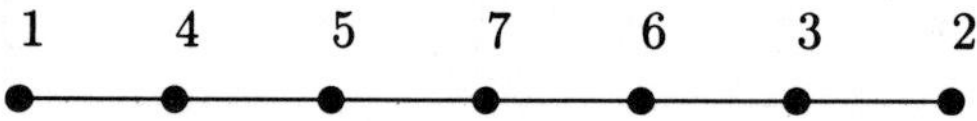

PROJECTIVES:

Nr.	1	2	3	4	5	6	7
CC	r	r	r	r	r	r	r
N&C	×	×	∘	∘	×	×	∘
$\chi_2 \otimes \chi_5$	1			1			
$\chi_2 \otimes \chi_{20}$		1	1				
$\chi_7 \otimes \chi_{10}$			1			1	
$\chi_4 \otimes \chi_5$				1	1		
$\chi_{16} \otimes \chi_5$					1		1
$\chi_{22} \otimes \chi_5$						1	1

Group: C2 Prime: 7 Block: 4

Nr.	CAS-Nr.	Degree	CC	N&C
1	4	275	r	×
2	15	23000	r	∘
3	22	91125	23	∘
3	23	91125	22	∘
4	24	113850	r	×

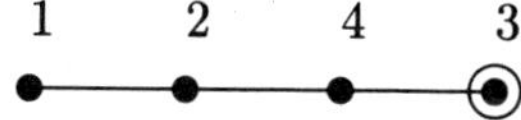

PROJECTIVES:

Nr.	1	2	3	4
CC	r	r	r	r
N&C	×	∘	∘	×
$\chi_8 \otimes \chi_5$	1	1		

Group: C2 Prime: 11 Block: 1

Nr.	CAS-Nr.	Degree	CC	N&C
1	1	1	r	×
2	2	23	r	×
3	8	4025	r	∘
4	15	23000	r	∘
5	22	91125	23	×
6	23	91125	22	×
7	36	312984	r	×
8	44	558900	r	×
9	50	1288000	r	∘
10	54	1835008	r	∘
11	60	2095875	r	×

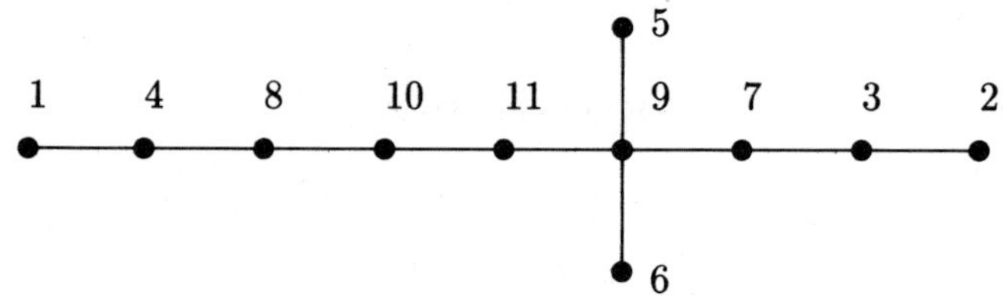

PROJECTIVES:

Nr.	1	2	3	4	5	6	7	8	9	10	11
CC	r	r	r	r	6	5	r	r	r	r	r
N&C	×	×	∘	∘	×	×	×	×	∘	∘	×
$\chi_3 \otimes \chi_3$	1			1							
χ_{10}^{2+}				1	1	1		2	2	2	1
$\chi_{15} \otimes \chi_3$				1				1			
$\chi_{22} \otimes \chi_3$						1			1	1	1
$\chi_{20} \otimes \chi_3$							1		1		
$\chi_{10} \otimes \chi_7$								1		1	
$\chi_2 \otimes \chi_{34}$									1		1
$\chi_{27} \otimes \chi_3$										1	1

Group: C2 Prime: 23 Block: 1

Nr.	CAS-Nr.	Degree	CC	N&C
1	1	1	r	×
2	4	275	r	∘
3	10	9625	11	∘
3	11	9625	10	∘
4	12	10395	13	∘
5	13	10395	12	∘
6	19	37422	r	×
7	22	91125	23	∘
8	23	91125	22	∘
9	42	462000	r	∘
10	43	467775	r	×
11	54	1835008	r	∘
12	57	2004750	r	×

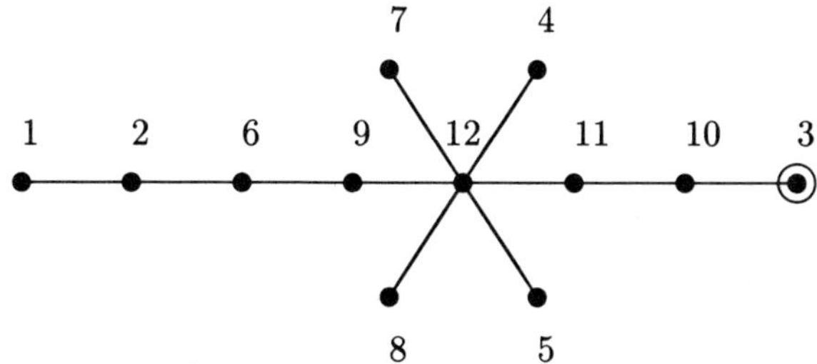

PROJECTIVES:

Nr.	1	2	3	4	5	6	7	8	9	10	11	12
CC	r	r	r	5	4	r	8	7	r	r	r	r
N&C	×	∘	∘	∘	∘	×	∘	∘	∘	×	∘	×
$\chi_2 \otimes \chi_2$	1	1										
$\chi_7 \otimes \chi_2$		1				1						
$\chi_{32} \otimes \chi_2$					1							1
$\chi_{20} \otimes \chi_2$						1			1			
$\chi_{22} \otimes \chi_3$								1			1	2
$\chi_{39} \otimes \chi_2$									1			1

The planar embedding follows from $\chi_{12}^{2-} \approx 2\chi_{42}+2\chi_{54}+5\chi_{57}$ and the fact that the two pairs of complex conjugate characters have independent irrationalities.

6.17 The Covering Group of the Fischer Group F22

Group: 6F22 Prime: 2 Block: 2

Nr.	CAS-Nr.	Degree	CC	N&C
1	56	1441792	r	×
2	104	1441792	r	∘

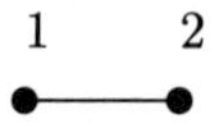

Group: F22 Prime: 2 Block: 3

Nr.	CAS-Nr.	Degree	CC	N&C
1	63	2555904	r	×
2	64	2555904	r	∘

Remark: This is a block of defect 1 in the simple group F22. The inverse image of its defect group in 2F22 is a Klein four group. There are no other 2-blocks with a non-trivial cyclic defect group in 3F22.

Group: F22 Prime: 3 Block: 2

Nr.	CAS-Nr.	Degree	CC	N&C
1	29	360855	r	×
2	38	577368	r	×
3	49	938223	r	∘

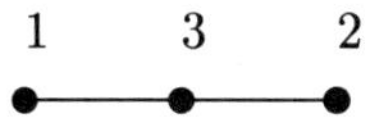

1 3 2

Group: F22 Prime: 3 Block: 3

Nr.	CAS-Nr.	Degree	CC	N&C
1	48	852930	r	×
2	58	1876446	r	×
3	65	2729376	r	∘

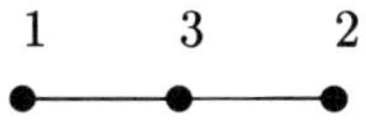

1 3 2

Remark: The Blocks 2 and 3 are blocks of defect 1 in the simple group F22. The inverse image of their defect group (which is contained in class 3A in ATLAS notation) in 3F22 is elementary abelian. There are no other 3-blocks with a non-trivial cyclic defect group in 2F22.

Group: 6F22 Prime: 3 Block: 4

Nr.	CAS-Nr.	Degree	CC	N&C
1	57	1791153	r	×
2	201	1791153	202	∘
2	202	1791153	201	∘

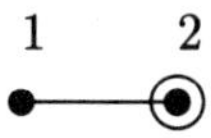

1 2

Group: 6F22 Prime: 5 Block: 2

Nr.	CAS-Nr.	Degree	CC	N&C
1	5	1430	r	×
2	7	3080	r	×
3	13	45045	r	∘
4	28	320320	r	∘
5	29	360855	r	×

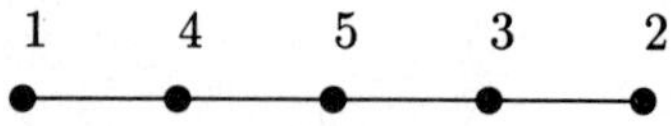

1 — 4 — 5 — 3 — 2

PROJECTIVES:

Nr.	1	2	3	4	5
CC	r	r	r	r	r
N&C	×	×	∘	∘	×
$\chi_3 \otimes \chi_{27}$	1			2	1
$\chi_2 \otimes \chi_{22}$			1		1

Group: 6F22 Prime: 5 Block: 3

Nr.	CAS-Nr.	Degree	CC	N&C
1	10	30030	r	×
2	12	43680	r	×
3	25	205920	r	∘
4	46	720720	r	∘
5	48	852930	r	×

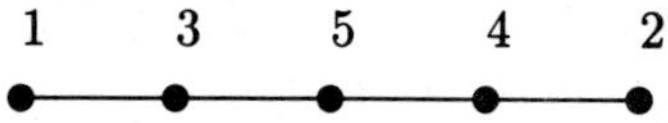

1 — 3 — 5 — 4 — 2

PROJECTIVES:

Nr.	1	2	3	4	5
CC	r	r	r	r	r
N&C	×	×	∘	∘	×
$\chi_2 \otimes \chi_8$	1		1		
$\chi_2 \otimes \chi_{17}$		1		2	1
$\chi_2 \otimes \chi_{36}$			1		1

Group: 6F22 Prime: 5 Block: 5

Nr.	CAS-Nr.	Degree	CC	N&C
1	67	2080	r	×
2	87	320320	r	∘
3	94	480480	r	×
4	107	2196480	r	×
5	110	2358720	r	∘

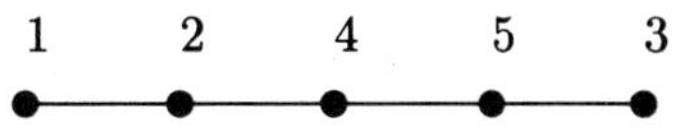

1 — 2 — 4 — 5 — 3

PROJECTIVES:

Nr.	1	2	3	4	5
CC	r	r	r	r	r
N&C	×	∘	×	×	∘
$\chi_2 \otimes \chi_{90}$		1		1	
$\chi_2 \otimes \chi_{80}$			1		1
$\chi_2 \otimes \chi_{84}$				1	1

Group: 6F22 Prime: 5 Block: 6

Nr.	CAS-Nr.	Degree	CC	N&C
1	68	2080	r	×
2	88	320320	r	∘
3	93	480480	r	×
4	108	2196480	r	×
5	109	2358720	r	∘

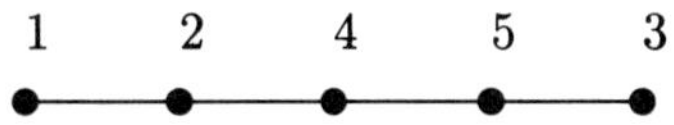

1 — 2 — 4 — 5 — 3

PROJECTIVES:

Nr.	1	2	3	4	5
CC	r	r	r	r	r
N&C	×	∘	×	×	∘
$\chi_2 \otimes \chi_{89}$		1		1	
$\chi_2 \otimes \chi_{80}$			1		1
$\chi_2 \otimes \chi_{84}$				1	1

Group: 6F22 Prime: 5 Block: 9

Nr.	CAS-Nr.	Degree	CC	AC	N&C
1	123	19305	124	ar	×
2	129	42120	130	ar	∘
3	155	405405	157	ar	×
4	173	852930	174	ar	×
5	179	1235520	180	ar	∘

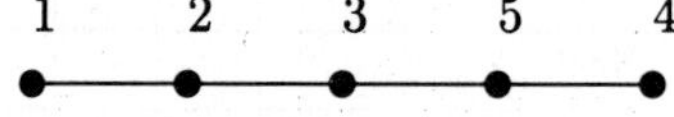

PROJECTIVES:

Nr.	1	2	3	4	5
AC	ar	ar	ar	ar	ar
N&C	×	∘	×	×	∘
$\chi_2 \otimes \chi_{145}$	1	1			
$\chi_2 \otimes \chi_{168}$		1	1		
$\chi_2 \otimes \chi_{167}$			1	3	4

Group: 6F22 Prime: 5 Block: 10

Complex conjugate to Block 9.

Group: 6F22 Prime: 5 Block: 11

Nr.	CAS-Nr.	Degree	CC	AC	N&C
1	147	360855	148	ar	×
2	156	405405	158	ar	×
3	177	997920	178	ar	∘
4	193	1621620	194	ar	∘
5	203	1853280	204	ar	×

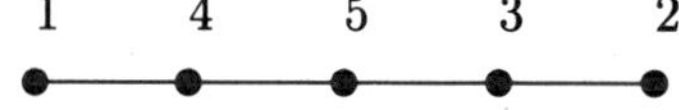

PROJECTIVES:

Nr.	1	2	3	4	5
AC	ar	ar	ar	ar	ar
N&C	×	×	∘	∘	×
$\chi_2 \otimes \chi_{145}$	2			3	1
$\chi_2 \otimes \chi_{143}$		1	1		
$\chi_2 \otimes \chi_{137}$			1		1

Group: 6F22 Prime: 5 Block: 12

Complex conjugate to Block 11.

Group: 6F22 Prime: 5 Block: 15

Nr.	CAS-Nr.	Degree	CC	AC	A	N&C
1	223	71280	225	224	226	×
2	227	112320	229	228	230	∘
3	234	308880	236	233	235	×
4	254	967680	256	253	255	×
5	262	1235520	264	261	263	∘

1 —— 2 —— 4 —— 5 —— 3

PROJECTIVES:

Nr.	1	2	3	4	5
N&C	×	∘	×	×	∘
$\chi_4 \otimes \chi_{258}$	9	11		6	4
$\chi_2 \otimes \chi_{257}$			2	4	6

The column 'A' in the above block information table lists the characters which are conjugate under the outer automorphism to the characters of the second column.

Group: 6F22 Prime: 5 Block: 16

Conjugate to Block 15 under the composition of the outer automorphism with complex conjugation.

Group: 6F22 Prime: 5 Block: 17

Complex conjugate to Block 15.

Group: 6F22 Prime: 5 Block: 18

Complex conjugate to Block 16.

Group: 6F22 Prime: 7 Block: 1

Nr.	CAS-Nr.	Degree	CC	N&C
1	1	1	r	×
2	2	78	r	×
3	21	114400	r	∘
4	26	289575	r	∘
5	38	577368	r	×
6	63	2555904	r	×
7	65	2729376	r	∘

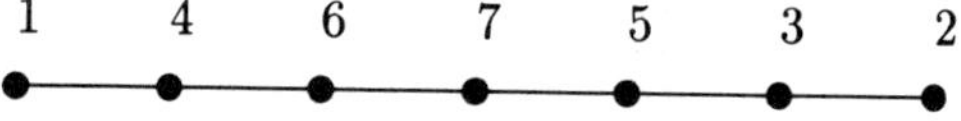

PROJECTIVES:

Nr.	1	2	3	4	5	6	7
CC	r	r	r	r	r	r	r
N&C	×	×	∘	∘	×	×	∘
$\chi_2 \otimes \chi_6$		1	1				
$\chi_2 \otimes \chi_{42}$			1		2		1
$\chi_2 \otimes \chi_{22}$				1		1	
$\chi_2 \otimes \chi_{27}$						1	1

Group: 6F22 Prime: 7 Block: 2

Nr.	CAS-Nr.	Degree	CC	N&C
1	3	429	r	×
2	5	1430	r	×
3	29	360855	r	∘
4	39	579150	r	∘
5	55	1372800	r	×
6	56	1441792	r	×
7	58	1876446	r	∘

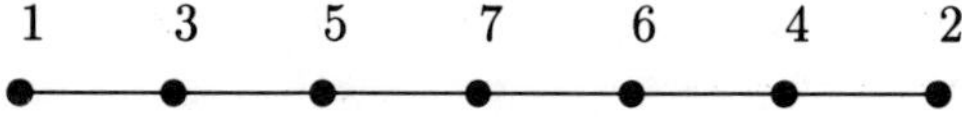

PROJECTIVES:

Nr.	1	2	3	4	5	6	7
CC	r	r	r	r	r	r	r
N&C	×	×	∘	∘	×	×	∘
$\chi_2 \otimes \chi_{11}$	1		1				
$\chi_2 \otimes \chi_{10}$		1		1			
$\chi_2 \otimes \chi_{27}$			1	1	1	1	
$\chi_2 \otimes \chi_{12}$			1		1		
$\chi_2 \otimes \chi_{28}$				1		1	
$\chi_2 \otimes \chi_{40}$					2	1	3

Group: 6F22 Prime: 7 Block: 3

Nr.	CAS-Nr.	Degree	CC	N&C
1	8	10725	r	×
2	25	205920	r	×
3	30	370656	r	∘
4	48	852930	r	×
5	49	938223	r	∘
6	61	2316600	r	∘
7	64	2555904	r	×

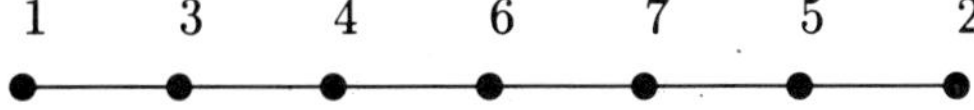

PROJECTIVES:

Nr.	1	2	3	4	5	6	7
CC	r	r	r	r	r	r	r
N&C	×	×	∘	×	∘	∘	×
$\chi_2 \otimes \chi_{10}$	1		1				
$\chi_2 \otimes \chi_{18}$		1			1		
$\chi_2 \otimes \chi_{35}$			1	3		4	2
$\chi_2 \otimes \chi_{31}$			1	1	1	1	2
$\chi_2 \otimes \chi_{17}$			1	1			
$\chi_2 \otimes \chi_{24}$				1		1	
$\chi_2 \otimes \chi_{47}$					1	1	2

Group: 6F22 Prime: 7 Block: 4

Nr.	CAS-Nr.	Degree	CC	N&C
1	66	352	r	×
2	75	27456	r	×
3	78	105600	79	∘
4	79	105600	78	∘
5	84	228800	r	∘
6	99	1029600	r	∘
7	104	1441792	r	×

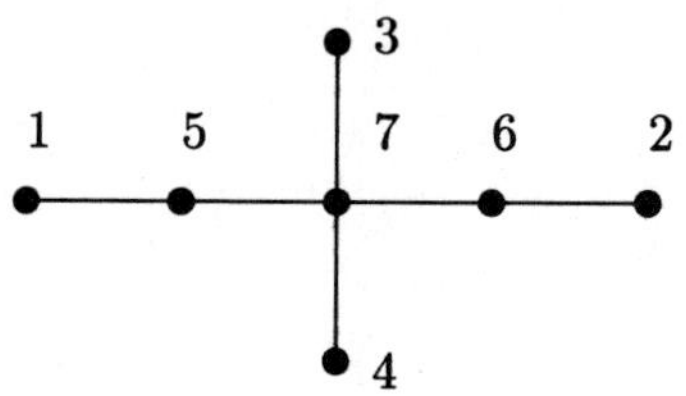

PROJECTIVES:

Nr.	1	2	3	4	5	6	7
CC	r	r	4	3	r	r	r
N&C	×	×	∘	∘	∘	∘	×
$\chi_2 \otimes \chi_{80}$		1				1	
$\chi_2 \otimes \chi_{105}$			1	1		1	3
$\chi_2 \otimes \chi_{102}$					1	1	2

Group: 6F22 Prime: 7 Block: 5

Nr.	CAS-Nr.	Degree	CC	N&C
1	67	2080	r	×
2	74	13728	r	×
3	82	146432	r	∘
4	95	686400	r	×
5	98	915200	r	∘
6	107	2196480	r	∘
7	113	2555904	r	×

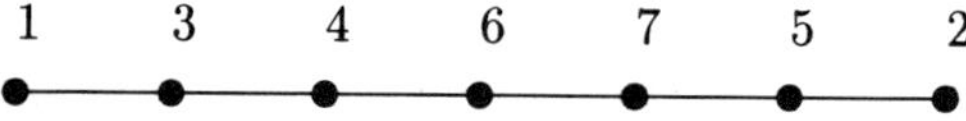

PROJECTIVES:

Nr.	1	2	3	4	5	6	7
CC	r	r	r	r	r	r	r
N&C	×	×	∘	×	∘	∘	×
$\chi_3 \otimes \chi_{85}$	1		3	4		4	2
$\chi_2 \otimes \chi_{85}$		1			2		1
$\chi_2 \otimes \chi_{77}$			1	1			
$\chi_2 \otimes \chi_{88}$				1		1	
$\chi_2 \otimes \chi_{81}$						1	1

Group: 6F22 Prime: 7 Block: 6

Nr.	CAS-Nr.	Degree	CC	N&C
1	68	2080	r	×
2	73	13728	r	×
3	83	146432	r	∘
4	96	686400	r	×
5	97	915200	r	∘
6	108	2196480	r	∘
7	114	2555904	r	×

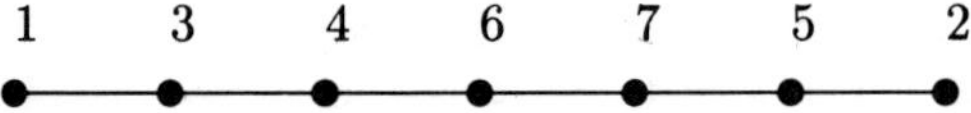

PROJECTIVES:

Nr.	1	2	3	4	5	6	7
CC	r	r	r	r	r	r	r
N&C	×	×	∘	×	∘	∘	×
$\chi_3 \otimes \chi_{86}$	1		3	4		4	2
$\chi_2 \otimes \chi_{86}$		1			2		1
$\chi_2 \otimes \chi_{76}$			1	1			
$\chi_2 \otimes \chi_{87}$				1		1	
$\chi_2 \otimes \chi_{81}$						1	1

Group: 6F22 Prime: 7 Block: 7

Nr.	CAS-Nr.	Degree	CC	AC	N&C
1	115	351	116	ar	×
2	117	7722	118	ar	×
3	123	19305	124	ar	∘
4	145	289575	146	ar	∘
5	153	386100	154	ar	×
6	173	852930	174	ar	×
7	175	938223	176	ar	∘

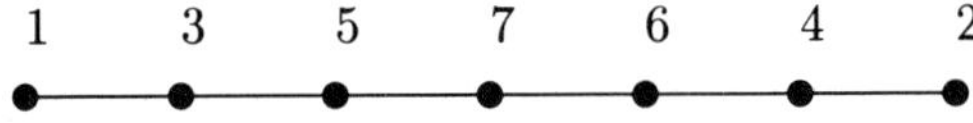

PROJECTIVES:

Nr.	1	2	3	4	5	6	7
AC	ar	ar	ar	ar	ar	ar	ar
N&C	×	×	∘	∘	×	×	∘
$\chi_2 \otimes \chi_{125}$	1		1				
$\chi_2 \otimes \chi_{135}$		1		2		1	
$\chi_2 \otimes \chi_{156}$			1		2		1
$\chi_2 \otimes \chi_{181}$					1	2	3

Group: 6F22 Prime: 7 Block: 8

Complex conjugate to Block 7.

Group: 6F22 Prime: 7 Block: 9

Nr.	CAS-Nr.	Degree	CC	AC	N&C
1	129	42120	130	ar	×
2	143	237600	144	ar	∘
3	161	494208	162	ar	×
4	163	577368	164	ar	×
5	179	1235520	180	ar	∘
6	211	2729376	212	ar	∘
7	213	3088800	214	ar	×

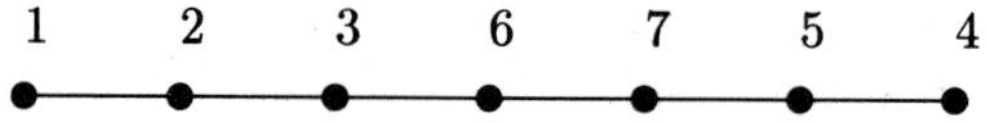

PROJECTIVES:

Nr.	1	2	3	4	5	6	7
AC	ar	ar	ar	ar	ar	ar	ar
N&C	×	∘	×	×	∘	∘	×
$\chi_2 \otimes \chi_{149}$	1	2	2			1	
$\chi_3 \otimes \chi_{127}$	1	1					
$\chi_2 \otimes \chi_{156}$		1	2			1	
$\chi_2 \otimes \chi_{167}$				2	4	1	3
$\chi_2 \otimes \chi_{155}$						1	1

Group: 6F22 Prime: 7 Block: 10

Complex conjugate to Block 9.

Group: 6F22 Prime: 7 Block: 11

Nr.	CAS-Nr.	Degree	CC	AC	N&C
1	137	96525	138	ar	×
2	139	123552	140	ar	×
3	147	360855	148	ar	∘
4	165	608256	166	ar	∘
5	171	772200	172	ar	×
6	203	1853280	204	ar	×
7	205	1876446	206	ar	∘

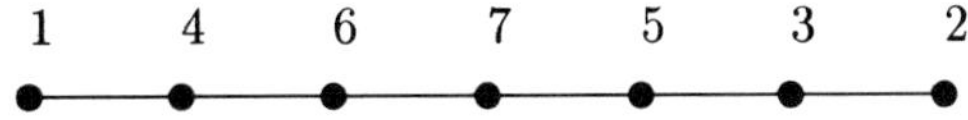

PROJECTIVES:

Nr.	1	2	3	4	5	6	7
AC	ar	ar	ar	ar	ar	ar	ar
N&C	×	×	∘	∘	×	×	∘
$\chi_2 \otimes \chi_{181}$	1			2		2	1
$\chi_2 \otimes \chi_{127}$	1			1			
$\chi_2 \otimes \chi_{156}$		2	2	1		2	1
$\chi_2 \otimes \chi_{149}$		1	2		2	1	2
$\chi_2 \otimes \chi_{131}$					1		1
$\chi_2 \otimes \chi_{155}$						2	2

Group: 6F22 Prime: 7 Block: 12

Complex conjugate to Block 11.

Group: 6F22 Prime: 7 Block: 13

Nr.	CAS-Nr.	Degree	CC	AC	A	N&C
1	215	1728	217	216	218	○
2	219	61776	221	220	222	×
3	224	71280	226	223	225	○
4	258	1123200	260	257	259	×
5	261	1235520	263	262	264	○
6	276	2965248	278	275	277	○
7	279	3088800	281	280	282	×

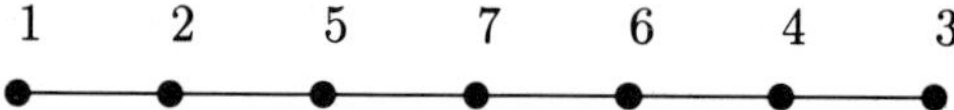

PROJECTIVES:

Nr.	1	2	3	4	5	6	7
N&C	○	×	○	×	○	○	×
$\chi_{220} \otimes \chi_4$	1	5			6		2
$\chi_3 \otimes \chi_{249}$	1	2		1	6	6	10
$\chi_{216} \otimes \chi_4$	1	1					
$\chi_2 \otimes \chi_{249}$			1	2		2	1
$\chi_{215} \otimes \chi_4$			1	1			
$\chi_2 \otimes \chi_{237}$				2		3	1
$\chi_2 \otimes \chi_{239}$					1		1

The column 'A' in the above block information table lists the characters which are conjugate under the outer automorphism to the characters of the second column.

Group: 6F22 Prime: 7 Block: 14

Conjugate to Block 13 under the composition of the outer automorphism with complex conjugation.

Group: 6F22 Prime: 7 Block: 15

Complex conjugate to Block 13.

Group: 6F22 Prime: 7 Block: 16

Complex conjugate to Block 14.

Group: 6F22 Prime: 7 Block: 17

Nr.	CAS-Nr.	Degree	CC	AC	N&C
1	227	112320	229	ar	∘
2	228	112320	230	ar	∘
3	231	123552	232	ar	×
4	233	308880	235	ar	∘
5	234	308880	236	ar	∘
6	245	359424	247	ar	×
7	246	359424	248	ar	×

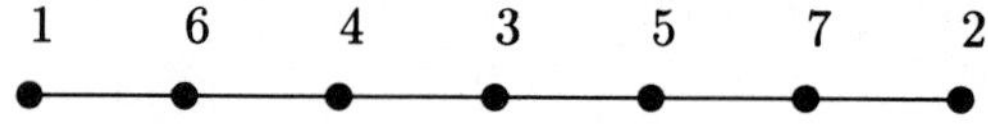

PROJECTIVES:

Nr.	1	2	3	4	5	6	7
AC	ar	ar	ar	ar	ar	ar	ar
N&C	∘	∘	×	∘	∘	×	×
$\chi_2 \otimes \chi_{239}$	1					1	
$\chi_2 \otimes \chi_{237}$		1					1
$\chi_2 \otimes \chi_{250}$			1	2		1	
$\chi_2 \otimes \chi_{249}$			1		2		1

Group: 6F22 Prime: 7 Block: 18

Complex conjugate to Block 17.

Group: 6F22 Prime: 11 Block: 1

Nr.	CAS-Nr.	Degree	CC	N&C
1	1	1	r	×
2	9	13650	r	∘
3	40	582400	41	∘
3	41	582400	40	∘
4	54	1360800	r	×
5	57	1791153	r	×
6	63	2555904	r	∘

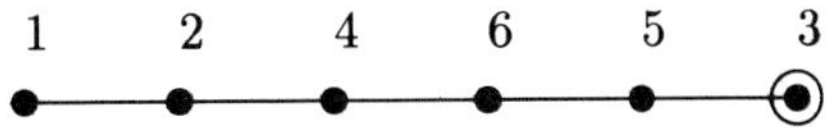

1 — 2 — 4 — 6 — 5 — 3 (◉)

PROJECTIVES:

Nr.	1	2	3	4	5	6
CC	r	r	r	r	r	r
N&C	×	∘	∘	×	×	∘
$\chi_3 \otimes \chi_3$	1	1				
$\chi_2 \otimes \chi_{27}$		1		2		1

Group: 6F22 Prime: 11 Block: 2

Nr.	CAS-Nr.	Degree	CC	N&C
1	2	78	r	×
2	12	43680	r	∘
3	48	852930	r	×
4	51	982800	52	∘
4	52	982800	51	∘
5	64	2555904	r	∘
6	65	2729376	r	×

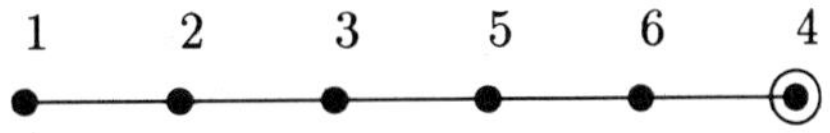

1 — 2 — 3 — 5 — 6 — 4 (◉)

PROJECTIVES:

Nr.	1	2	3	4	5	6
CC	r	r	r	r	r	r
N&C	×	∘	×	∘	∘	×
$\chi_2 \otimes \chi_6$	1	1				
$\chi_2 \otimes \chi_{17}$		1	1			
$\chi_2 \otimes \chi_{26}$			1		2	1

Group: 6F22 Prime: 11 Block: 3

Nr.	CAS-Nr.	Degree	CC	N&C
1	67	2080	r	×
2	71	5824	72	∘
2	72	5824	71	∘
3	85	235872	r	∘
4	92	436800	r	×
5	110	2358720	r	×
6	113	2555904	r	∘

PROJECTIVES:

Nr.	1	2	3	4	5	6
CC	r	r	r	r	r	r
N&C	×	∘	∘	×	×	∘
$\chi_2 \otimes \chi_{74}$	1		1			
$\chi_2 \otimes \chi_{76}$		1		1		
$\chi_2 \otimes \chi_{80}$			1		1	
$\chi_2 \otimes \chi_{84}$					1	1

Group: 6F22 Prime: 11 Block: 4

Nr.	CAS-Nr.	Degree	CC	N&C
1	68	2080	r	×
2	69	5824	70	∘
2	70	5824	69	∘
3	86	235872	r	∘
4	91	436800	r	×
5	109	2358720	r	×
6	114	2555904	r	∘

PROJECTIVES:

Nr.	1	2	3	4	5	6
CC	r	r	r	r	r	r
N&C	×	∘	∘	×	×	∘
$\chi_2 \otimes \chi_{73}$	1		1			
$\chi_2 \otimes \chi_{77}$		1		1		
$\chi_2 \otimes \chi_{80}$			1		1	
$\chi_2 \otimes \chi_{84}$					1	1

Group: 6F22 Prime: 11 Block: 5

Nr.	CAS-Nr.	Degree	CC	AC	N&C
1	115	351	116	ar	∘
2	129	42120	130	ar	×
3	173	852930	174	ar	×
4	195	1658475	197	ar	∘
4	196	1658475	198	ar	∘
5	208	1965600	210	ar	∘
6	211	2729376	212	ar	×

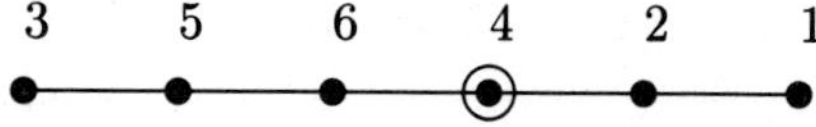

PROJECTIVES:

Nr.	1	2	3	4	5	6
AC	ar	ar	ar	ar	ar	ar
N&C	∘	×	×	∘	∘	×
$\chi_2 \otimes \chi_{117}$	1	1				
$\chi_2 \otimes \chi_{153}$			1		2	1
$\chi_2 \otimes \chi_{155}$				1		1

Group: 6F22 Prime: 11 Block: 6

Complex conjugate to Block 5.

Group: 6F22 Prime: 11 Block: 7

Nr.	CAS-Nr.	Degree	CC	AC	N&C
1	125	19656	126	ar	○
2	135	61425	136	ar	×
3	141	235872	142	ar	○
4	149	368550	151	ar	×
4	150	368550	152	ar	×
5	201	1791153	202	ar	×
6	207	1965600	209	ar	○

PROJECTIVES:

Nr.	1	2	3	4	5	6
AC	ar	ar	ar	ar	ar	ar
N&C	○	×	○	×	×	○
$\chi_2 \otimes \chi_{123}$	1			1		
$\chi_2 \otimes \chi_{117}$		1	1			
$\chi_2 \otimes \chi_{155}$			1		2	1

Group: 6F22 Prime: 11 Block: 8

Complex conjugate to Block 7.

Group: 6F22 Prime: 11 Block: 9

Nr.	CAS-Nr.	Degree	CC	AC	A	N&C
1	215	1728	217	216	218	×
2	227	112320	229	228	230	∘
3	237	314496	241	239	243	×
3	238	314496	242	240	244	×
4	246	359424	248	245	247	∘
5	254	967680	256	253	255	∘
6	258	1123200	260	257	259	×

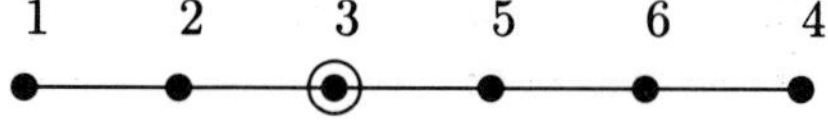

PROJECTIVES:

Nr.	1	2	3	4	5	6
N&C	×	∘	×	∘	∘	×
$\chi_2 \otimes \chi_{220}$	1	1				
$\chi_2 \otimes \chi_{262}$		1	1			
$\chi_2 \otimes \chi_{269}$			1		3	2

The column 'A' in the above block information table lists the characters which are conjugate under the outer automorphism to the characters of the second column.

Group: 6F22 Prime: 11 Block: 10

Conjugate to Block 9 under the composition of the outer automorphism with complex conjugation.

Group: 6F22 Prime: 11 Block: 11

Complex conjugate to Block 9.

Group: 6F22 Prime: 11 Block: 12

Complex conjugate to Block 10.

Group: 6F22 Prime: 13 Block: 1

Nr.	CAS-Nr.	Degree	CC	N&C
1	1	1	r	×
2	7	3080	r	∘
3	22	138600	r	×
3	23	138600	r	×
4	29	360855	r	×
5	38	577368	r	∘
6	54	1360800	r	∘
7	56	1441792	r	×

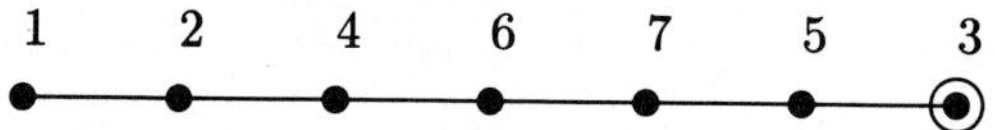

PROJECTIVES:

Nr.	1	2	3	4	5	6	7
CC	r	r	r	r	r	r	r
N&C	×	∘	×	×	∘	∘	×
$\chi_2 \otimes \chi_2$	1	1					
$\chi_{11} \otimes \chi_2$		1		1			
$\chi_{27} \otimes \chi_2$				1		2	1

Group: 6F22 Prime: 13 Block: 2

Nr.	CAS-Nr.	Degree	CC	N&C
1	66	352	r	×
2	78	105600	79	×
3	79	105600	78	×
4	80	123200	r	∘
5	81	133056	r	×
6	104	1441792	r	×
7	105	1663200	r	∘
7	106	1663200	r	∘

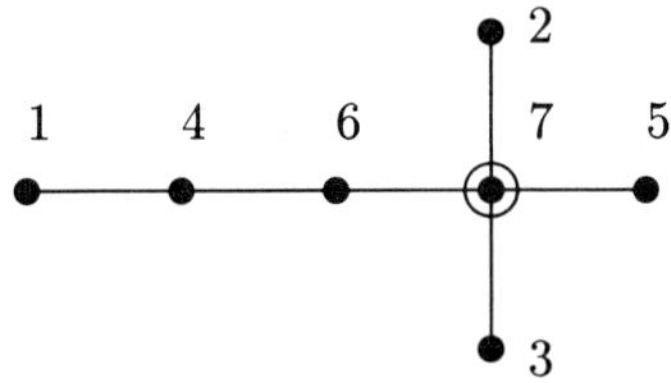

PROJECTIVES:

Nr.	1	2	3	4	5	6	7
CC	r	3	2	r	r	r	r
N&C	×	×	×	∘	×	×	∘
$\chi_{75} \otimes \chi_2$	1			1			
$\chi_{78} \otimes \chi_2$			1				1
$\chi_{85} \otimes \chi_2$				1		1	
$\chi_{91} \otimes \chi_2$					1		1
$\chi_{87} \otimes \chi_2$						1	1

Group: 6F22 Prime: 13 Block: 3

Nr.	CAS-Nr.	Degree	CC	AC	N&C
1	119	12474	121	120	×
1	120	12474	122	119	×
2	131	51975	132	ar	×
3	143	237600	144	ar	○
4	147	360855	148	ar	×
5	163	577368	164	ar	○
6	165	608256	166	ar	○
7	177	997920	178	ar	×

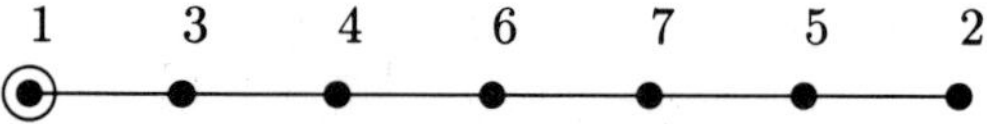

PROJECTIVES:

Nr.	1	2	3	4	5	6	7
AC	ar	ar	ar	ar	ar	ar	ar
N&C	×	×	○	×	○	○	×
$\chi_{117} \otimes \chi_2$	1		1				
$\chi_{135} \otimes \chi_2$		1			1		
$\chi_{156} \otimes \chi_2$			1	2		1	
$\chi_{129} \otimes \chi_2$					1		1
$\chi_{127} \otimes \chi_2$						1	1

Group: 6F22 Prime: 13 Block: 4

Complex conjugate to Block 3.

Group: 6F22 Prime: 13 Block: 5

Nr.	CAS-Nr.	Degree	CC	AC	N&C
1	215	1728	217	216	∘
2	216	1728	218	215	∘
3	223	71280	225	224	×
4	224	71280	226	223	×
5	253	967680	255	254	∘
6	254	967680	256	253	∘
7	269	1796256	271	270	×
7	270	1796256	272	269	×

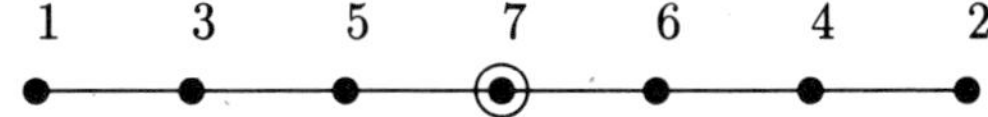

PROJECTIVES:

Nr.	1	2	3	4	5	6	7
AC	2	1	4	3	6	5	ar
N&C	∘	∘	×	×	∘	∘	×
$\chi_{223} \otimes \chi_2$		1		2		1	
$\chi_{237} \otimes \chi_2$						1	1

Group: 6F22 Prime: 13 Block: 6

Complex conjugate to Block 5.

6.18 The Harada–Norton Group

Group: HA Prime: 3 Block: 3

Nr.	CAS-Nr.	Degree	CC	N&C
1	23	406296	r	×
2	38	1625184	r	×
3	39	2031480	r	∘

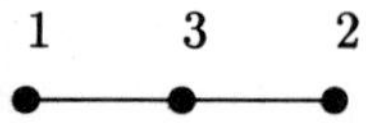

1 — 3 — 2

Group: HA Prime: 5 Block: 2

Nr.	CAS-Nr.	Degree	CC	N&C
1	24	653125	r	×
2	43	2784375	r	∘
3	45	3200000	r	×
4	50	4809375	r	×
5	54	5878125	r	∘

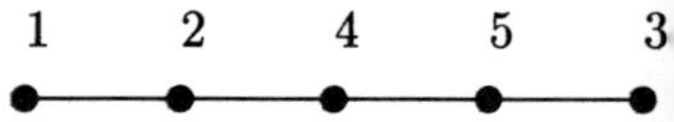

1 — 2 — 4 — 5 — 3

PROJECTIVES:

Nr.:	1	2	3	4	5
CC:	r	r	r	r	r
N&C:	×	∘	×	×	∘
$\chi_2 \otimes \chi_{25}$			2	1	3
$\chi_2 \otimes (\chi_{24} + \chi_{43})$	3	8	3	8	6

The first tensor product shows that node 1 is joined to 2, since node 5 is in the middle of 3 and 4.

Group: HA | Prime: 7 | Block: 1

Nr.	CAS-Nr.	Degree	CC	N&C
1	1	1	r	×
2	8	8910	r	∘
3	18	267520	r	×
4	38	1625184	r	×
5	45	3200000	r	∘
6	49	4561920	r	∘
7	54	5878125	r	×

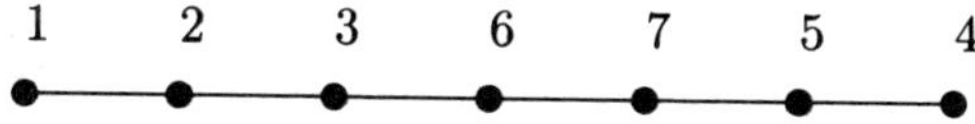

PROJECTIVES:

Nr.:	1	2	3	4	5	6	7
CC:	r	r	r	r	r	r	r
N&C:	×	∘	×	×	∘	∘	×
$\chi_2 \otimes \chi_2$	1	1					
$\chi_{11} \otimes \chi_2$		1	1				
$\chi_{18} \otimes \chi_2$			1			1	
$\chi_{19} \otimes \chi_2$						1	1

Group: HA Prime: 7 Block: 2

Nr.	CAS-Nr.	Degree	CC	N&C
1	4	760	r	×
2	9	9405	r	×
3	10	16929	r	∘
4	24	653125	r	×
5	39	2031480	r	∘
6	46	3424256	r	∘
7	50	4809375	r	×

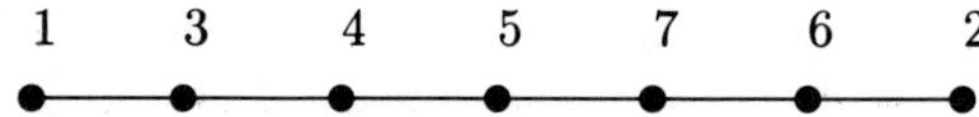

PROJECTIVES:

Nr.:	1	2	3	4	5	6	7
CC:	r	r	r	r	r	r	r
N&C:	×	×	∘	×	∘	∘	×
$\chi_3 \otimes \chi_2$	1		1				
$\chi_{13} \otimes \chi_2$		1				1	
$\chi_{20} \otimes \chi_2$			1	1		1	1
$\chi_{23} \otimes \chi_2$				1	2		1
$\chi_{21} \otimes \chi_2$						1	1

Group: HA Prime: 7 Block: 3

Nr.	CAS-Nr.	Degree	CC	N&C
1	5	3344	r	×
2	17	214016	r	×
3	23	406296	r	∘
4	33	1354320	r	∘
5	37	1575936	r	×
6	40	2375000	r	×
7	41	2407680	r	∘

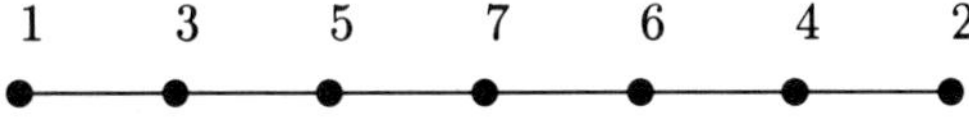

PROJECTIVES:

Nr.:	1	2	3	4	5	6	7
CC:	r	r	r	r	r	r	r
N&C:	×	×	∘	∘	×	×	∘
$\chi_5 \otimes \chi_2$	1		1				
$\chi_{12} \otimes \chi_2$		1	1	1	1		
$\chi_{14} \otimes \chi_2$		1		1	1		1
$\chi_{17} \otimes \chi_2$			1		1		
$\chi_6 \otimes \chi_6$				1		1	
$\chi_{29} \otimes \chi_2$					1		1
$\chi_{25} \otimes \chi_2$						1	1

Group: HA Prime: 7 Block: 4

Nr.	CAS-Nr.	Degree	CC	N&C
1	15	69255	16	∘
1	16	69255	15	∘
2	19	270864	r	×
3	43	2784375	r	×
4	44	2985984	r	∘

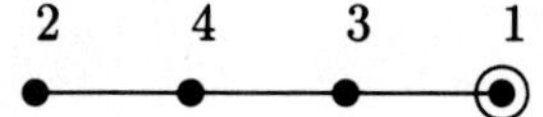

PROJECTIVES:

Nr.:	1	2	3	4
CC:	r	r	r	r
N&C:	∘	×	×	∘
$\chi_{47} \otimes \chi_2$	1		5	4

Group: HA | Prime: 11 | Block: 1

Nr.	CAS-Nr.	Degree	CC	N&C
1	1	1	r	×
2	4	760	r	×
3	34	1361920	r	∘
4	37	1575936	r	∘
5	40	2375000	r	×
6	45	3200000	r	×
7	47	3878280	r	∘
8	48	4156250	r	∘
9	50	4809375	r	∘
10	51	5103000	r	×
11	52	5103000	r	×

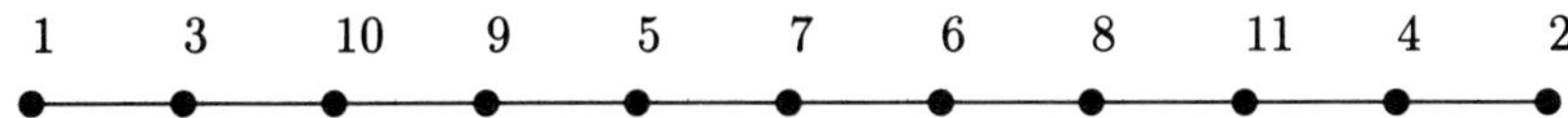

PROJECTIVES:

Nr.:	1	2	3	4	5	6	7	8	9	10	11
CC:	r	r	r	r	r	r	r	r	r	r	r
N&C:	×	×	∘	∘	×	×	∘	∘	∘	×	×
$\chi_8 \otimes \chi_8$	1		1	1	1		1		1	1	1
χ_5^{2+}	1		1								
$\chi_{10} \otimes \chi_8$		2	1	4	1			1	3	3	3
$\chi_4 \otimes \chi_5$		1		1							
$\chi_8 \otimes \chi_5$			1	1						1	1
$\chi_7 \otimes \chi_6$			1		1	1	1	1	1	1	1
$\chi_{15} \otimes \chi_5$				1	2	3	5	3	4	4	4
$\chi_6 \otimes \chi_5$				1					1	1	1
$\chi_2 \otimes \chi_{21}$						1	1	1	1	1	1
χ_6^{2-}								1	1	1	1

Without loss of generality we assume that of the pair of algebraically conjugate characters χ_{51} and χ_{52} (nodes 10 and 11), χ_{51} comes first on the tree. Then only the tree given above is consistent with all the projectives.

Group: HA Prime: 11 Block: 2

Nr.	CAS-Nr.	Degree	CC	N&C
1	2	133	r	×
2	3	133	r	×
3	15	69255	16	∘
4	16	69255	15	∘
5	25	656250	26	×
6	26	656250	25	×
7	27	718200	r	∘
8	28	718200	r	∘
9	35	1361920	36	∘
10	36	1361920	35	∘
11	44	2985984	r	×

PROJECTIVES:

Nr.:	1	2	3	4	5	6	7	8	9	10	11
CC:	r	r	4	3	6	5	r	r	10	9	r
N&C:	×	×	∘	∘	×	×	∘	∘	∘	∘	×
$\chi_2 \otimes \chi_6$	1						1				
$\chi_3 \otimes \chi_7$		1						1			
$\chi_9 \otimes \chi_9$			1	1	1	1					
$\chi_2 \otimes \chi_{38}$					1	1			2	2	2
$\chi_4 \otimes \chi_{18}$							1	1	1	1	4

Since all three pairs of complex conjugate characters have their values in fields which are pairwise disjoint, we have without loss of generality only one tree.

Group: HA Prime: 19 Block: 1

Nr.	CAS-Nr.	Degree	CC	N&C
1	1	1	r	×
2	8	8910	r	∘
3	25	656250	26	∘
3	26	656250	25	∘
4	42	2661120	r	∘
5	43	2784375	r	×
6	44	2985984	r	×
7	45	3200000	r	×
8	49	4561920	r	×
9	51	5103000	r	∘
10	52	5103000	r	∘

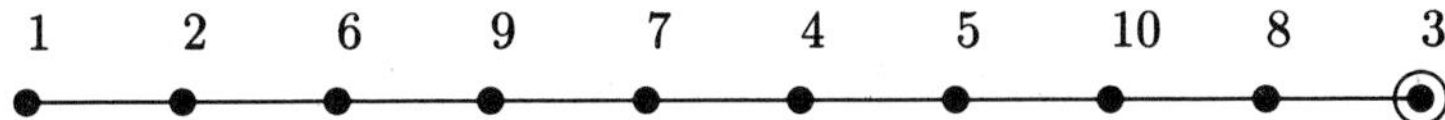

PROJECTIVES:

Nr.:	1	2	3	4	5	6	7	8	9	10
CC:	r	r	r	r	r	r	r	r	r	r
N&C:	×	∘	∘	∘	×	×	×	×	∘	∘
$\chi_2 \otimes \chi_2$	1	1								
$\chi_{20} \otimes \chi_2$		1		1	2	2			1	1
$\chi_{11} \otimes \chi_2$		1				1				
$\chi_{21} \otimes \chi_2$				1		1	1	1	1	1
$\chi_{17} \otimes \chi_2$					1	1			1	1
$\chi_{19} \otimes \chi_2$						1		1	1	1
$\chi_{22} \otimes \chi_2$							1	1	1	1

We have one pair of algebraically conjugate characters, namely χ_{51} and χ_{52} (nodes 9 and 10). Assuming as usual that χ_{51} comes first on the tree, the given projectives leave only the above tree.

6.19 The Lyons Group

Group: LY Prime: 3 Block: 2

Nr.	CAS-Nr.	Degree	CC	N&C
1	12	3028266	r	×
2	49	52994655	r	×
3	51	56022921	r	∘

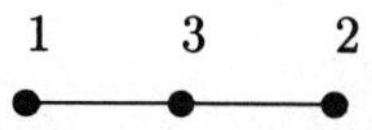

1 3 2

Group: LY Prime: 3 Block: 3

Nr.	CAS-Nr.	Degree	CC	N&C
1	20	18395586	r	∘
2	21	18395586	r	×
2	22	18395586	r	×

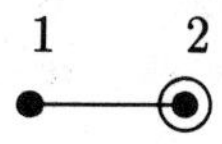

1 2

Group: LY Prime: 7 Block: 1

Nr.	CAS-Nr.	Degree	CC	N&C
1	1	1	r	×
2	13	3073960	r	×
3	20	18395586	r	∘
4	23	19212250	r	×
5	24	21312500	r	∘
6	25	21312500	r	∘
7	38	38734375	r	×

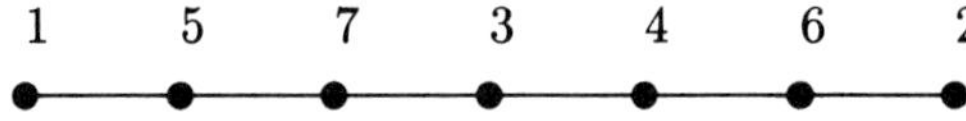

PROJECTIVES:

Nr.:	1	2	3	4	5	6	7
CC:	r	r	r	r	r	r	r
N&C:	×	×	∘	×	∘	∘	×
χ_5^{2-}		1			1	1	1
$\chi_7 \otimes \chi_5$			2	1	1	1	3
$\chi_3 \otimes \chi_7$			1	1			
$\chi_4 \otimes \chi_5$			1				1

We have two algebraically conjugate characters in this block, namely χ_{24} and χ_{25} (nodes 5 and 6). As usual we assume that χ_{24} comes first on the Brauer tree. Then only the one tree given above is consistent with all projectives.

Group: LY Prime: 7 Block: 2

Nr.	CAS-Nr.	Degree	CC	N&C
1	2	2480	3	×
2	3	2480	2	×
3	31	27252720	r	∘
4	32	27252720	r	∘
5	37	36887520	r	∘
6	47	45694000	r	×
7	48	45694000	r	×

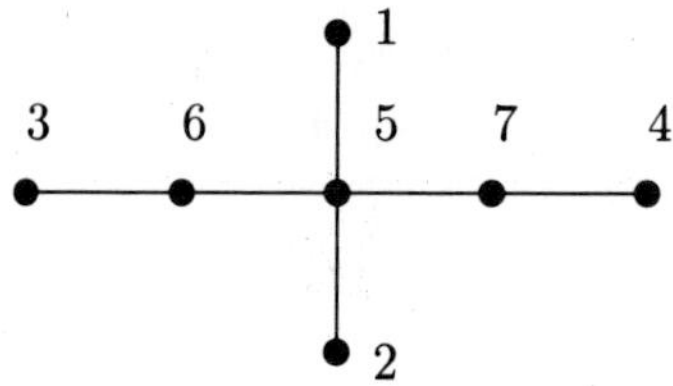

PROJECTIVES:

Nr.:	1	2	3	4	5	6	7
CC:	2	1	r	r	r	r	r
N&C:	×	×	∘	∘	∘	×	×
$\chi_2 \otimes \chi_5$		1			1		
χ_5^{2-}					2	1	1

We have two pairs of algebraically conjugate characters, namely χ_{31}, χ_{32} (nodes 3 and 4) and χ_{47}, χ_{48} (nodes 6 and 7). The characters in the two orbits generate disjoint fields, each of which in turn is disjoint from the field generated by the characters of the principal block. Hence we can assume without loss of generality that in each orbit the character with the smallest CAS-number comes first on the Brauer tree. Then only the tree given above is consistent with all the projectives.

Group: LY Prime: 7 Block: 3

Nr.	CAS-Nr.	Degree	CC	N&C
1	4	45694	r	○
2	11	1534500	r	×
3	19	16906780	r	×
4	21	18395586	r	○
4	22	18395586	r	○

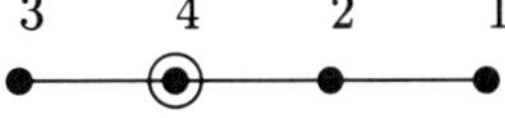

Without using any projectives at all, we only have the following two possible Brauer trees:

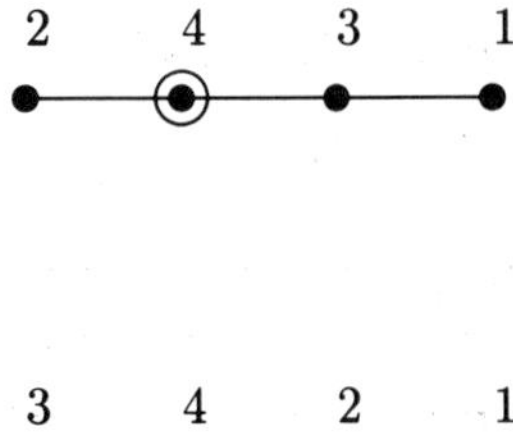

Consider the permutation character on the maximal subgroup $G_2(5)$. We have $\mathrm{Ind}_{G_2(5)}^{Ly}(\chi_1) \approx \chi_4 + \chi_{11}$, where χ_1 denotes the trivial character of $G_2(5)$ and $\approx$ as usual indicates the character restricted to the block. Every direct summand of a trivial source module is a trivial source module and hence is liftable. In a block with cyclic defect group every indecomposable trivial source module is either projective or lifts to an irreducible of type cross (or to the sum of the exceptionals, which must then be of type cross). It follows that $\chi_4 + \chi_{11}$ is projective, hence the second tree is correct.

Group: LY Prime: 7 Block: 4

Nr.	CAS-Nr.	Degree	CC	N&C
1	10	1152735	r	×
2	12	3028266	r	×
3	36	33813560	r	∘
4	45	45648306	r	∘
5	46	45648306	r	∘
6	51	56022921	r	×
7	52	64906250	r	×

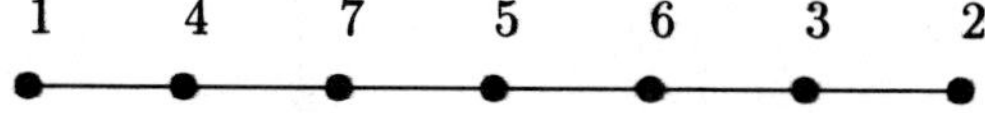

PROJECTIVES:

Nr.:	1	2	3	4	5	6	7
CC:	r	r	r	r	r	r	r
N&C:	×	×	∘	∘	∘	×	×
χ_5^{2+}		2	3			1	
χ_5^{2-}				2	1	1	2
$\chi_2 \otimes \chi_7$					1	1	
$\chi_3 \otimes \chi_7$					1		1

Group: LY Prime: 7 Block: 5

Nr.	CAS-Nr.	Degree	CC	N&C
1	14	4226695	r	×
2	29	27252720	r	∘
2	30	27252720	r	∘
3	35	30739600	r	∘
4	50	53765625	r	×

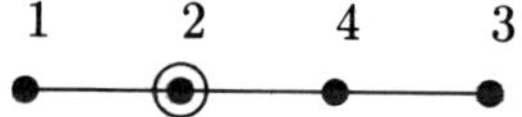

PROJECTIVES:

Nr.:	1	2	3	4
CC:	r	r	r	r
N&C:	×	∘	∘	×
χ_5^{2+}	1	2		1

Group: LY Prime: 11 Block: 1

Nr.	CAS-Nr.	Degree	CC	N&C
1	1	1	r	×
2	5	48174	6	○
2	6	48174	5	○
3	10	1152735	r	×
4	12	3028266	r	○
5	13	3073960	r	○
6	15	4997664	r	×

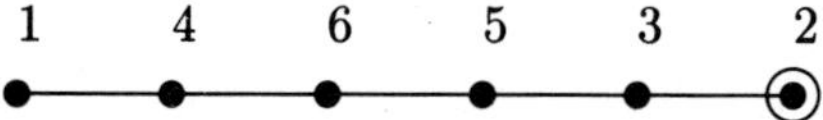

PROJECTIVES:

Nr.:	1	2	3	4	5	6
CC:	r	r	r	r	r	r
N&C:	×	○	×	○	○	×
χ_4^{2+}	1			2		1
$\chi_9 \otimes \chi_4$					1	1

Group: LY Prime: 11 Block: 2

Nr.	CAS-Nr.	Degree	CC	N&C
1	2	2480	3	∘
1	3	2480	2	∘
2	16	5379430	r	×
3	35	30739600	r	×
4	37	36887520	r	∘
5	49	52994655	r	∘
6	50	53765625	r	×

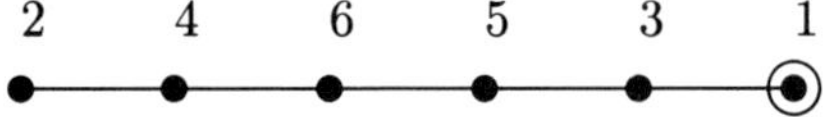

PROJECTIVES:

Nr.:	1	2	3	4	5	6
CC:	r	r	r	r	r	r
N&C:	∘	×	×	∘	∘	×
$\chi_5 \otimes \chi_4$			1	2	2	3
$\chi_2 \otimes \chi_4$			1		1	
$\chi_2 \otimes \chi_9$					1	1

There are two trees which are consistent with the above set of projectives.

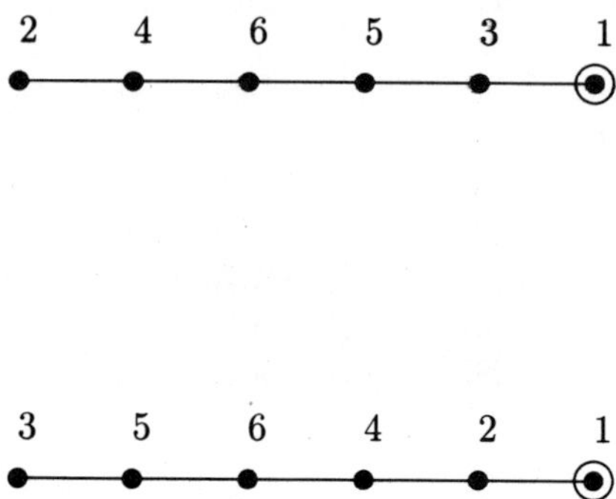

To rule out the second one we use the following argument. The centralizer of an element of order 11 in the Lyons group is the direct product of a cyclic group of order 11 with a symmetric group on 3 letters. In the the subgroup of the Lyons group isomorphic to the triple cover of the McLaughlin group (which is not maximal) this centralizer is cyclic of order 33. Observe, e.g. by inspecting the values of the characters on 11-elements, that the present block corresponds via Brauer correpondence to the alternating character of the symmetric group on three letters. By the generalized Green correspondence due to Burry and Carlson, the permutation module on the cosets of the triple cover of McLaughlin's group must have a non-projective direct summand in the present block. A non-projective trivial source module is liftable to an irreducible character which is of type cross (or to the sum of the exceptionals, which must then be of type cross). Since the dimension of the permutation module is smaller than the degree of χ_{35}, we see that the indecomposable non-projective component has Brauer character χ_{16}. Now the usual duality argument shows that the non-projective component in the present block must be irreducible. Hence χ_{16} occupies a leaf of the tree, ruling out the second possibility.

Group: LY Prime: 11 Block: 3

Nr.	CAS-Nr.	Degree	CC	N&C
1	7	120064	8	∘
1	8	120064	7	∘
2	17	10758860	r	×
3	23	19212250	r	×
4	38	38734375	r	∘
5	51	56022921	r	∘
6	52	64906250	r	×

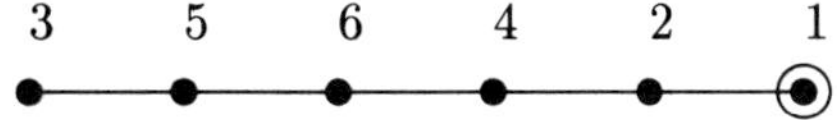

PROJECTIVES:

Nr.:	1	2	3	4	5	6
CC:	r	r	r	r	r	r
N&C:	∘	×	×	∘	∘	×
χ_4^{2+}		1		1		
$\chi_5 \otimes \chi_4$				1	2	3

We have the following two trees consistent with the above set of projectives:

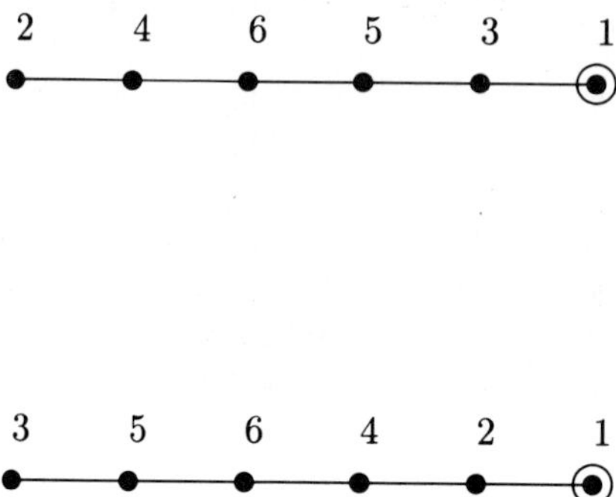

To rule out the first of these trees we consider the component of the permutation module on the cosets of the subgroup isomorphic to 2A11, the double cover of the alternating group on 11 letters. We have

$$\mathrm{Ind}_{2A11}^{Ly}(\chi_1) \approx \chi_{23} + \chi_{51} + \chi_{52}.$$

With the arguments already used above we find that one of χ_{23} or χ_{52} is the Brauer character of the non-projective indecomposable direct summand of the permutation module in this block. This character then must occupy an end node of the Brauer tree. Since in both possibilities the character χ_{52}, which corresponds to node 6, is located at the centre of the tree, we must have that character χ_{23} is at the end of the tree. This excludes the first possibility since χ_{23} is node 3.

Group: LY Prime: 31 Block: 1

Nr.	CAS-Nr.	Degree	CC	N&C
1	1	1	r	×
2	7	120064	8	×
3	8	120064	7	×
4	9	381766	r	×
5	15	4997664	r	∘
6	38	38734375	r	∘
7	39	43110144	r	×
7	40	43110144	r	×
7	41	43110144	r	×
7	42	43110144	r	×
7	43	43110144	r	×

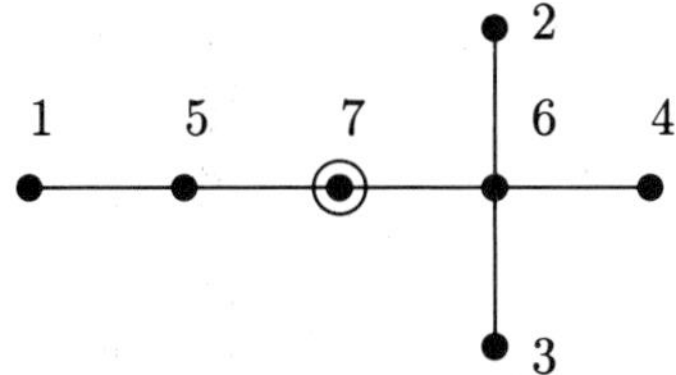

PROJECTIVES:

Nr.:	1	2	3	4	5	6	7
CC:	r	3	2	r	r	r	r
N&C:	×	×	×	×	∘	∘	×
$\chi_3 \otimes \chi_2$	1				1		
$\chi_7 \otimes \chi_2$			1			1	
χ_5^{2+}					1		1

Group: LY Prime: 37 Block: 1

Nr.	CAS-Nr.	Degree	CC	N&C
1	1	1	r	×
2	2	2480	3	×
3	3	2480	2	×
4	4	45694	r	∘
5	7	120064	8	∘
6	8	120064	7	∘
7	11	1534500	r	∘
8	12	3028266	r	×
9	24	21312500	r	×
9	25	21312500	r	×
10	33	28787220	r	∘
11	39	43110144	r	×
12	40	43110144	r	×
13	41	43110144	r	×
14	42	43110144	r	×
15	43	43110144	r	×
16	47	45694000	r	∘
17	48	45694000	r	∘
18	49	52994655	r	∘
19	52	64906250	r	∘

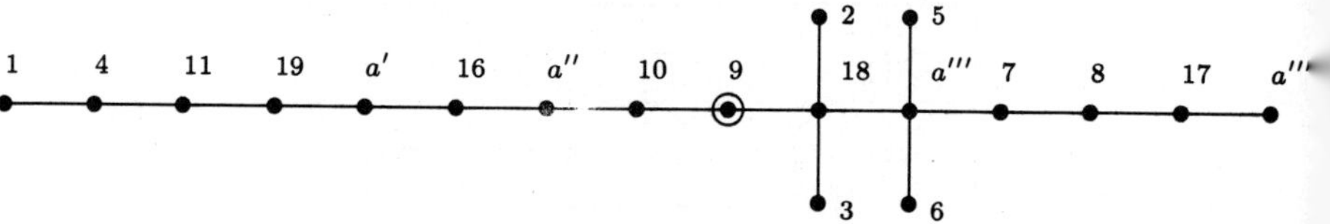

$$\{a', a'', a''', a''''\} = \{12, 13, 14, 15\}$$

PROJECTIVES:

Nr.:	1	2	3	4	5	6	7	8	9	10	11	12	13	14	15	16	17	18	19
CC:	r	3	2	r	6	5	r	r	r	r	r	r	r	r	r	r	r	r	r
N&C:	×	×	×	∘	∘	∘	∘	×	×	∘	×	×	×	×	×	∘	∘	∘	∘
$\mathrm{Ind}_{2.A_{11}}(\chi_1)$	1			1			2	2			1	1	1	1	1	2	2		1
$\chi_2 \otimes \chi_{10}$		1						1	1		2	2	2	2	2	3	3	4	3
$\chi_3 \otimes \chi_6$		1																1	
$\chi_4 \otimes \chi_{10}$				1			2	4	23	31	45	45	45	45	45	49	49	55	65
χ_5^{2+}				1			1	2			1	1	1	1	1	2	2	1	
$\chi_8 \otimes \chi_6$					1				2	4	5	5	5	5	5	5	5	5	7
χ_9^{2-}							1		36	46	61	61	61	61	61	61	61	79	93
$\chi_4 \otimes \chi_6$								1			2	2	2	2	2	3	3	2	3
$\chi_7 \otimes \chi_6$									2	4	5	5	5	5	5	5	5	5	8

Up to algebraic conjugacy, only ten trees are consistent with the projectives given above. We have two possible types of real stems and 5 possibilities for the complex conjugate pair of nodes 5 and 6. Here are the two possible types of trees with this pair removed:

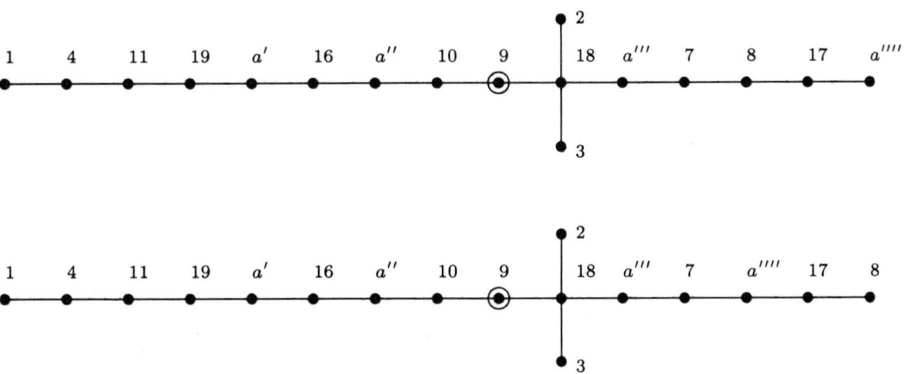

The complex conjugate pair of nodes can be located at nodes 9, 11, 12, 13 or 14. To complete the proof we use Green correspondence and the following tensor products:

$$\begin{array}{rcl} \chi_2 \otimes \chi_2 & \approx & \chi_{12} \\ \chi_2 \otimes \chi_7 & \approx & \chi_8 \end{array}$$

In terms of nodes these characters are as follows:

$$\begin{array}{rcl} 2 \otimes 2 & \approx & 8 \\ 2 \otimes 5 & \approx & 6 \end{array}$$

If the nodes 5 and 6 come out of the tree to the left of 2 and 3, then the irreducible module corresponding to the edge joining 2 with 18 has Green correspondent $1A6$. Otherwise the Green correspondent is $1A5$. If it is $1A6$, then the complex conjugate edge has $1A12$. Also, $2 \otimes 2$ has then Green correspondent $1A6 \otimes 1A6 = 1A12$. However, node 3 does not occur at all in $2 \otimes 2$. This contradiction shows that the nodes 5 and 6 come out right of 2 and 3. Furthermore, there must be an edge incident to node 8 with label $5 + 5 = 10$. This can happen only if 15 is the leaf. Finally the second tensor product shows that the edge joining nodes 5 and 14 has label 7, rather than 12, completing the proof of the planar embedding. Observe that all our arguments are independent of the actual location of a particular member of an orbit of algebraically conjugates, since the characters in an orbit all occur with equal multiplicity in the projectives we have used.

Group: LY Prime: 67 Block: 1

Nr.	CAS-Nr.	Degree	CC	N&C
1	1	1	r	×
2	2	2480	3	×
3	3	2480	2	×
4	5	48174	6	×
5	6	48174	5	×
6	11	1534500	r	○
7	20	18395586	r	○
8	21	18395586	r	○
9	22	18395586	r	○
10	24	21312500	r	×
11	25	21312500	r	×
12	26	22609664	r	×
12	27	22609664	r	×
12	28	22609664	r	×
13	29	27252720	r	×
14	30	27252720	r	×
15	31	27252720	r	×
16	32	27252720	r	×
17	39	43110144	r	○
18	40	43110144	r	○
19	41	43110144	r	○
20	42	43110144	r	○
21	43	43110144	r	○
22	44	44159500	r	×
23	50	53765625	r	×

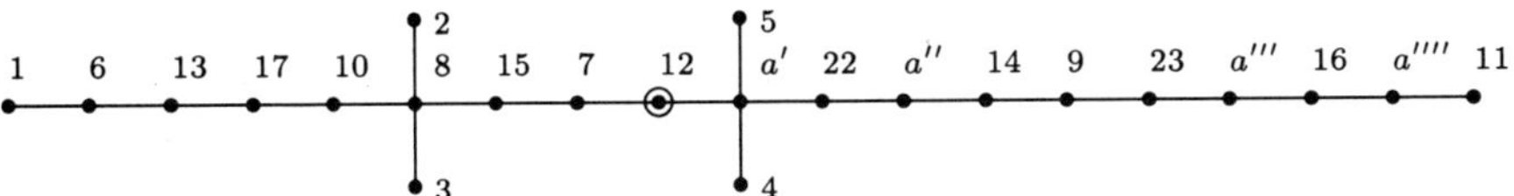

$$\{a', a'', a''', a''''\} = \{18, 19, 20, 21\}$$

PROJECTIVES:

Nr.:	1	2	3	4	5	6	7	8	9	10	11	12	13	14	15	16	17	18	19	20	21	22	23
CC:	r	3	2	5	4	r	r	r	r	r	r	r	r	r	r	r	r	r	r	r	r	r	r
N&C:	×	×	×	×	×	∘	∘	∘	∘	×	×	×	×	×	×	×	∘	∘	∘	∘	∘	×	×
χ_4^{2+}	1					2	1	1	1				2	2	2	2	1	1	1	1	1	1	
$\mathrm{Ind}_{G_2(5)}(\chi_1)$	1					1																	
$\chi_3 \otimes \chi_{10}$			1					2	2	1	1	1	1	1	1	1	2	2	2	2	2	2	4
$\chi_5 \otimes \chi_4$				1	1		1	1	1				2	2	2	2	2	2	2	2	2		3
χ_7^{2+}					1	2	1	5	5	2	2	2	7	7	5	5	6	6	6	6	6	4	8
$\mathrm{Ind}_{G_2(5)}(\chi_2)$							2					1	1	1	2	2	1	1	1	1	1		
$\chi_2 \otimes \chi_7$							1					1											
$\chi_2 \otimes \chi_9$							1						1	1	1	1	1	1	1	1	1	1	1
$\mathrm{Ind}_{G_2(5)}(\chi_3)$								1	1	2	2	1	1	1			2	2	2	2	2	2	3
χ_4^{2-}										1	1						1	1	1	1	1	2	1

We have three orbits of algebraically conjugate characters, generating disjoint fields. The first two orbits each has length two. The third orbit consists of the five characters χ_{40}–χ_{44}. We have four types of trees consistent with the above set of projectives:

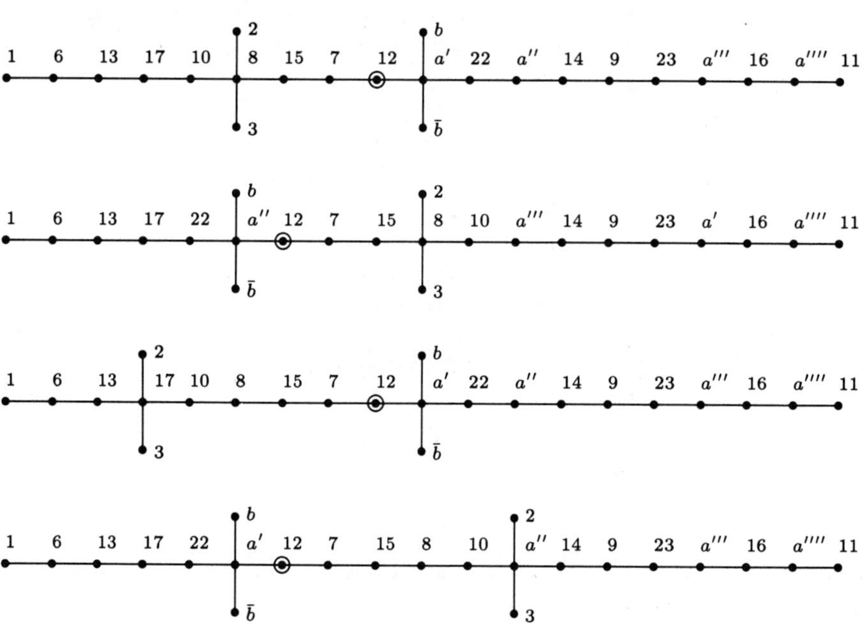

We have used the notation $\{b, \bar{b}\} = \{4, 5\}$. The correct tree (up to algebraic conjugacy) is derived using Green correspondence and the following tensor product $\chi_2 \otimes \chi_2 \approx \chi_6$, which is $2 \otimes 2 \approx 5$ in terms of nodes. The easy proof is left to the reader.

6.20 The Thompson Group

Group: TH Prime: 3 Block: 2

Nr.	CAS-Nr.	Degree	CC	N&C
1	25	4881384	r	×
2	43	76271625	r	×
3	45	81153009	r	○

1 — 3 — 2

Group: TH Prime: 7 Block: 2

Nr.	CAS-Nr.	Degree	CC	N&C
1	3	4123	r	×
2	20	2572752	r	×
3	24	4123000	r	○
4	32	11577384	r	×
5	39	40199250	r	○
6	45	81153009	r	○
7	47	111321000	r	×

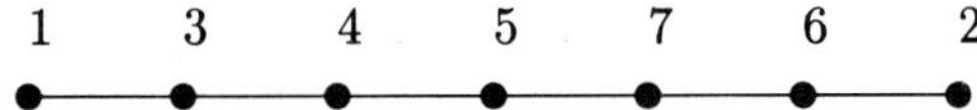

PROJECTIVES:

Nr.:	1	2	3	4	5	6	7
CC:	r	r	r	r	r	r	r
N&C:	×	×	○	×	○	○	×
$\chi_2 \otimes \chi_{21}$		1				1	
$\chi_{11} \otimes \chi_9$			1	2	5	11	15
$\chi_6 \otimes \chi_9$				1	2	2	3
$\chi_4 \otimes \chi_9$					1	2	3

Group: TH Prime: 13 Block: 1

Nr.	CAS-Nr.	Degree	CC	N&C
1	1	1	r	×
2	2	248	r	×
3	4	27000	5	∘
4	5	27000	4	∘
5	11	147250	r	∘
6	22	4096000	23	∘
7	23	4096000	22	∘
8	25	4881384	r	×
9	29	6696000	30	∘
10	30	6696000	29	∘
11	37	28861000	r	∘
12	38	30507008	r	∘
13	43	76271625	r	×

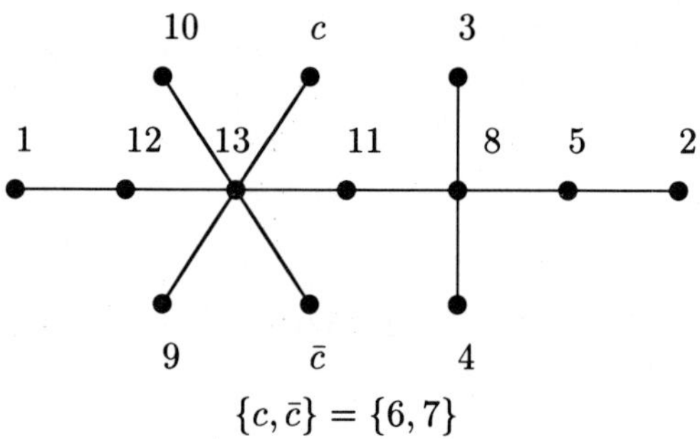

$\{c, \bar{c}\} = \{6, 7\}$

PROJECTIVES:

Nr.:	1	2	3	4	5	6	7	8	9	10	11	12	13
CC:	r	r	4	3	r	7	6	r	10	9	r	r	r
N&C:	×	×	∘	∘	∘	∘	∘	×	∘	∘	∘	∘	×
χ_6^{2+}	1							1			1	1	
$\chi_5 \otimes \chi_{19}$				1		3	3	5	4	5	21	21	53
$\chi_2 \otimes \chi_{39}$						1	1		1	1	2	3	9
$\chi_3 \otimes \chi_7$								1			1		

We have two possible trees consistent with all the projectives given above.

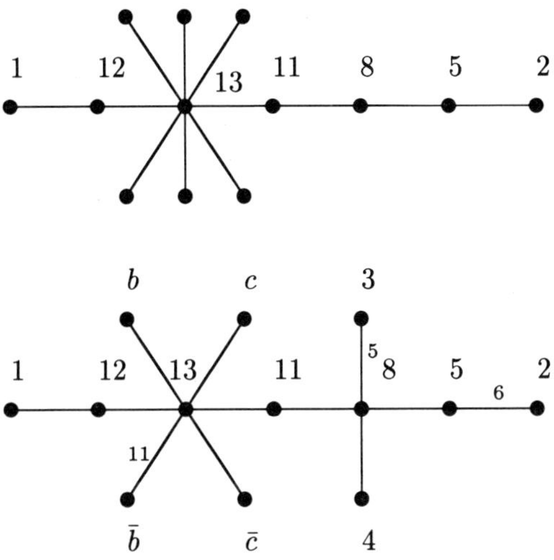

We have used the notation $\{b, \bar{b}, c, \bar{c}\} = \{6, 7, 9, 10\}$ to denote some of the nodes. We have also labelled some edges of the second tree according to Green correspondence. The numbers indicate the socles of the Green correspondents of the irreducible modules corresponding to the edges. We use the following symmetrization and tensor product:

$$\begin{aligned} \chi_4^{2+} &\approx \chi_5 + \chi_{25} \\ \chi_2 \otimes \chi_4 &\approx \chi_{29} \end{aligned}$$

(Here $\approx$ denotes the character restricted to the block.) In terms of nodes these characters are as follows:

$$\begin{aligned} 3^{2+} &\approx 4 + 8 \\ 2 \otimes 3 &\approx 9 \end{aligned}$$

By Lemma 4.3.5 we know that χ_4^{2+} must be the character of a projective module, showing that node 4 is joined to node 8. The Green correspondent of the module sitting at edge 5 is $12A5$. The Green correspondent sitting at edge 6 is $1A6$. Hence the Green correspondent of $3 \otimes 2$ must be $12A5 \otimes 1A6 = 12A11$. Since $3 \otimes 2 \approx 9$ this implies that node 9 must sit on the edge with label 11. This completes the proof of the planar embedding as far as we know it. Unfortunately we do not know which of c, $\bar{c}$ is 6.

Group: TH Prime: 13 Block: 2

Nr.	CAS-Nr.	Degree	CC	N&C
1	3	4123	r	×
2	14	779247	15	×
2	15	779247	14	×
3	24	4123000	r	∘
4	31	10822875	r	∘
5	40	44330496	r	×
6	45	81153009	r	×
7	47	111321000	r	∘

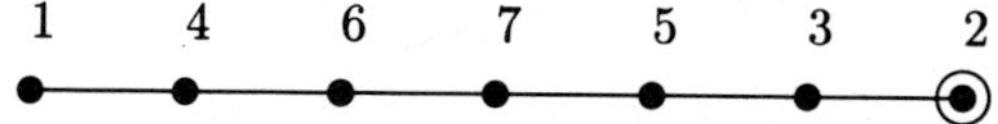

PROJECTIVES:

Nr.:	1	2	3	4	5	6	7
CC:	r	r	r	r	r	r	r
N&C:	×	×	∘	∘	×	×	∘
$\chi_3 \otimes \chi_6$	1			1			
$\chi_2 \otimes \chi_{20}$			1		1		
$\chi_7 \otimes \chi_6$				1		1	
$\chi_2 \otimes \chi_{17}$					1		1
$\chi_4 \otimes \chi_6$						1	1

Group: TH Prime: 19 Block: 1

Nr.	CAS-Nr.	Degree	CC	N&C
1	1	1	r	×
2	2	248	r	×
3	4	27000	5	×
4	5	27000	4	×
5	9	85995	10	×
6	10	85995	9	×
7	12	767637	13	∘
8	13	767637	12	∘
9	22	4096000	23	∘
10	23	4096000	22	∘
11	25	4881384	r	∘
12	26	4936750	r	∘
13	29	6696000	30	×
14	30	6696000	29	×
15	35	21326760	36	×
16	36	21326760	35	×
17	43	76271625	r	×
18	44	77376000	r	×
19	48	190373976	r	∘

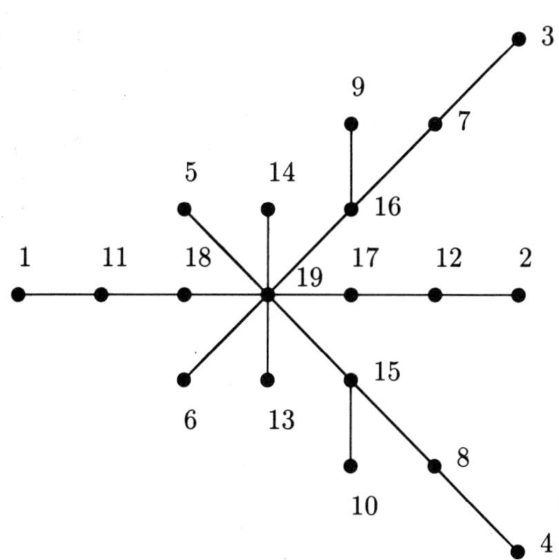

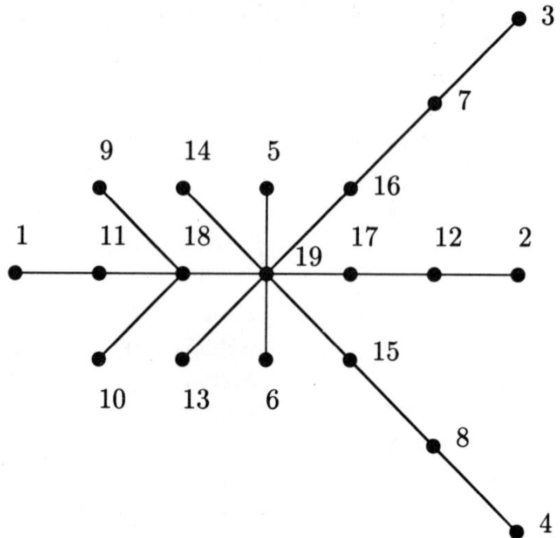

Warning: We have not been able to decide which of the two trees above is correct. Both are consistent with Green correspondence and tensor products.

PROJECTIVES:

Nr.:	1	2	3	4	5	6	7	8	9	10	11	12	13	14	15	16	17	18	19
CC:	r	r	4	3	6	5	8	7	10	9	r	r	14	13	16	15	r	r	r
N&C:	×	×	×	×	×	×	∘	∘	∘	∘	∘	∘	×	×	×	×	×	×	∘
χ_3^{2+}	1										1								
$\chi_{10} \otimes \chi_8$						1									2	1	4	4	12
$\chi_{12} \otimes \chi_7$								1	1	1	2	1	1	1	6	6	18	22	48
$\chi_2 \otimes \chi_{37}$									1	1	1	1			2	2	7	8	15
$\chi_7 \otimes \chi_3$											1							1	
$\chi_5 \otimes \chi_6$														1			1		2

The above set of projectives determines the real stem of the Brauer tree and shows that the nodes $\{5, 6, 13, 14, 15, 16\}$ are joined to node 19. The real stem of the tree is as follows:

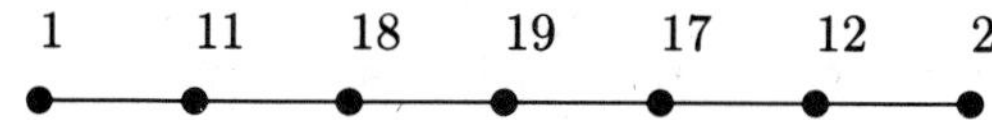

It remains to determine where the nodes $\{3, 4, 7, 8, 9, 10\}$ are joined to, and the planar embedding. The projectives show that the nodes 3, 7 and 9 are joined to one of the nodes in the following table:

3	$7, 8, 9, 10, 11, 12, 19$
7	$13, 14, 15, 16, 17, 18$
9	$15, 16, 17, 18$

Observe that the characters in the four sets $\{3,4,13,14\}$, $\{5,6,15,16\}$, $\{7,8\}$ and $\{9,10\}$ generate pairwise disjoint fields. Hence we can assume without loss of generality that the nodes $\{3,5,7,9\}$ lie in the upper half of the tree.

We shall use the following tensor products and symmetrizations for the proof:

$$\begin{array}{ccl}
\chi_2 \otimes \chi_4 & \approx & \chi_{13} \\
\chi_2 \otimes \chi_9 & \approx & \chi_{35} \\
\chi_2 \otimes \chi_{12} & \approx & \chi_{48} \\
\chi_2 \otimes \chi_{22} & \approx & \chi_{23} + \chi_{43} + \chi_{44} + 2\chi_{48} \\
\chi_2 \otimes \chi_{29} & \approx & \chi_4 + \chi_{29} + \chi_{43} + \chi_{44} + 3\chi_{48} \\
\chi_2 \otimes \chi_{35} & \approx & \chi_9 + 2\chi_{35} + 2\chi_{36} + 4\chi_{43} + 4\chi_{44} + 12\chi_{48} \\
\chi_4^{2+} & \approx & \chi_5 + \chi_{25} + \chi_{44} \\
\chi_4 \otimes \chi_9 & \approx & \chi_{35} + \chi_{36} + 2\chi_{43} + 2\chi_{44} + 5\chi_{48} \\
\chi_4 \otimes \chi_{10} & \approx & \chi_{35} + \chi_{36} + 2\chi_{43} + 2\chi_{44} + 5\chi_{48} \\
\chi_9^{2+} & \approx & \chi_{25} + \chi_{35} + \chi_{36} + 2\chi_{43} + 4\chi_{44} + 6\chi_{48}
\end{array}$$

We note that $\chi_{12} \otimes \chi_{12}$ contains none of χ_9, χ_{10} as constituent:

In terms of nodes the above characters are as follows.

$$\begin{array}{ccl}
2 \otimes 3 & \approx & 13 \\
2 \otimes 5 & \approx & 15 \\
2 \otimes 7 & \approx & 19 \\
2 \otimes 9 & \approx & 10 + 17 + 18 + 19^2 \\
2 \otimes 13 & \approx & 3 + 13 + 17 + 18 + 19^3 \\
2 \otimes 15 & \approx & 5 + 15^2 + 16^2 + 17^4 + 18^4 + 19^{12} \\
3^{2+} & \approx & 4 + 11 + 18 \\
3 \otimes 5 & \approx & 15 + 16 + 17^2 + 18^2 + 19^5 \\
3 \otimes 6 & \approx & 15 + 16 + 17^2 + 18^2 + 19^5 \\
5^{2+} & \approx & 11 + 15 + 16 + 17^2 + 18^4 + 19^6
\end{array}$$

(Here, as usual, the exponents on the nodes denote their multiplicity.) We note that $7 \otimes 7$ does contain none of 5 and 6.

From the above projectives it follows that node 2, which is χ_2, is irreducible modulo 19, i.e. is a leaf on the real stem. Since it is of type cross half way round the tree, the edge incident to it has label 9 in the labelling according to Green correspondence. Also, the edge incident to node 1, which corresponds to the trivial character, has label 0. Now consider the symmetric square of node 3. By the general theory we have to subtract a cross from $4 + 11 + 18$ to obtain a projective character. The crosses in $\{4, 11, 18\}$ are 4 and 18. Suppose first that $4 + 11$ is projective. Then 4 is joined to 11. Since 11 is joined to 1 on the real stem, the edge incident to 3 receives label 1. Now the Green correspondent of the irreducible module corresponding to node 3 is $1A1$. Hence $2 \otimes 3 \approx 13$ has Green correspondent $1A9 \otimes 1A1 = 1A10$. Since 13 is a node joined to 19, which is located at the centre of the real stem, there is no way that it can be incident to an edge with label 10. This contradiction shows that $11 + 18$ is projective, in other words that 4 is the character of the Green correspondent of 3^{2+}.

Certainly, 3 and 4 are leaves. Let $1Ax$ be the Green correspondent of the irreducible module with Brauer character 3. Then $1A(18 - x)$ is the Green correspondent of the edge incident to the leaf 4. From the discussion above we obtain $1A2x = 1A(18 - x)$ since $1A2x$ is the Green correspondent of 3^{2+}. By our convention that 3 should be on the upper half tree, x is less than 9. The congruence $2x \equiv 18 - x (\mathrm{mod} 18)$ then has the unique solution $x = 6$. Hence the edge incident to node 3 has label 6 and the edge incident to node 4 has label 12.

From $2 \otimes 3 \approx 13$ and $2 \otimes 4 \approx 14$ we conclude that node 13 must be incident to an edge with label 15 and node 14 to an edge with label 3.

It is clear that the nodes 5 and 6 are leaves. Let x be the label on the edge joining nodes 5 and 19. We now distinguish cases for the possibilities of x. By our convention x is smaller than 9 and 0, 3 and 6 are already excluded as possibilities for x. If x equals 8, then the edge joining node 6 with node 19 has label 10. But we have already seen that there can be no edge with label 10 incident to 19. So we have only four possibilities for x, namely 2, 4, 5 and 7.

CASE I. $x = 2$

This assumption leads to the following labelled planar embedded tree, which gives our first possibility above:

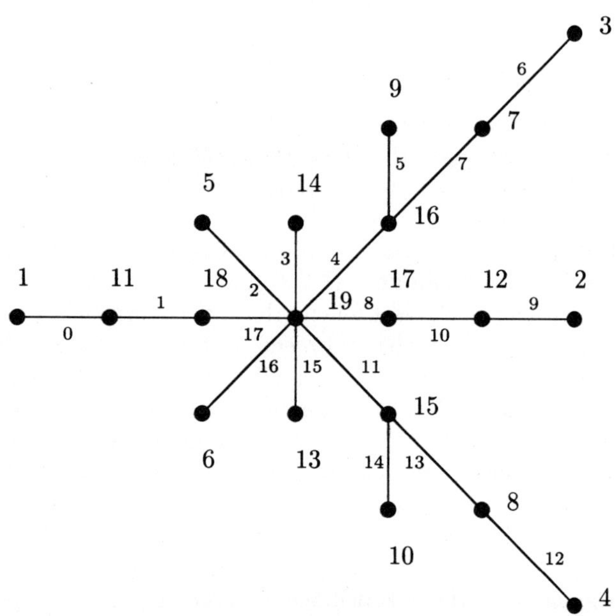

We briefly comment on how this tree is forced by the assumption above and some of the tensor products. Since the edge joining node 5 with 19 has label 2, there can be no arms going off to the left of 19. Thus the labels 1, 2, 16 and 17 are forced. Node 5 is a leaf, forcing the next edge coming from 19 to have label 3. By the discussion above this must therefore join 19 and 14. Now 5^{2+} has Green correspondent $1A4$, which implies that there must be an edge with label 4 incident to one of the nodes 15, 16, 17 or 18. This forces 14 to be a leaf and the next edge joining 19 with either 15 or 16. Furthermore the tensor product $2 \otimes 5 \approx 15$ implies that there is an edge incident to node 15 with label $9 + 2 = 11$. Hence there is no arm going off right of 19. The rest of the argument is fairly straightforward and is left to the reader. One has to use the tensor product $2 \otimes 7 \approx 19$ to show that node 7 cannot have an edge with label 5.

CASE II. $x = 4$

This implies the following labelled planar embedded tree which is the second of the above possibilities:

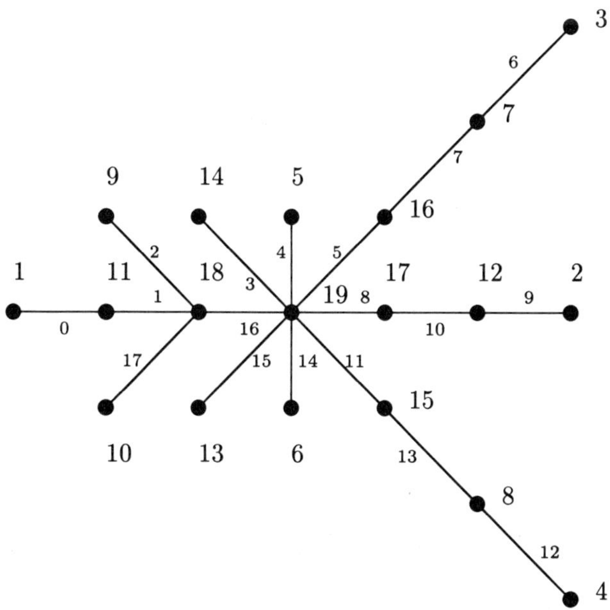

Again we remark on how this tree is derived. As we have already seen that node 3 is not joined to 11, we can put the labels 0, 1, 9 in their position. Then by assumption we can label 4 and 14, then 15, 3 and 5. The tensor product $3 \otimes 6$ implies that one of the nodes 15, 16, 17 or 18 must have an edge with label $6 + 14 = 2$. This forces an edge meeting node 18 to have label 2. The node on the other end of this edge is 7 or 9. In any case the tensor products $2 \otimes 7$ and $2 \otimes 9$ show that there is an edge with label $9 + 2 = 11$ incident to 19. Since this edge cannot join 19 with 4, we are done, as the reader easily verifies.

CASE III. $x = 5$

This leads to the following labelled planar embedded tree which, however, contradicts the tensor product $7 \otimes 7$:

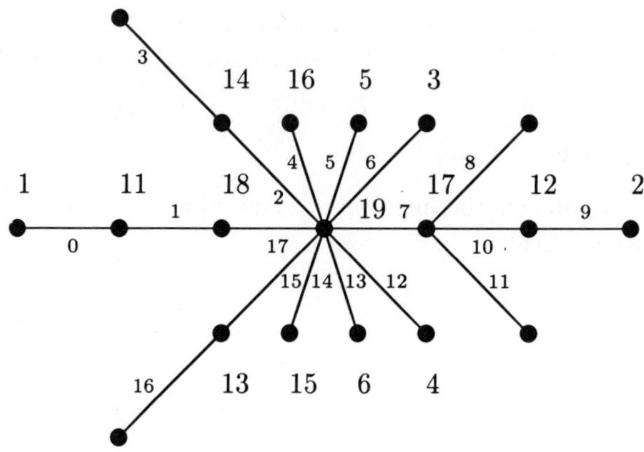

Let us again give a couple of remarks on how this tree is forced upon us by the above assumption. First we can put the labels 0, 1, 9, 5 and 13 in their place. Since 5 is a leaf we are forced by the fact that node 3 is a leaf with label 6 to put 6 on an edge joining 19 with 3. Then we know where to put label 12. Since 6 is a leaf we know where to put label 14. Since node 13 is incident to an edge with label 15, this implies that the node incident to edge 14 different from 19 must be a leaf, and that it must be one of 15 or 16. We now have 10 edges coming out of node 19. Since there can be no more, we know where to put label 7. We now see that there must be an arm going off node 17 with label 8, and we can then put the labels 10 and 11 in place. As above the tensor products $2 \otimes 7$ and $2 \otimes 9$ imply that there is an edge with label $9 + 11 = 2$ incident to node 19. This implies the result.

CASE IV. $x = 7$

This leads to the following planar embedded tree which, however, contradicts the symmetric square of the module corresponding to node 5:

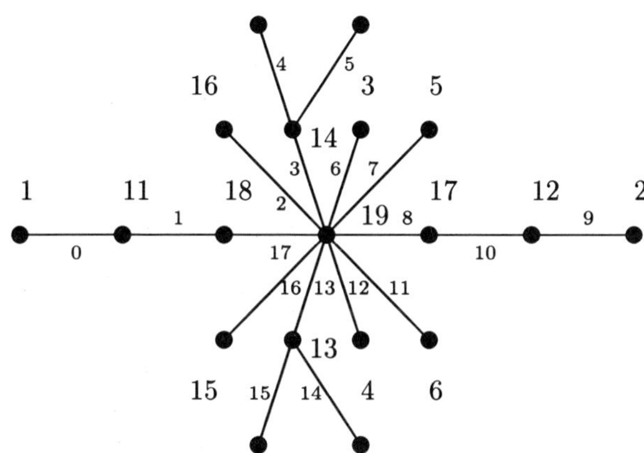

This last tree is easily derived from the above assumption as follows. First we can put labels 0, 1, 9, 7 and 11 in their position. This in turn forces the position of the labels 10, 8, 12, 6 and finally 13. The tensor product $2 \otimes 5 = 15$ shows that there must be an edge incident to node 15 which has label 16. This implies the tree as it is given.

Group: TH Prime: 31 Block: 1

Nr.	CAS-Nr.	Degree	CC	N&C
1	1	1	r	×
2	4	27000	5	∘
3	5	27000	4	∘
4	7	30875	r	∘
5	9	85995	10	×
6	10	85995	9	×
7	12	767637	13	∘
7	13	767637	12	∘
8	17	1707264	18	×
9	18	1707264	17	×
10	22	4096000	23	×
11	23	4096000	22	×
12	27	6669000	28	×
13	28	6669000	27	×
14	34	18154500	r	×
15	38	30507008	r	×
16	42	72925515	r	∘

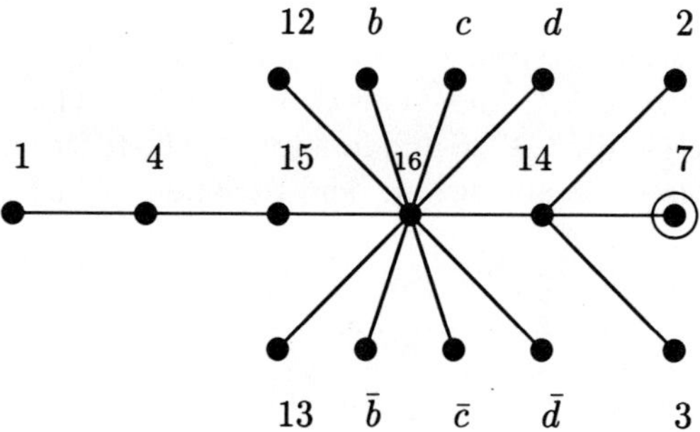

Here, $\{b, \bar{b}, c, \bar{c}, d, \bar{d}\} = \{5, 6, 8, 9, 10, 11\}$, and $d \notin \{5, 6\}$.

PROJECTIVES:

Nr.:	1	2	3	4	5	6	7	8	9	10	11	12	13	14	15	16
CC:	r	3	2	r	6	5	r	9	8	11	10	13	12	r	r	r
N&C:	×	○	○	○	×	×	○	×	×	×	×	×	×	×	×	○
$\chi_2 \otimes \chi_2$	1			1												
$\chi_5 \otimes \Phi_1$			1											1		
$\chi_{11} \otimes \chi_2$				1											1	
$\chi_{10} \otimes \Phi_1$						1									1	2
$\chi_{17} \otimes \chi_3$									1			1	1	1	2	6
$\chi_{22} \otimes \chi_2$											1					1

Note that $\Phi_1 = \chi_1 + \chi_7$. It remains to prove the information on the planar embedding of the tree we have given. For this we use the following tensor product and symmetrization:

$$\begin{aligned} \chi_4 \otimes \chi_4 &\approx \chi_5 + \chi_{28} + \chi_{34} \\ \chi_{10}^{2+} &\approx 2\chi_{34} + \chi_{38} + 2\chi_{42} \end{aligned}$$

In terms of nodes these characters are as follows:

$$\begin{aligned} 2 \otimes 2 &\approx 3 + 13 + 14 \\ 6^{2+} &\approx 14^2 + 15 + 16^2 \end{aligned}$$

The label on the edge of node 2 is 7. By Lemma 4.3.2 the Green correspondent of $2 \otimes 2$ is $30A7 \otimes 30A7 = 1A13$. There is no edge meeting node 14 which has label 13. Hence node 13 must have an edge with label 13. Thus nodes 12 and 13 have the position as given above.

It remains to show that d cannot be 5 or 6. The label on the edge joining node d with 16 is 5. Hence the Green correspondent of d^{2+} is $1A5^{2+} = 1A10$. It follows that node $\bar{d}$ must occur in d^{2+}. But 6 does not occur in 5^{2+} nor does 5 in 6^{2+}. We have not been able to determine the full planar embedding of this Brauer tree.

6.21 The Fischer Group F23

Group: F23 Prime: 2 Block: 2

Nr.	CAS-Nr.	Degree	CC	N&C
1	56	73531392	r	×
2	57	73531392	r	∘

1 —— 2

Group: F23 Prime: 3 Block: 2

Nr.	CAS-Nr.	Degree	CC	N&C
1	69	207793431	r	×
2	82	289103904	r	×
3	93	496897335	r	∘

1 —— 3 —— 2

Group: F23 Prime: 5 Block: 3

Nr.	CAS-Nr.	Degree	CC	N&C
1	11	279565	r	×
2	19	1677390	r	×
3	25	6709560	r	∘
4	34	17892160	r	∘
5	39	22644765	r	×

1 —— 3 —— 5 —— 4 —— 2

PROJECTIVES:

Nr.:	1	2	3	4	5
CC:	r	r	r	r	r
N&C:	×	×	∘	∘	×
$\chi_3 \otimes \chi_{38}$	1		1		
$\chi_2 \otimes \chi_{38}$		1		1	
$\chi_2 \otimes \chi_{66}$			1		1
$\chi_2 \otimes \chi_{47}$				1	1

Group: F23 Prime: 5 Block: 4

Nr.	CAS-Nr.	Degree	CC	N&C
1	13	789360	r	×
2	24	5533110	r	×
3	40	26838240	r	∘
4	43	35225190	r	∘
5	49	55740960	r	×

1 — 4 — 5 — 3 — 2

PROJECTIVES:

Nr.:	1	2	3	4	5
CC:	r	r	r	r	r
N&C:	×	×	∘	∘	×
$\chi_3 \otimes \chi_{31}$	1			1	
$\chi_2 \otimes \chi_{53}$		1	1		
$\chi_2 \otimes \chi_{36}$			1		1
$\chi_2 \otimes \chi_{12}$				1	1

Group: F23 Prime: 5 Block: 5

Nr.	CAS-Nr.	Degree	CC	N&C
1	21	2236520	r	×
2	55	65875680	r	∘
3	60	97976320	r	×
4	79	287721720	r	×
5	86	322058880	r	∘

1 — 2 — 4 — 5 — 3

PROJECTIVES:

Nr.:	1	2	3	4	5
CC:	r	r	r	r	r
N&C:	×	∘	×	×	∘
$\chi_3 \otimes \chi_{12}$		1		1	
$\chi_2 \otimes \chi_{41}$			1		1
$\chi_2 \otimes \chi_{45}$				1	1

Group: F23 Prime: 5 Block: 6

Nr.	CAS-Nr.	Degree	CC	N&C
1	22	2322540	r	×
2	68	203802885	r	∘
3	87	336061440	r	×
4	91	362316240	r	×
5	93	496897335	r	∘

1 — 2 — 3 — 5 — 4

PROJECTIVES:

Nr.:	1	2	3	4	5
CC:	r	r	r	r	r
N&C:	×	∘	×	×	∘
$\chi_3 \otimes \chi_{17}$		1	1		
$\chi_2 \otimes \chi_{36}$			1	1	2

Group: F23 Prime: 5 Block: 7

Nr.	CAS-Nr.	Degree	CC	N&C
1	23	3913910	r	×
2	33	15096510	r	×
3	59	93933840	r	∘
4	70	211351140	r	∘
5	78	286274560	r	×

1 — 4 — 5 — 3 — 2

PROJECTIVES:

Nr.:	1	2	3	4	5
CC:	r	r	r	r	r
N&C:	×	×	∘	∘	×
$\chi_2 \otimes \chi_{31}$	1			1	
$\chi_4 \otimes \chi_{12}$		1	2		1
$\chi_2 \otimes \chi_{12}$				1	1

Group: F23 Prime: 7 Block: 1

Nr.	CAS-Nr.	Degree	CC	N&C
1	1	1	r	×
2	19	1677390	r	×
3	27	8783424	r	∘
4	64	166559744	r	×
5	69	207793431	r	∘
6	79	287721720	r	∘
7	87	336061440	r	×

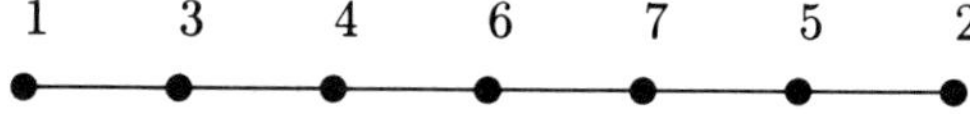

PROJECTIVES:

Nr.:	1	2	3	4	5	6	7
CC:	r	r	r	r	r	r	r
N&C:	×	×	∘	×	∘	∘	×
χ_8^{2+}	1		3	3		1	
$\chi_4 \otimes \chi_8$			1	1			
$\chi_2 \otimes \chi_{28}$				1		1	
$\chi_2 \otimes \chi_{45}$					2	1	3

Group: F23 Prime: 7 Block: 2

Nr.	CAS-Nr.	Degree	CC	N&C
1	2	782	r	×
2	13	789360	r	×
3	33	15096510	r	○
4	48	48308832	r	×
5	49	55740960	r	○
6	77	264536064	r	○
7	78	286274560	r	×

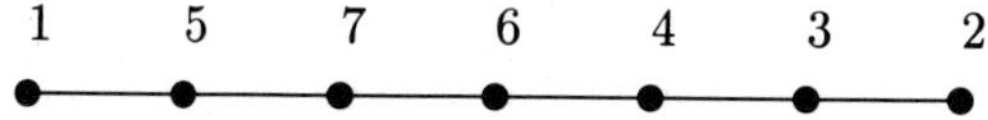

PROJECTIVES:

Nr.:	1	2	3	4	5	6	7
CC:	r	r	r	r	r	r	r
N&C:	×	×	○	×	○	○	×
$\chi_4 \otimes \chi_{23}$	1				1		
$\chi_2 \otimes \chi_{23}$		1	1				
$\chi_2 \otimes \chi_{30}$			1	1			
$\chi_2 \otimes \chi_{44}$				1		1	
$\chi_2 \otimes \chi_{12}$					1		1
$\chi_2 \otimes \chi_{45}$						1	1

Group: F23 Prime: 7 Block: 3

Nr.	CAS-Nr.	Degree	CC	N&C
1	3	3588	r	×
2	6	30888	r	×
3	22	2322540	r	∘
4	25	6709560	r	×
5	32	12077208	r	∘
6	55	65875680	r	∘
7	56	73531392	r	×

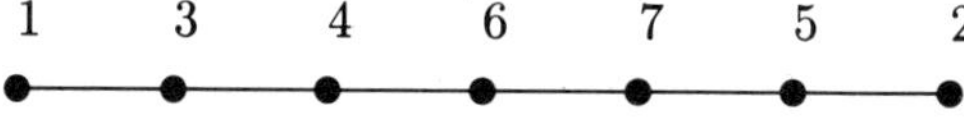

PROJECTIVES:

Nr.:	1	2	3	4	5	6	7
CC:	r	r	r	r	r	r	r
N&C:	×	×	∘	×	∘	∘	×
$\chi_5 \otimes \chi_{14}$	1		1				
$\chi_4 \otimes \chi_{23}$		1			1		
$\chi_2 \otimes \chi_{41}$			1	1			
$\chi_4 \otimes \chi_{45}$				1		1	
$\chi_2 \otimes \chi_{42}$					1		1
$\chi_2 \otimes \chi_{36}$						1	1

Group: F23 Prime: 7 Block: 4

Nr.	CAS-Nr.	Degree	CC	N&C
1	4	5083	r	×
2	9	111826	r	×
3	21	2236520	r	∘
4	34	17892160	r	∘
5	52	57254912	r	×
6	65	166559744	r	×
7	68	203802885	r	∘

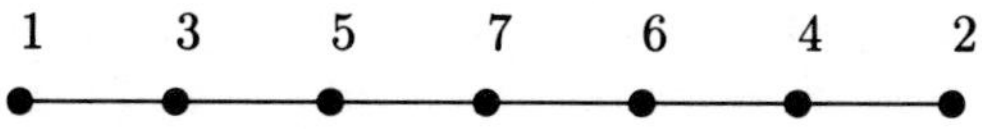

1 — 3 — 5 — 7 — 6 — 4 — 2

PROJECTIVES:

Nr.:	1	2	3	4	5	6	7
CC:	r	r	r	r	r	r	r
N&C:	×	×	∘	∘	×	×	∘
$\chi_2 \otimes \chi_{23}$	1		1				
$\chi_3 \otimes \chi_{14}$		1		1		1	1
$\chi_4 \otimes \chi_{23}$			1		1		
$\chi_3 \otimes \chi_8$				1		1	
$\chi_2 \otimes \chi_{44}$					1		1
$\chi_2 \otimes \chi_{37}$						1	1

Group: F23 Prime: 7 Block: 5

Nr.	CAS-Nr.	Degree	CC	N&C
1	5	25806	r	×
2	35	18812574	r	×
3	39	22644765	r	∘
4	57	73531392	r	×
5	73	244563462	r	∘
6	86	322058880	r	∘
7	93	496897335	r	×

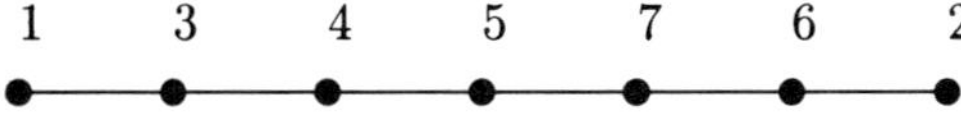

PROJECTIVES:

Nr.:	1	2	3	4	5	6	7
CC:	r	r	r	r	r	r	r
N&C:	×	×	∘	×	∘	∘	×
$\chi_3 \otimes \chi_{14}$	1		1				
$\chi_2 \otimes \chi_{14}$		1				1	
$\chi_3 \otimes \chi_{28}$			2	4	5		3
$\chi_3 \otimes \chi_{17}$			1	1			
$\chi_2 \otimes \chi_{45}$				1	3	1	3
$\chi_5 \otimes \chi_{15}$					1	1	2

Group: F23 Prime: 7 Block: 6

Nr.	CAS-Nr.	Degree	CC	N&C
1	7	60996	r	×
2	10	274482	r	×
3	24	5533110	r	∘
4	26	7468032	r	×
5	40	26838240	r	∘
6	76	264536064	r	∘
7	82	289103904	r	×

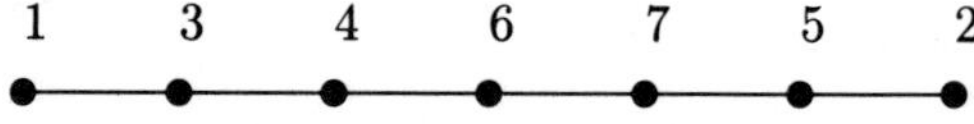

PROJECTIVES:

Nr.:	1	2	3	4	5	6	7
CC:	r	r	r	r	r	r	r
N&C:	×	×	∘	×	∘	∘	×
$\chi_2 \otimes \chi_{14}$	1		1				
$\chi_2 \otimes \chi_{30}$		2	2	2	2		
$\chi_2 \otimes \chi_8$		1			1		
$\chi_4 \otimes \chi_{14}$			1	1		1	1
$\chi_2 \otimes \chi_{44}$				1	1	3	3
$\chi_2 \otimes \chi_{74}$				1		18	17
$\chi_2 \otimes \chi_{36}$					1	1	2

Group: F23 Prime: 7 Block: 7

Nr.	CAS-Nr.	Degree	CC	N&C
1	11	279565	r	×
2	20	1951872	r	×
3	60	97976320	r	∘
4	61	133398252	r	∘
5	89	343529472	r	×
6	91	362316240	r	×
7	92	476702577	r	∘

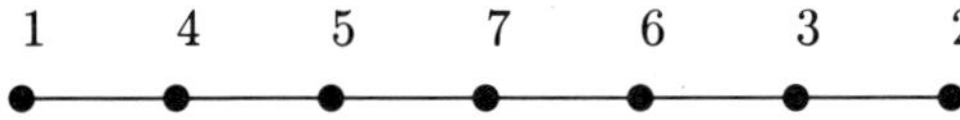

PROJECTIVES:

Nr.:	1	2	3	4	5	6	7
CC:	r	r	r	r	r	r	r
N&C:	×	×	∘	∘	×	×	∘
$\chi_4 \otimes \chi_{14}$	1			1			
$\chi_2 \otimes \chi_{23}$		1	2			1	
$\chi_3 \otimes \chi_{12}$				1	1		
$\chi_2 \otimes \chi_{28}$					1		1
$\chi_3 \otimes \chi_{28}$						1	1

Group: F23 Prime: 11 Block: 1

Nr.	CAS-Nr.	Degree	CC	N&C
1	1	1	r	×
2	2	782	r	×
3	8	106743	r	∘
4	10	274482	r	∘
5	27	8783424	r	×
6	41	28464800	r	×
7	64	166559744	r	∘
8	69	207793431	r	∘
9	71	216154575	r	∘
10	77	264536064	r	×
11	82	289103904	r	×

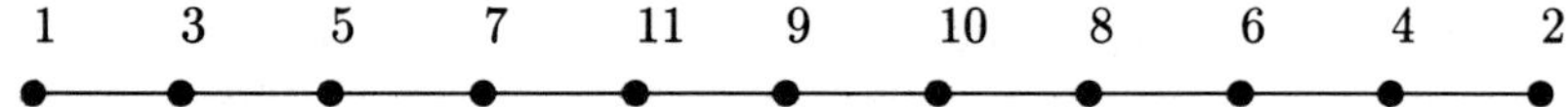

PROJECTIVES:

Nr.:	1	2	3	4	5	6	7	8	9	10	11
CC:	r	r	r	r	r	r	r	r	r	r	r
N&C:	×	×	∘	∘	×	×	∘	∘	∘	×	×
$\chi_5 \otimes \chi_5$	1	1	1	1							
$\chi_{12} \otimes \chi_{12}$	1		6		24		98	2	28	6	103
$\chi_5 \otimes \chi_9$			1		1		1				1
$\chi_3 \otimes \chi_9$					1		1				
$\chi_2 \otimes \chi_{19}$						1		1			
$\chi_2 \otimes \chi_{44}$							1		3	1	3
$\chi_2 \otimes \chi_{32}$							1				1
$\chi_2 \otimes \chi_{45}$								2	3	4	1
$\chi_2 \otimes \chi_{37}$								2	1	2	1

Group: F23 Prime: 11 Block: 2

Nr.	CAS-Nr.	Degree	CC	N&C
1	3	3588	r	×
2	28	9108736	29	∘
2	29	9108736	28	∘
3	38	21348600	r	∘
4	55	65875680	r	×
5	60	97976320	r	×
6	61	133398252	r	∘

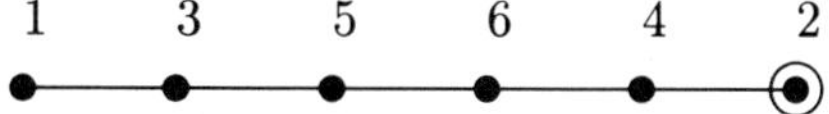

PROJECTIVES:

Nr.:	1	2	3	4	5	6
CC:	r	r	r	r	r	r
N&C:	×	∘	∘	×	×	∘
$\chi_3 \otimes \chi_6$	1		1			
$\chi_4 \otimes \chi_{17}$		1		1		
$\chi_2 \otimes \chi_{14}$			1		1	
$\chi_2 \otimes \chi_{21}$					1	1

Group: F23 Prime: 11 Block: 3

Nr.	CAS-Nr.	Degree	CC	N&C
1	4	5083	r	×
2	7	60996	r	×
3	15	837200	16	×
4	16	837200	15	×
5	20	1951872	r	∘
6	31	10674300	r	∘
7	47	48034350	r	×
8	65	166559744	r	∘
9	76	264536064	r	×
10	88	341577600	r	×
11	92	476702577	r	∘

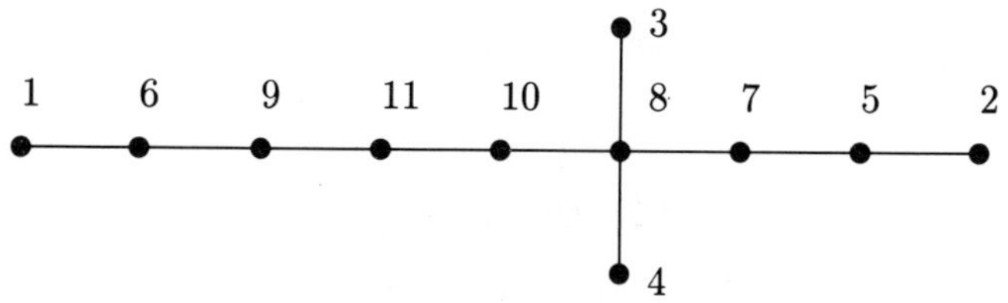

PROJECTIVES:

Nr.:	1	2	3	4	5	6	7	8	9	10	11
CC:	r	r	4	3	r	r	r	r	r	r	r
N&C:	×	×	×	×	∘	∘	×	∘	×	×	∘
$\chi_4 \otimes \chi_6$	1					1					
$\chi_2 \otimes \chi_{14}$		1			1						
$\chi_4 \otimes \chi_{12}$			1	1				3		1	
$\chi_3 \otimes \chi_6$					1		1				
$\chi_3 \otimes \chi_{12}$						1			1		
$\chi_4 \otimes \chi_9$							1	1			
$\chi_2 \otimes \chi_{36}$									1	1	2

Group: F23 Prime: 13 Block: 1

Nr.	CAS-Nr.	Degree	CC	N&C
1	1	1	r	×
2	12	752675	r	×
3	14	812889	r	∘
4	35	18812574	r	×
5	57	73531392	r	∘
6	62	153014400	r	∘
6	63	153014400	r	∘
7	69	207793431	r	×

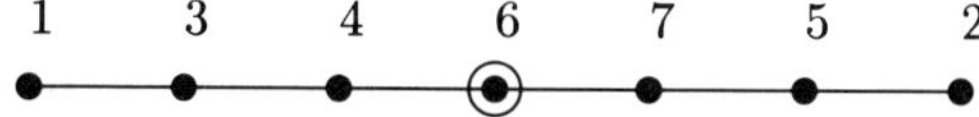

PROJECTIVES:

Nr.:	1	2	3	4	5	6	7
CC:	r	r	r	r	r	r	r
N&C:	×	×	∘	×	∘	∘	×
$\chi_3 \otimes \chi_3$	1		1				
$\chi_2 \otimes \chi_7$			1	1			
$\chi_2 \otimes \chi_{13}$				1		1	
$\chi_2 \otimes \chi_{19}$						1	1

Group: F23 Prime: 13 Block: 2

Nr.	CAS-Nr.	Degree	CC	N&C
1	2	782	r	×
2	24	5533110	r	∘
3	49	55740960	r	×
4	72	216770400	r	×
5	82	289103904	r	∘
6	94	504627200	r	∘
6	95	504627200	r	∘
7	96	526752072	r	×

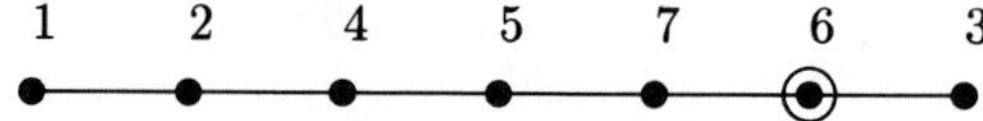

PROJECTIVES:

Nr.:	1	2	3	4	5	6	7
CC:	r	r	r	r	r	r	r
N&C:	×	∘	×	×	∘	∘	×
$\chi_2 \otimes \chi_6$	1	1					
$\chi_2 \otimes \chi_{21}$		1		1			
$\chi_2 \otimes \chi_{39}$			1			1	
$\chi_4 \otimes \chi_{17}$				1	1		
$\chi_3 \otimes \chi_{15}$					1		1
$\chi_3 \otimes \chi_{21}$						1	1

Group: F23 Prime: 13 Block: 3

Nr.	CAS-Nr.	Degree	CC	N&C
1	5	25806	r	×
2	22	2322540	r	∘
3	36	20322225	r	×
4	56	73531392	r	∘
5	74	263376036	r	×
6	80	289027200	r	×
6	81	289027200	r	×
7	93	496897335	r	∘

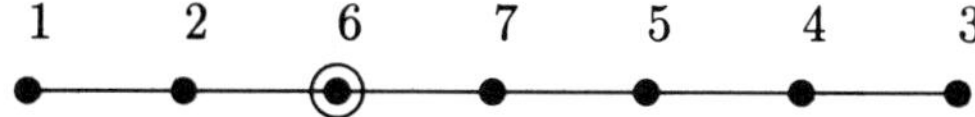

PROJECTIVES:

Nr.:	1	2	3	4	5	6	7
CC:	r	r	r	r	r	r	r
N&C:	×	∘	×	∘	×	×	∘
$\chi_2 \otimes \chi_{10}$	1	1					
$\chi_3 \otimes \chi_{21}$		1				1	
$\chi_3 \otimes \chi_{15}$			1	1			
$\chi_3 \otimes \chi_{19}$				1	2		1
$\chi_2 \otimes \chi_{27}$						1	1

Group: F23 Prime: 17 Block: 1

Nr.	CAS-Nr.	Degree	CC	N&C
1	1	1	r	×
2	3	3588	r	×
3	6	30888	r	∘
4	13	789360	r	∘
5	15	837200	16	×
6	16	837200	15	×
7	24	5533110	r	×
8	60	97976320	r	∘
9	62	153014400	r	×
10	63	153014400	r	×
11	76	264536064	r	∘
12	77	264536064	r	∘
13	79	287721720	r	×
14	92	476702577	r	×
15	94	504627200	r	∘
16	95	504627200	r	∘
17	98	559458900	r	×

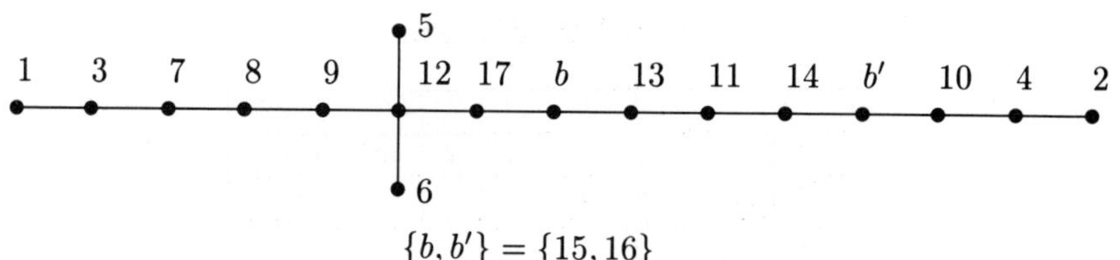

$\{b, b'\} = \{15, 16\}$

PROJECTIVES:

Nr.:	1	2	3	4	5	6	7	8	9	10	11	12	13	14	15	16	17
CC:	r	r	r	r	6	5	r	r	r	r	r	r	r	r	r	r	r
N&C:	×	×	∘	∘	×	×	×	∘	×	×	∘	∘	×	×	∘	∘	×
$\chi_2 \otimes \chi_2$	1		1														
$\chi_2 \otimes \chi_{10}$			1				1										
$\chi_2 \otimes \chi_{19}$				1				1	1	1							
$\chi_4 \otimes \chi_{19}$				1					1	1		3					2
$\chi_4 \otimes \chi_{12}$					1	1						2					
$\chi_2 \otimes \chi_{14}$							1	1									
$\chi_2 \otimes \chi_{27}$								1	1	1			1		1	1	
$\chi_5 \otimes \chi_{17}$									1	1		4		1	2	2	5
$\chi_4 \otimes \chi_{17}$											2		1	1			
$\chi_2 \otimes \chi_{36}$											1		3	2	2	2	

Group: F23 Prime: 23 Block: 1

Nr.	CAS-Nr.	Degree	CC	N&C
1	1	1	r	×
2	6	30888	r	○
3	17	850850	18	○
3	18	850850	17	○
4	26	7468032	r	×
5	45	40840800	r	○
6	56	73531392	r	×
7	57	73531392	r	×
8	82	289103904	r	○
9	87	336061440	r	○
10	93	496897335	r	○
11	94	504627200	r	×
12	95	504627200	r	×

PROJECTIVES:

Nr.:	1	2	3	4	5	6	7	8	9	10	11	12
CC:	r	r	r	r	r	r	r	r	r	r	r	r
N&C:	×	○	○	×	○	×	×	○	○	○	×	×
$\chi_2 \otimes \chi_2$	1	1										
$\chi_2 \otimes \chi_{10}$		1		1								
$\chi_2 \otimes \chi_{25}$				1						1		
$\chi_4 \otimes \chi_9$					1		1					
$\chi_3 \otimes \chi_{21}$					1				1		1	1
$\chi_5 \otimes \chi_8$							1		1			
$\chi_2 \otimes \chi_{27}$								1		1	1	1
$\chi_2 \otimes \chi_{33}$									1	1	1	1

Without loss of generality we assume that of the two algebraic conjugate characters χ_{94} and χ_{95} (nodes 11 and 12), χ_{94} comes first on the tree. Then only the above tree is consistent with all the projectives.

6.22 The Double Cover of the Conway Group C1

Group: 2C1 Prime: 5 Block: 2

Nr.	CAS-Nr.	Degree	CC	N&C
1	6	17250	r	×
2	58	66602250	r	×
3	78	241741500	r	∘
4	89	326956500	r	∘
5	97	502078500	r	×

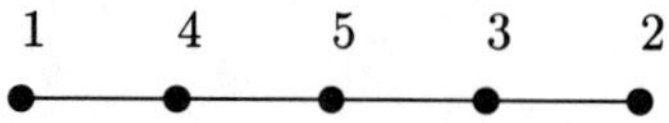

1 — 4 — 5 — 3 — 2

PROJECTIVES:

Nr.	1	2	3	4	5
CC	r	r	r	r	r
N&C	×	×	∘	∘	×
$\chi_2 \otimes \chi_{25}$		1	1		
$\chi_2 \otimes \chi_{63}$			1	4	5

Group: 2C1 Prime: 5 Block: 4

Nr.	CAS-Nr.	Degree	CC	N&C
1	11	94875	r	×
2	41	21528000	r	×
3	73	205395750	r	∘
4	86	299710125	r	∘
5	96	483483000	r	×

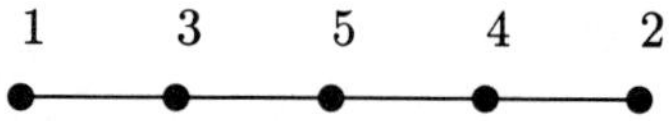

1 — 3 — 5 — 4 — 2

PROJECTIVES:

Nr.	1	2	3	4	5
CC	r	r	r	r	r
N&C	×	×	∘	∘	×
$\chi_2 \otimes \chi_{63}$		4		8	4
$\chi_2 \otimes \chi_{45}$			1		1

Group: 2C1 Prime: 5 Block: 5

Nr.	CAS-Nr.	Degree	CC	N&C
1	15	483000	r	×
2	40	21049875	r	×
3	59	77702625	r	∘
4	67	163478250	r	∘
5	77	219648000	r	×

1 — 3 — 5 — 4 — 2

PROJECTIVES:

Nr.	1	2	3	4	5
CC	r	r	r	r	r
N&C	×	×	∘	∘	×
$\chi_2 \otimes \chi_{63}$		2		6	4
$\chi_2 \otimes \chi_{79}$			2	2	4

Group: 2C1 Prime: 5 Block: 6

Nr.	CAS-Nr.	Degree	CC	N&C
1	19	822250	r	×
2	23	1771000	r	×
3	47	25900875	r	∘
4	68	184184000	r	∘
5	74	207491625	r	×

1 — 4 — 5 — 3 — 2

PROJECTIVES:

Nr.	1	2	3	4	5
CC	r	r	r	r	r
N&C	×	×	∘	∘	×
$\chi_6 \otimes \chi_{45}$	1			28	27
$\chi_2 \otimes \chi_{79}$			1	2	3

Group: 2C1 Prime: 5 Block: 7

Nr.	CAS-Nr.	Degree	CC	N&C
1	30	2877875	r	×
2	32	5494125	r	×
3	46	24794000	r	∘
4	51	44013375	r	∘
5	56	60435375	r	×

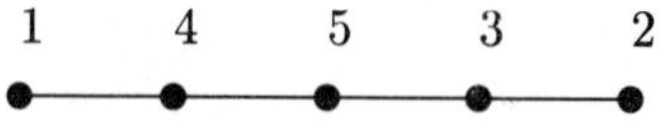

1 4 5 3 2

PROJECTIVES:

Nr.	1	2	3	4	5
CC	r	r	r	r	r
N&C	×	×	∘	∘	×
$\chi_{102} \otimes \chi_{134}$	1			1	
$\chi_{102} \otimes \chi_{146}$		1	1		
$\chi_{2} \otimes \chi_{63}$			2	3	5

Group: 2C1 Prime: 5 Block: 9

Nr.	CAS-Nr.	Degree	CC	N&C
1	109	299000	r	∘
2	133	19734000	r	∘
3	149	161161000	r	×
4	151	210496000	r	×
5	159	351624000	r	∘

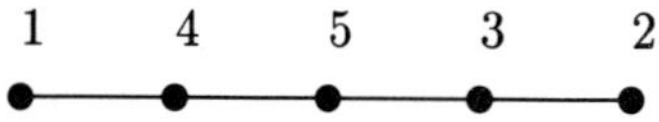

1 4 5 3 2

PROJECTIVES:

Nr.	1	2	3	4	5
CC	r	r	r	r	r
N&C	∘	∘	×	×	∘
$\chi_{2} \otimes \chi_{134}$		3	4		1
$\chi_{2} \otimes \chi_{128}$				1	1

Group: 2C1 Prime: 5 Block: 10

Nr.	CAS-Nr.	Degree	CC	N&C
1	116	4004000	r	∘
2	121	9152000	122	∘
2	122	9152000	121	∘
3	124	13156000	r	×

1 3 2

Group: 2C1 Prime: 7 Block: 2

Nr.	CAS-Nr.	Degree	CC	N&C
1	4	1771	r	×
2	12	313950	r	×
3	37	12432420	r	∘
4	59	77702625	r	∘
5	68	184184000	r	×
6	85	292953024	r	×
7	91	387317700	r	∘

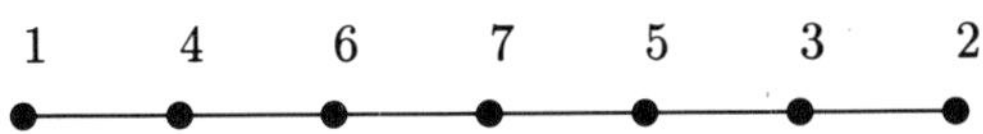

PROJECTIVES:

Nr.	1	2	3	4	5	6	7
CC	r	r	r	r	r	r	r
N&C	×	×	∘	∘	×	×	∘
$\chi_{102} \otimes \chi_{119}$		1	1				
$\chi_{2} \otimes \chi_{63}$			1		1		
$\chi_{2} \otimes \chi_{33}$				1		1	
$\chi_{6} \otimes \chi_{17}$					1		1
$\chi_{2} \otimes \chi_{27}$						1	1

Group: 2C1 Prime: 7 Block: 3

Nr.	CAS-Nr.	Degree	CC	N&C
1	5	8855	r	×
2	21	1434510	r	×
3	40	21049875	r	∘
4	47	25900875	r	∘
5	52	46621575	r	×
6	98	503513010	r	×
7	99	504627200	r	∘

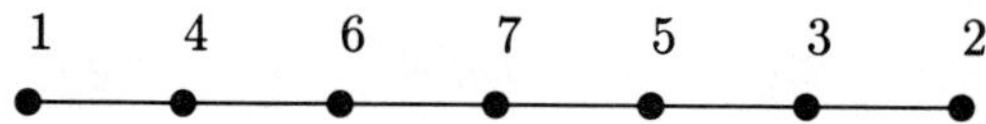

PROJECTIVES:

Nr.	1	2	3	4	5	6	7
CC	r	r	r	r	r	r	r
N&C	×	×	∘	∘	×	×	∘
$\chi_{102} \otimes \chi_{119}$		1	1				
$\chi_{102} \otimes \chi_{132}$			1		1		
$\chi_{2} \otimes \chi_{39}$				1		1	
$\chi_{6} \otimes \chi_{17}$					1	1	2

Group: 2C1 Prime: 7 Block: 4

Nr.	CAS-Nr.	Degree	CC	N&C
1	7	27300	r	×
2	15	483000	r	×
3	23	1771000	r	∘
4	29	2816856	r	∘
5	50	40370176	r	×
6	54	55255200	r	×
7	62	91547820	r	∘

PROJECTIVES:

Nr.	1	2	3	4	5	6	7
CC	r	r	r	r	r	r	r
N&C	×	×	∘	∘	×	×	∘
$\chi_2 \otimes \chi_{22}$	1			1			
$\chi_2 \otimes \chi_{36}$		1	1				
$\chi_2 \otimes \chi_{49}$			1		1		
$\chi_{102} \otimes \chi_{127}$				1		1	
$\chi_2 \otimes \chi_{70}$					1		1
$\chi_2 \otimes \chi_{26}$						1	1

Group: 2C1 Prime: 7 Block: 5

Nr.	CAS-Nr.	Degree	CC	N&C
1	8	37674	r	×
2	13	345345	r	×
3	30	2877875	r	∘
4	51	44013375	r	∘
5	80	251756505	r	×
6	82	259008750	r	×
7	93	464257024	r	∘

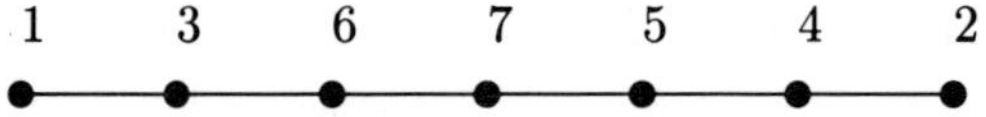

PROJECTIVES:

Nr.	1	2	3	4	5	6	7
CC	r	r	r	r	r	r	r
N&C	×	×	∘	∘	×	×	∘
$\chi_{102} \otimes \chi_{112}$	1		1				
$\chi_{2} \otimes \chi_{36}$		1		2	1		
$\chi_{6} \otimes \chi_{16}$			1			2	1
$\chi_{2} \otimes \chi_{53}$					1		1

Group: 2C1 Prime: 7 Block: 6

Nr.	CAS-Nr.	Degree	CC	N&C
1	9	44275	r	×
2	14	376740	r	×
3	31	4100096	r	∘
4	34	7628985	r	×
5	60	83720000	r	∘
6	64	106142400	r	∘
7	69	185912496	r	×

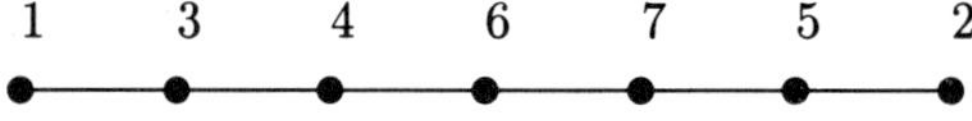

PROJECTIVES:

Nr.	1	2	3	4	5	6	7
CC	r	r	r	r	r	r	r
N&C	×	×	∘	×	∘	∘	×
$\chi_{102} \otimes \chi_{111}$	1		1				
$\chi_2 \otimes \chi_{39}$		1			1		
$\chi_2 \otimes \chi_{49}$			2	3		1	
$\chi_6 \otimes \chi_{43}$				1	6	12	17
$\chi_2 \otimes \chi_{70}$					1		1
$\chi_2 \otimes \chi_{44}$						1	1

Group: 2C1 Prime: 7 Block: 8

Nr.	CAS-Nr.	Degree	CC	N&C
1	104	2576	r	×
2	114	1841840	r	×
3	123	11051040	r	∘
4	144	62790000	r	×
5	146	106260000	r	∘
6	153	230230000	r	∘
7	155	282906624	r	×

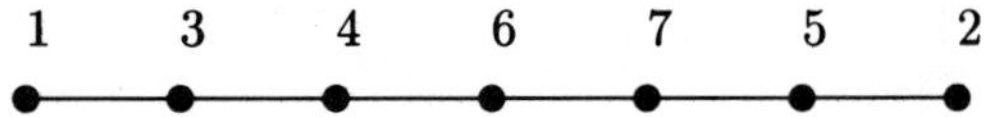

PROJECTIVES:

Nr.	1	2	3	4	5	6	7
CC	r	r	r	r	r	r	r
N&C	×	×	∘	×	∘	∘	×
$\chi_2 \otimes \chi_{112}$	1		1				
$\chi_{102} \otimes \chi_{32}$		1			1		
$\chi_2 \otimes \chi_{119}$			1	1			
$\chi_{102} \otimes \chi_{56}$				1		1	
$\chi_2 \otimes \chi_{118}$					1		1
$\chi_{102} \otimes \chi_{44}$						1	1

Group: 2C1 Prime: 7 Block: 9

Nr.	CAS-Nr.	Degree	CC	N&C
1	108	170016	r	∘
2	116	4004000	r	×
3	135	40370176	136	×
3	136	40370176	135	×
4	137	44204160	r	∘

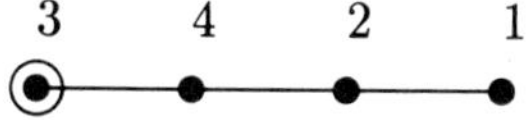

PROJECTIVES:

Nr.	1	2	3	4
CC	r	r	r	r
N&C	∘	×	×	∘
$\chi_2 \otimes \chi_{112}$	1	1		

Group: 2C1 Prime: 11 Block: 1

Nr.	CAS-Nr.	Degree	CC	N&C
1	1	1	r	×
2	12	313950	r	∘
3	14	376740	r	×
4	34	7628985	r	×
5	38	16347825	r	∘
6	41	21528000	r	∘
7	60	83720000	r	×
8	64	106142400	r	×
9	73	205395750	r	∘
10	79	247235625	r	∘
11	85	292953024	r	×

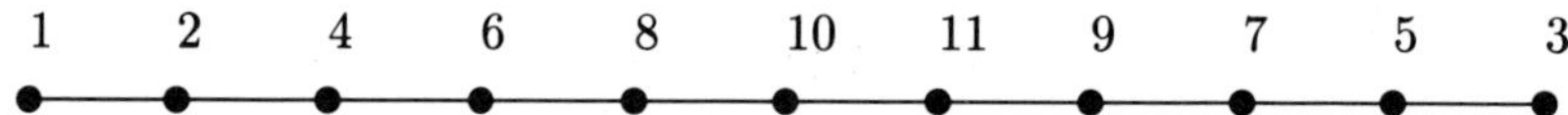

PROJECTIVES:

Nr.	1	2	3	4	5	6	7	8	9	10	11
CC	r	r	r	r	r	r	r	r	r	r	r
N&C	×	∘	×	×	∘	∘	×	×	∘	∘	×
$\chi_2 \otimes \chi_{11}$		1		1							
$\chi_2 \otimes \chi_{24}$			1		1						
$\chi_2 \otimes \chi_{22}$				1		1					
$\chi_2 \otimes \chi_{54}$					1		1				
$\chi_2 \otimes \chi_{44}$						1		1			
$\chi_2 \otimes \chi_{45}$							1		1		
$\chi_2 \otimes \chi_{50}$								1		1	
$\chi_2 \otimes \chi_{33}$									1		1
$\chi_2 \otimes \chi_{43}$										1	1

Group: 2C1 Prime: 11 Block: 2

Nr.	CAS-Nr.	Degree	CC	N&C
1	2	276	r	×
2	8	37674	r	∘
3	10	80730	r	×
4	15	483000	r	×
5	27	2464749	28	×
6	28	2464749	27	×
7	32	5494125	r	∘
8	51	44013375	r	∘
9	62	91547820	r	×
10	67	163478250	r	×
11	75	210974400	r	∘

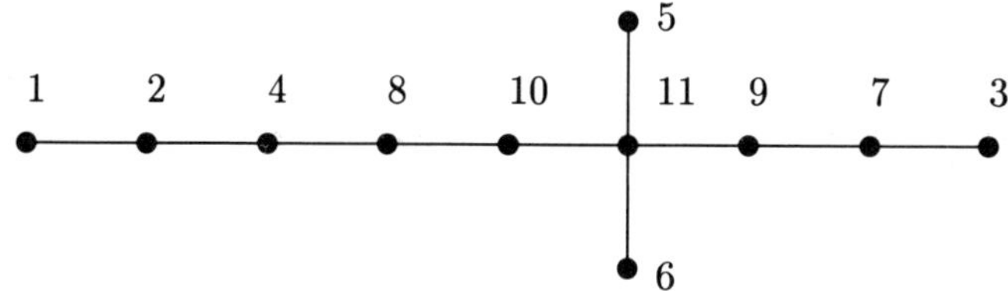

PROJECTIVES:

Nr.	1	2	3	4	5	6	7	8	9	10	11
CC	r	r	r	r	6	5	r	r	r	r	r
N&C	×	∘	×	×	×	×	∘	∘	×	×	∘
$\chi_2 \otimes \chi_4$	1	1									
$\chi_2 \otimes \chi_9$		1		1							
$\chi_2 \otimes \chi_{24}$			1				1				
$\chi_6 \otimes \chi_5$				1				1			
$\chi_2 \otimes \chi_{91}$					1	1			1	1	4
$\chi_2 \otimes \chi_{53}$							1		1		
$\chi_2 \otimes \chi_{47}$								1		1	

Group: 2C1 Prime: 11 Block: 3

Nr.	CAS-Nr.	Degree	CC	N&C
1	3	299	r	×
2	6	17250	r	×
3	7	27300	r	∘
4	20	871884	r	×
5	29	2816856	r	∘
6	65	109882500	r	×
7	66	150732800	r	∘
8	71	191102976	r	∘
9	83	267014475	r	×
10	89	326956500	r	×
11	90	360062976	r	∘

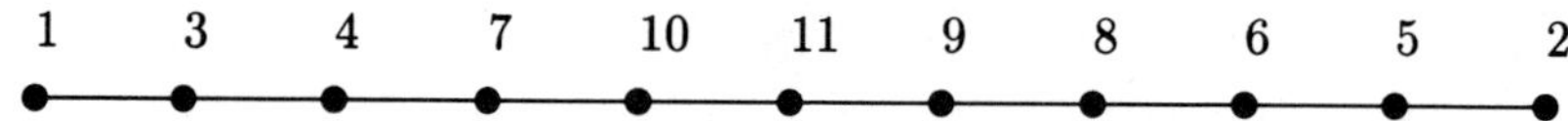

PROJECTIVES:

Nr.	1	2	3	4	5	6	7	8	9	10	11
CC	r	r	r	r	r	r	r	r	r	r	r
N&C	×	×	∘	×	∘	×	∘	∘	×	×	∘
$\chi_{102} \otimes \chi_{105}$	1		1								
$\chi_{2} \otimes \chi_{24}$		1			1						
$\chi_{2} \otimes \chi_{11}$			1	1							
$\chi_{2} \otimes \chi_{36}$				1		1	1	1			
$\chi_{102} \otimes \chi_{127}$					1	1					
$\chi_{2} \otimes \chi_{52}$						3	2	3		2	
$\chi_{2} \otimes \chi_{35}$						1	1	1		1	
$\chi_{2} \otimes \chi_{47}$							1	1	1	1	
$\chi_{6} \otimes \chi_{17}$							1			1	
$\chi_{2} \otimes \chi_{50}$								1	1	1	1
$\chi_{2} \otimes \chi_{43}$									1		1
$\chi_{2} \otimes \chi_{42}$										1	1

Group: 2C1 Prime: 11 Block: 4

Nr.	CAS-Nr.	Degree	CC	N&C
1	102	24	r	×
2	107	95680	r	×
3	109	299000	r	∘
4	115	1937520	r	×
5	131	17310720	r	∘
6	142	59153976	143	×
7	143	59153976	142	×
8	147	112519680	r	∘
9	157	313524224	r	×
10	159	351624000	r	×
11	165	655360000	r	∘

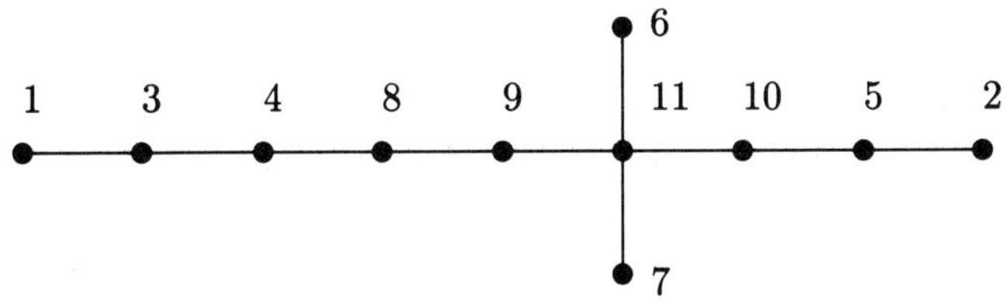

PROJECTIVES:

Nr.	1	2	3	4	5	6	7	8	9	10	11
CC	r	r	r	r	r	7	6	r	r	r	r
N&C	×	×	∘	×	∘	×	×	∘	×	×	∘
$\chi_2 \otimes \chi_{103}$	1		1								
$\chi_2 \otimes \chi_{110}$		1			1						
$\chi_2 \otimes \chi_{106}$			1	1							
$\chi_2 \otimes \chi_{116}$				1				1			
$\chi_2 \otimes \chi_{118}$					1					1	
$\chi_{102} \otimes \chi_{91}$						1	1				2
$\chi_2 \otimes \chi_{124}$								1	1		
$\chi_{102} \otimes \chi_{50}$									1		1
$\chi_2 \otimes \chi_{128}$										1	1

The consistency with Block 2 follows from $\chi_{102} \otimes \chi_{102} \approx_B \chi_2$ $(1_D \otimes 1_D \approx_B 1_B)$ and $\chi_{102} \otimes \chi_{142} \approx_B \chi_{27}$ $(1_D \otimes 6_D \approx_B 5_B)$.

Group: 2C1 Prime: 11 Block: 5

Nr.	CAS-Nr.	Degree	CC	N&C
1	104	2576	r	×
2	111	351624	r	∘
3	125	15002624	126	∘
3	126	15002624	125	∘
4	132	18987696	r	×
5	141	59153976	r	×
6	144	62790000	r	∘

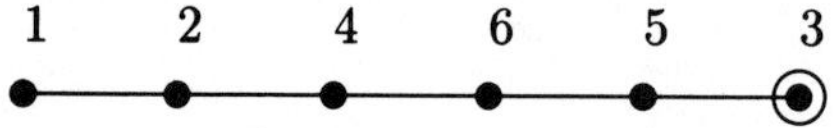

PROJECTIVES:

Nr.	1	2	3	4	5	6
CC	r	r	r	r	r	r
N&C	×	∘	∘	×	×	∘
$\chi_2 \otimes \chi_{105}$	1	1				
$\chi_6 \otimes \chi_{118}$		1		1		
$\chi_2 \otimes \chi_{152}$			1		1	
$\chi_2 \otimes \chi_{113}$				1		1

Group: 2C1 Prime: 13 Block: 1

Nr.	CAS-Nr.	Degree	CC	N&C
1	1	1	r	×
2	6	17250	r	∘
3	11	94875	r	×
4	17	673750	18	∘
5	18	673750	17	∘
6	21	1434510	r	∘
7	24	1821600	r	×
8	35	9221850	r	×
9	39	20083140	r	∘
10	71	191102976	r	∘
11	73	205395750	r	×
12	97	502078500	r	×
13	99	504627200	r	∘

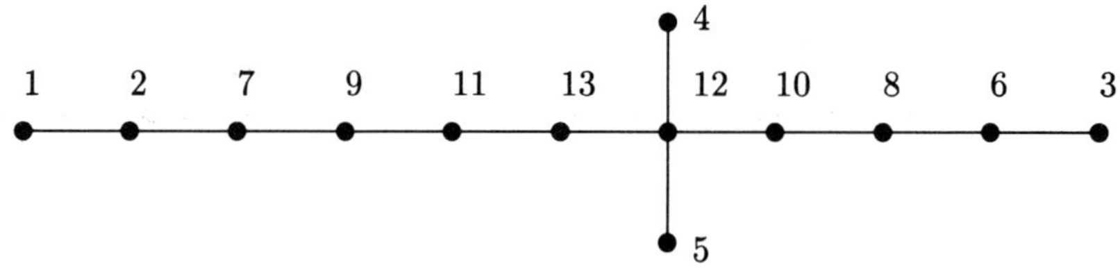

PROJECTIVES:

Nr.	1	2	3	4	5	6	7	8	9	10	11	12	13
CC	r	r	r	5	4	r	r	r	r	r	r	r	r
N&C	×	∘	×	∘	∘	∘	×	×	∘	∘	×	×	∘
$\chi_2 \otimes \chi_{30}$		1					1						
$\chi_2 \otimes \chi_7$			1			1							
$\chi_6 \otimes \chi_{25}$				1	1		1		2	1	2	5	3
$\chi_{102} \otimes \chi_{113}$						1		1					
$\chi_2 \otimes \chi_{10}$							1		1				
$\chi_6 \otimes \chi_{12}$								1		1			
$\chi_2 \otimes \chi_{33}$									1		1		
$\chi_2 \otimes \chi_{44}$										1		1	
$\chi_{102} \otimes \chi_{154}$											1		1
$\chi_2 \otimes \chi_{42}$												1	1

Group: 2C1 Prime: 13 Block: 2

Nr.	CAS-Nr.	Degree	CC	N&C
1	2	276	r	×
2	4	1771	r	×
3	9	44275	r	∘
4	23	1771000	r	∘
5	29	2816856	r	×
6	46	24794000	r	∘
7	51	44013375	r	×
8	66	150732800	r	∘
9	69	185912496	r	×
10	70	185955000	r	∘
11	74	207491625	r	×
12	91	387317700	r	×
13	93	464257024	r	∘

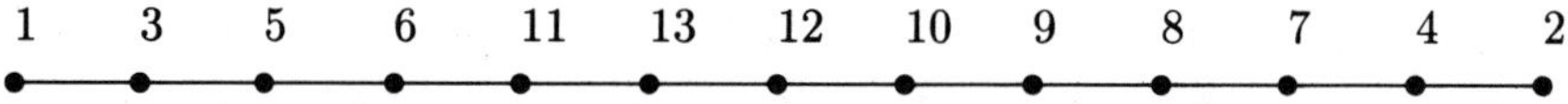

1 — 3 — 5 — 6 — 11 — 13 — 12 — 10 — 9 — 8 — 7 — 4 — 2

PROJECTIVES:

Nr.	1	2	3	4	5	6	7	8	9	10	11	12	13
CC	r	r	r	r	r	r	r	r	r	r	r	r	r
N&C	×	×	∘	∘	×	∘	×	∘	×	∘	×	×	∘
$\chi_2 \otimes \chi_3$	1		1										
$\chi_2 \otimes \chi_{13}$		1		1									
$\chi_6 \otimes \chi_3$			1		1								
$\chi_{102} \otimes \chi_{116}$				1			1						
$\chi_2 \otimes \chi_{22}$					1	1							
$\chi_2 \otimes \chi_{44}$						1		1	1		1		
$\chi_2 \otimes \chi_{37}$							1	1					
$\chi_{102} \otimes \chi_{160}$								1	1		1		1
$\chi_{102} \otimes \chi_{159}$									1	1	1		1
$\chi_{102} \otimes \chi_{151}$										1		1	
$\chi_{102} \otimes \chi_{163}$											1	1	2

Group: 2C1 Prime: 13 Block: 3

Nr.	CAS-Nr.	Degree	CC	N&C
1	5	8855	r	×
2	15	483000	r	∘
3	27	2464749	28	×
3	28	2464749	27	×
4	40	21049875	r	×
5	43	23244375	r	∘
6	49	40166280	r	∘
7	50	40370176	r	×

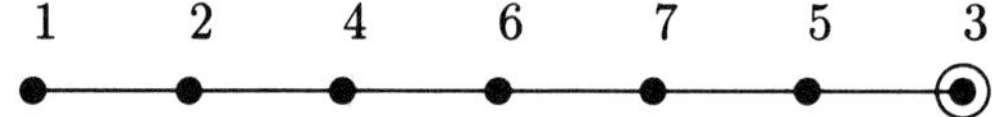

PROJECTIVES:

Nr.	1	2	3	4	5	6	7
CC	r	r	r	r	r	r	r
N&C	×	∘	×	×	∘	∘	×
$\chi_2 \otimes \chi_8$	1	1					
$\chi_2 \otimes \chi_{30}$		1		1			
$\chi_2 \otimes \chi_{85}$			1		1		
$\chi_2 \otimes \chi_{12}$				1		1	
$\chi_2 \otimes \chi_{78}$					1		1
$\chi_2 \otimes \chi_{64}$						1	1

Group: 2C1 Prime: 13 Block: 4

Nr.	CAS-Nr.	Degree	CC	N&C
1	102	24	r	∘
2	104	2576	r	×
3	106	40480	r	∘
4	108	170016	r	×
5	128	16170000	129	×
6	129	16170000	128	×
7	131	17310720	r	∘
8	135	40370176	136	×
9	136	40370176	135	×
10	141	59153976	r	∘
11	146	106260000	r	×
12	158	342752256	r	×
13	162	485760000	r	∘

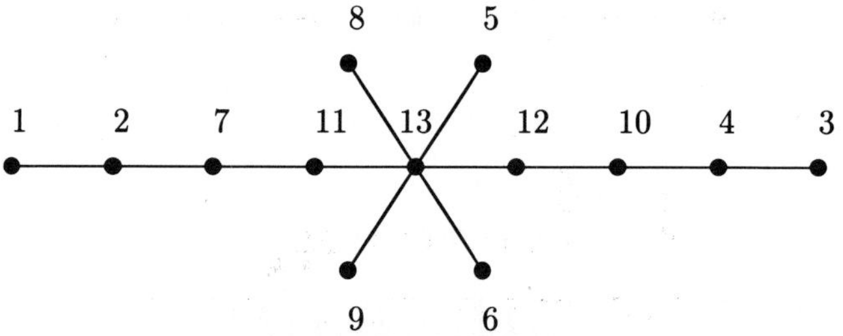

8 5

1 2 7 11 13 12 10 4 3

9 6

PROJECTIVES:

Nr.	1	2	3	4	5	6	7	8	9	10	11	12	13
CC	r	r	r	r	6	5	r	9	8	r	r	r	r
N&C	∘	×	∘	×	×	×	∘	×	×	∘	×	×	∘
$\chi_{102} \otimes \chi_3$	1	1											
$\chi_2 \otimes \chi_{110}$		1					1						
$\chi_2 \otimes \chi_{109}$			1	1									
$\chi_2 \otimes \chi_{116}$				1						1			
$\chi_{102} \otimes \chi_{72}$					1	1							2
$\chi_2 \otimes \chi_{114}$							1				1		
$\chi_2 \otimes \chi_{154}$								1	1		1	4	7
$\chi_2 \otimes \chi_{123}$										1		1	

The planar embedding and the consistency with Block 1 follow from $\chi_{102} \otimes \chi_{128} \approx_A \chi_{17}$ $(1_B \otimes 5_B \approx_A 4_A)$ and $\chi_{102} \otimes \chi_{135} \approx_A \chi_{99}$ $(1_B \otimes 8_B \approx_A 13_A)$.

Group: 2C1 Prime: 13 Block: 5

Nr.	CAS-Nr.	Degree	CC	N&C
1	103	2024	r	∘
2	112	388080	r	×
3	134	34155000	r	∘
4	142	59153976	143	∘
4	143	59153976	142	∘
5	150	190417920	r	×
6	164	557865000	r	×
7	165	655360000	r	∘

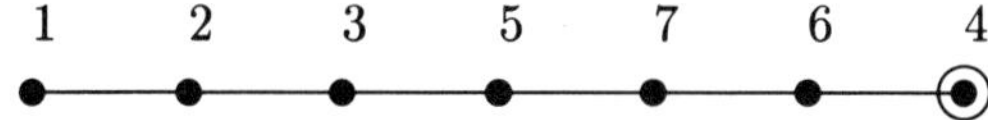

PROJECTIVES:

Nr.	1	2	3	4	5	6	7
CC	r	r	r	r	r	r	r
N&C	∘	×	∘	∘	×	×	∘
$\chi_2 \otimes \chi_{105}$	1	1					
$\chi_2 \otimes \chi_{110}$		1	1				
$\chi_2 \otimes \chi_{127}$			1		1		
$\chi_{102} \otimes \chi_{83}$				1		2	1
$\chi_{102} \otimes \chi_{68}$					1		1

Group: 2C1 Prime: 23 Block: 1

Nr.	CAS-Nr.	Degree	CC	N&C
1	1	1	r	×
2	7	27300	r	∘
3	17	673750	18	∘
3	18	673750	17	∘
4	31	4100096	r	×
5	40	21049875	r	∘
6	50	40370176	r	×
7	64	106142400	r	∘
8	71	191102976	r	×
9	77	219648000	r	×
10	84	280280000	r	∘
11	99	504627200	r	×
12	101	551675124	r	∘

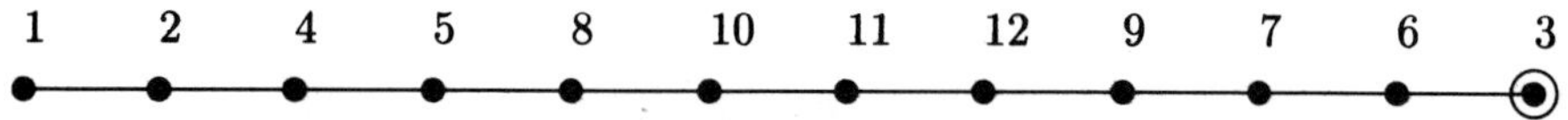

PROJECTIVES:

Nr.	1	2	3	4	5	6	7	8	9	10	11	12
CC	r	r	r	r	r	r	r	r	r	r	r	r
N&C	×	∘	∘	×	∘	×	∘	×	×	∘	×	∘
$\chi_2 \otimes \chi_2$	1	1										
$\chi_2 \otimes \chi_8$		1		1								
$\chi_2 \otimes \chi_{12}$				1	1							
$\chi_2 \otimes \chi_{47}$					1			1				
$\chi_2 \otimes \chi_{44}$							1	1	2	1		1
$\chi_6 \otimes \chi_{16}$								1		1		
$\chi_{102} \otimes \chi_{152}$									1			1
$\chi_2 \otimes \chi_{57}$										1	2	1

Group: 2C1 Prime: 23 Block: 2

Nr.	CAS-Nr.	Degree	CC	N&C
1	102	24	r	×
2	105	4576	r	∘
3	112	388080	r	×
4	116	4004000	r	∘
5	121	9152000	122	×
6	122	9152000	121	×
7	128	16170000	129	∘
7	129	16170000	128	∘
8	135	40370176	136	×
9	136	40370176	135	×
10	156	287006720	r	×
11	165	655360000	r	×
12	167	1021620600	r	∘

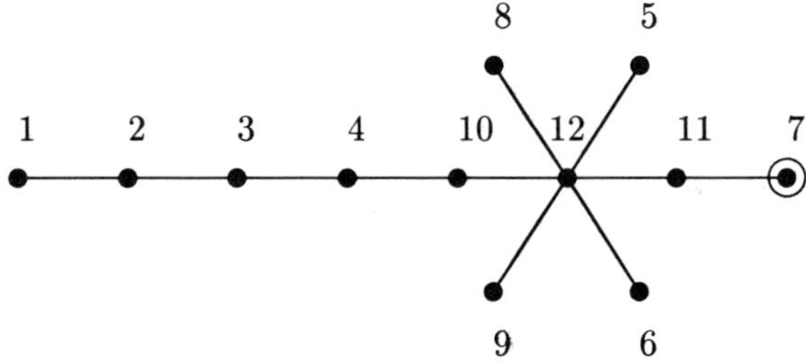

PROJECTIVES:

Nr.	1	2	3	4	5	6	7	8	9	10	11	12
CC	r	r	r	r	6	5	r	9	8	r	r	r
N&C	×	∘	×	∘	×	×	∘	×	×	×	×	∘
$\chi_2 \otimes \chi_{103}$	1	1										
$\chi_2 \otimes \chi_{104}$		1	1									
$\chi_2 \otimes \chi_{110}$			1	1								
$\chi_{102} \otimes \chi_{47}$				1						1		
$\chi_2 \otimes \chi_{153}$					1	1				7	7	16
$\chi_{102} \otimes \chi_{70}$							1				1	
$\chi_{102} \otimes \chi_{93}$								1	1		1	3

The planar embedding follows from $\chi_{102} \otimes \chi_{121} \approx_A \chi_{77}$ $(1_B \otimes 5_B \approx_A 9_A)$.

6.23 The Janko Group J4

We shall make extensive use of the character tables of the maximal subgroups of J4, which were computed by Fischer (1986).

Group: J4 Prime: 3 Block: 3

Nr.	CAS-Nr.	Degree	CC	N&C
1	4	299367	5	×
2	6	887778	7	×
3	9	1187145	10	○

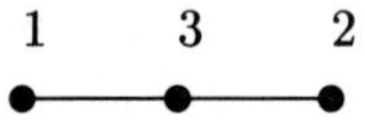

1 — 3 — 2

Group: J4 Prime: 3 Block: 4

Complex conjugate to Block 3.

Group: J4 Prime: 3 Block: 5

Nr.	CAS-Nr.	Degree	CC	N&C
1	11	1776888	r	×
2	42	1182518964	r	×
3	45	1184295852	r	○

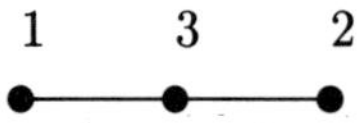

1 — 3 — 2

Group: J4 Prime: 3 Block: 7

Nr.	CAS-Nr.	Degree	CC	N&C
1	32	786127419	r	×
2	33	786127419	r	○
2	34	786127419	r	○

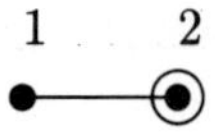

1 — 2 (exceptional vertex)

Group: J4 Prime: 5 Block: 1

Nr.	CAS-Nr.	Degree	CC	N&C
1	1	1	r	×
2	27	393877506	r	×
3	28	394765284	r	∘
4	42	1182518964	r	∘
5	44	1183406741	r	×

1 — 3 — 5 — 4 — 2

PROJECTIVES:

Nr.:	1	2	3	4	5
CC:	r	r	r	r	r
N&C:	×	×	∘	∘	×
$\chi_6 \otimes \chi_9$		5	4	15	14
$\chi_4 \otimes \chi_9$		1	2	4	5

Group: J4 Prime: 5 Block: 2

Nr.	CAS-Nr.	Degree	CC	N&C
1	2	1333	3	×
2	4	299367	5	∘
3	6	887778	7	×
4	15	32307363	16	×
5	17	32897107	18	∘

1 — 2 — 4 — 5 — 3

PROJECTIVES:

Nr.:	1	2	3	4	5
CC:	r	r	r	r	r
N&C:	×	∘	×	×	∘
$\chi_{14} \otimes \chi_9$		1		2	1
$\chi_{19} \otimes \chi_9$			1	13	14

Group: J4 Prime: 5 Block: 3

Complex conjugate to Block 2.

Group: J4 Prime: 5 Block: 4

Nr.	CAS-Nr.	Degree	CC	N&C
1	8	889111	r	×
2	26	366159104	r	∘
3	40	1085604531	r	×
4	43	1183406741	r	×
5	52	1903741279	r	∘

1 2 3 5 4

PROJECTIVES:

Nr.:	1	2	3	4	5
CC:	r	r	r	r	r
N&C:	×	∘	×	×	∘
$\chi_4 \otimes (\chi_2 + \chi_4)$	1	3	2	1	1
$\chi_4 \otimes \chi_9$		3	5	4	6

Observe that $\chi_2 + \chi_4$ is a projective in Block 2.

Group: J4 Prime: 5 Block: 5

Nr.	CAS-Nr.	Degree	CC	N&C
1	11	1776888	r	×
2	14	4290927	r	∘
3	35	789530568	r	×
4	59	2267824128	r	×
5	62	3054840657	r	∘

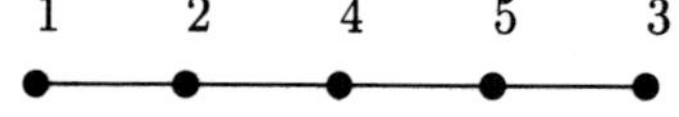

PROJECTIVES:

Nr.:	1	2	3	4	5
CC:	r	r	r	r	r
N&C:	×	∘	×	×	∘
$\chi_5 \otimes (\chi_2 + \chi_4)$	1	2		3	2
$\chi_4 \otimes \chi_9$			3	9	12

Observe that $\chi_2 + \chi_4$ is a projective in Block 2.

Group: J4 Prime: 5 Block: 6

Nr.	CAS-Nr.	Degree	CC	N&C
1	12	3403149	13	×
2	13	3403149	12	×
3	33	786127419	r	×
4	34	786127419	r	×
5	50	1579061136	r	∘

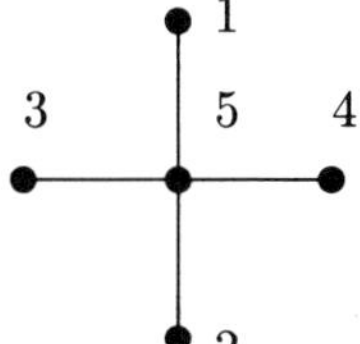

Group: J4 Prime: 5 Block: 7

Nr.	CAS-Nr.	Degree	CC	N&C
1	21	95288172	r	×
2	29	460559498	r	∘
3	31	690839247	r	×
4	45	1184295852	r	×
5	49	1509863773	r	∘

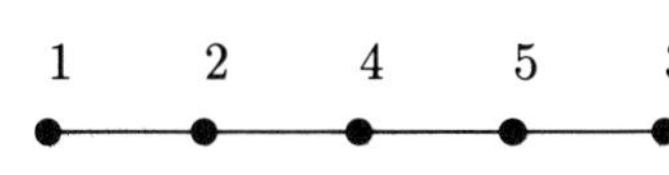

PROJECTIVES:

Nr.:	1	2	3	4	5
CC:	r	r	r	r	r
N&C:	×	∘	×	×	∘
$\chi_8 \otimes (\chi_2 + \chi_4)$	1	3	1	5	4
$\chi_4 \otimes \chi_9$		1	3	4	6

Observe that $\chi_2 + \chi_4$ is a projective in Block 2.

Group: J4 Prime: 5 Block: 8

Nr.	CAS-Nr.	Degree	CC	N&C
1	22	230279749	r	×
2	32	786127419	r	×
3	38	1016407168	r	∘
3	39	1016407168	r	∘

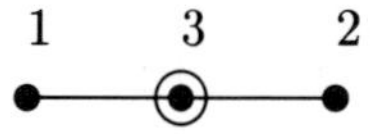

1 3 2

Group: J4 Prime: 5 Block: 9

Nr.	CAS-Nr.	Degree	CC	N&C
1	36	885257856	r	×
1	37	885257856	r	×
2	51	1842237992	r	×
3	61	2727495848	r	∘

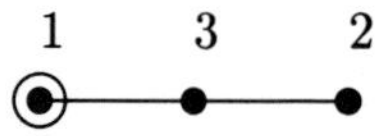

1 3 2

Group: J4 Prime: 7 Block: 1

Nr.	CAS-Nr.	Degree	CC	N&C
1	1	1	r	×
2	6	887778	7	∘
2	7	887778	6	∘
3	8	889111	r	∘
4	11	1776888	r	×

Only the following two trees are possible. The first of these is ruled out by considering $(887777^{2+}, \Phi_{887777}) = -1$.

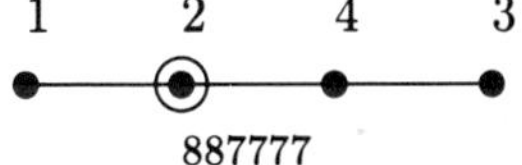

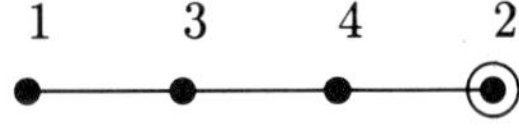

Group: J4 Prime: 7 Block: 2

Nr.	CAS-Nr.	Degree	CC	N&C
1	2	1333	3	∘
1	3	1333	2	∘
2	27	393877506	r	×
3	35	789530568	r	×
4	43	1183406741	r	∘

PROJECTIVES:

Nr.:	1	2	3	4
CC:	r	r	r	r
N&C:	∘	×	×	∘
$\chi_2 \otimes (\chi_1 + \chi_8)$	1		1	

Observe that $\chi_1 + \chi_8$ is a projective in the principal block.

Group: J4 Prime: 7 Block: 3

Nr.	CAS-Nr.	Degree	CC	N&C
1	4	299367	5	∘
1	5	299367	4	∘
2	40	1085604531	r	×
3	42	1182518964	r	×
4	59	2267824128	r	∘

2 4 3 1

Without using any projectives at all we have only the following two possible trees:

Let 1^- denote the alternating character of the maximal subgroup $2_+^{1+12} \cdot (3M_{22} : 2)$. This character lies in a 7-block of defect 1 and is irreducible modulo 7. Let M be a module over a field of characteristic 7 for 1^-. Then M is a trivial source module and is selfdual. Hence every direct summand of $\mathrm{Ind}_{2_+^{1+12} \cdot (3M_{22}:2)}(M)$ is liftable. Now $\mathrm{Ind}_{2_+^{1+12} \cdot (3M_{22}:2)}(1^-) \approx \chi_{40}$. Therefore $\mathrm{Ind}_{2_+^{1+12} \cdot (3M_{22}:2)}(M)$ restricted to the block is indecomposable with Brauer character χ_{40}. It follows that this module must be selfdual. If the second tree is correct then there is no selfdual indecomposable module with Brauer character χ_{40}. This shows that the first possibility is correct.

Group: J4 Prime: 7 Block: 4

Nr.	CAS-Nr.	Degree	CC	N&C
1	9	1187145	10	∘
1	10	1187145	9	∘
2	45	1184295852	r	×
3	49	1509863773	r	×
4	60	2692972480	r	∘

2 4 3 1

Without using any projectives at all, we have only the two possible trees below:

Let 23 denote the ordinary irreducible character of the subgroup isomorphic to $2^{11} : M_{24}$. Then the reduction of 23 modulo 7 is an irreducible trivial source module. Now $\mathrm{Ind}_{2^{11}:M_{24}}(23) \approx \chi_{45}$. The same argument as for Block 3 shows that the first tree is correct.

Group: J4 Prime: 7 Block: 5

Nr.	CAS-Nr.	Degree	CC	N&C
1	12	3403149	13	∘
1	13	3403149	12	∘
2	30	493456605	r	×
3	41	1089007680	r	×
4	50	1579061136	r	∘

Without using any projectives at all we have only the following two possible trees:

Let the notation be as in the proof for Block 4. We have $\mathrm{Ind}_{2^{11}:M_{24}}(23) \approx \chi_{30}$, showing that the first tree is correct.

Group: J4 Prime: 7 Block: 6

Nr.	CAS-Nr.	Degree	CC	N&C
1	14	4290927	r	×
2	17	32897107	18	∘
2	18	32897107	17	∘
3	26	366159104	r	∘
4	28	394765284	r	×

PROJECTIVES:

Nr.:	1	2	3	4
CC:	r	r	r	r
N&C:	×	∘	∘	×
$\chi_2 \otimes \chi_{25}$	1		3	2

Group: J4 Prime: 7 Block: 7

Nr.	CAS-Nr.	Degree	CC	N&C
1	15	32307363	16	∘
1	16	32307363	15	∘
2	44	1183406741	r	×
3	52	1903741279	r	×
4	62	3054840657	r	∘

Without using any projectives at all, we have only the following two possible trees:

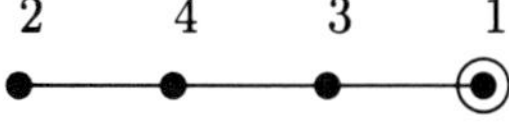

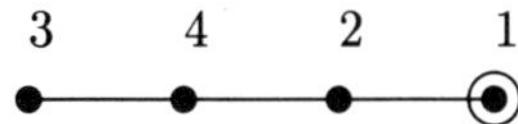

Let the notation be as in the proof for Block 3. We have $\mathrm{Ind}_{2_+^{1+12}\cdot(3M_{22}:2)}(1^-) \approx \chi_{52}$, showing that the second tree is correct.

Group: J4 Prime: 23 Block: 1

Nr.	CAS-Nr.	Degree	CC	N&C
1	1	1	r	×
2	2	1333	3	∘
3	3	1333	2	∘
4	4	299367	5	∘
5	5	299367	4	∘
6	6	887778	7	×
7	7	887778	6	×
8	14	4290927	r	×
9	15	32307363	16	∘
10	16	32307363	15	∘
11	23	259775040	r	∘
12	24	259775040	r	∘
13	25	300364890	r	×
14	26	366159104	r	×
15	27	393877506	r	∘
16	52	1903741279	r	∘
17	53	1981808640	r	×
18	54	1981808640	r	×
19	55	1981808640	r	×
20	56	2001151845	r	∘
21	57	2001151845	r	∘
22	58	2001151845	r	∘
23	59	2267824128	r	×

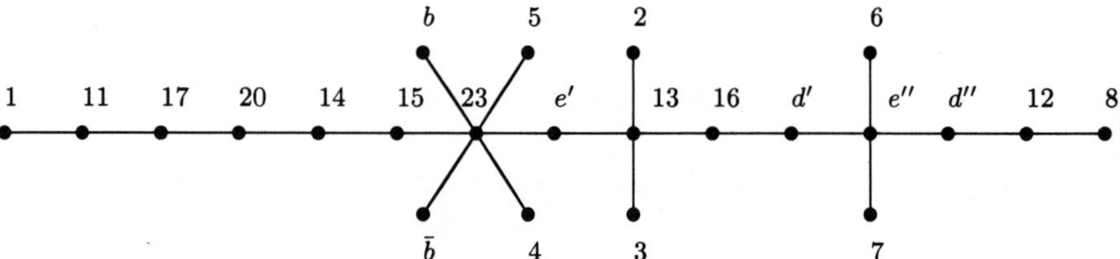

$\{b, \bar{b}\} = \{9, 10\}$, $\{d', d''\} = \{18, 19\}$ and $\{e', e''\} = \{21, 22\}$

PROJECTIVES:

Nr.:	1	2	3	4	5	6	7	8	9	10	11	12
CC:	r	3	2	5	4	7	6	r	10	9	r	r
N&C:	×	∘	∘	∘	∘	×	×	×	∘	∘	∘	∘
$\mathrm{Ind}_{2^{10}:L_5(2)}(\chi_1)$	1							1			1	1
$\mathrm{Ind}_{2^{3+12}.(S_5\times L_3(2))}(\chi_6)$			1								1	1
$\mathrm{Ind}_{2_+^{1+12}.(3M_{22}:2)}(\chi_8)$					1					1	1	1
$\mathrm{Ind}_{2_+^{1+12}.(3M_{22}:2)}(\chi_6)$							1					
$\mathrm{Ind}_{2^{10}:L_5(2)}(\chi_2)$								2			5	5
$\mathrm{Ind}_{2_+^{1+12}.(3M_{22}:2)}(\chi_3)$								1			2	2
$\mathrm{Ind}_{2_+^{1+12}.(3M_{22}:2)}(\chi_{13})$										1	1	1
$\mathrm{Ind}_{2_+^{1+12}.(3M_{22}:2)}(\chi_4)$											1	1
$\mathrm{Ind}_{2_+^{1+12}.(3M_{22}:2)}(\chi_{11})$												
$\mathrm{Ind}_{2_+^{1+12}.(3M_{22}:2)}(\chi_2)$												

PROJECTIVES (continued):

Nr.:	13	14	15	16	17	18	19	20	21	22	23
CC:	r	r	r	r	r	r	r	r	r	r	r
N&C:	×	×	∘	∘	×	×	×	∘	∘	∘	×
$\mathrm{Ind}_{2^{10}:L_5(2)}(\chi_1)$											
$\mathrm{Ind}_{2^{3+12}.(S_5\times L_3(2))}(\chi_6)$	6	1		11	9	9	9	9	9	9	7
$\mathrm{Ind}_{2_+^{1+12}.(3M_{22}:2)}(\chi_8)$			1	4	4	4	4	3	3	3	6
$\mathrm{Ind}_{2_+^{1+12}.(3M_{22}:2)}(\chi_6)$	1	2		3	4	4	4	6	6	6	5
$\mathrm{Ind}_{2^{10}:L_5(2)}(\chi_2)$	2	2		7	6	6	6	3	3	3	2
$\mathrm{Ind}_{2_+^{1+12}.(3M_{22}:2)}(\chi_3)$	2	1		3	2	2	2	1	1	1	
$\mathrm{Ind}_{2_+^{1+12}.(3M_{22}:2)}(\chi_{13})$	1	2	1	7	8	8	8	9	9	9	11
$\mathrm{Ind}_{2_+^{1+12}.(3M_{22}:2)}(\chi_4)$		1	2	2	2	2	2	1	1	1	2
$\mathrm{Ind}_{2_+^{1+12}.(3M_{22}:2)}(\chi_{11})$	2	1	1	5	5	5	5	6	6	6	6
$\mathrm{Ind}_{2_+^{1+12}.(3M_{22}:2)}(\chi_2)$	1			1							

We have three orbits of algebraically conjugate characters. They are $\{\chi_{23}, \chi_{24}\}$, $\{\chi_{53}, \chi_{54}, \chi_{55}\}$ and $\{\chi_{56}, \chi_{57}, \chi_{58}\}$. They correspond to sets of nodes $\mathcal{O}_1 = \{11, 12\}$, $\mathcal{O}_2 = \{17, 18, 19\}$ and $\mathcal{O}_3 = \{20, 21, 22\}$. These orbits generate pairwise disjoint fields. Let S be the subgroup of the symmetric group on 23 letters permuting only the points inside each of these orbits. Then S acts on the set of all Brauer trees which are consistent with the above projectives. The reason is that the characters of an orbit occur with equal multiplicity in each of the projectives. There are 12 orbits under this action and we give a representative for each orbit:

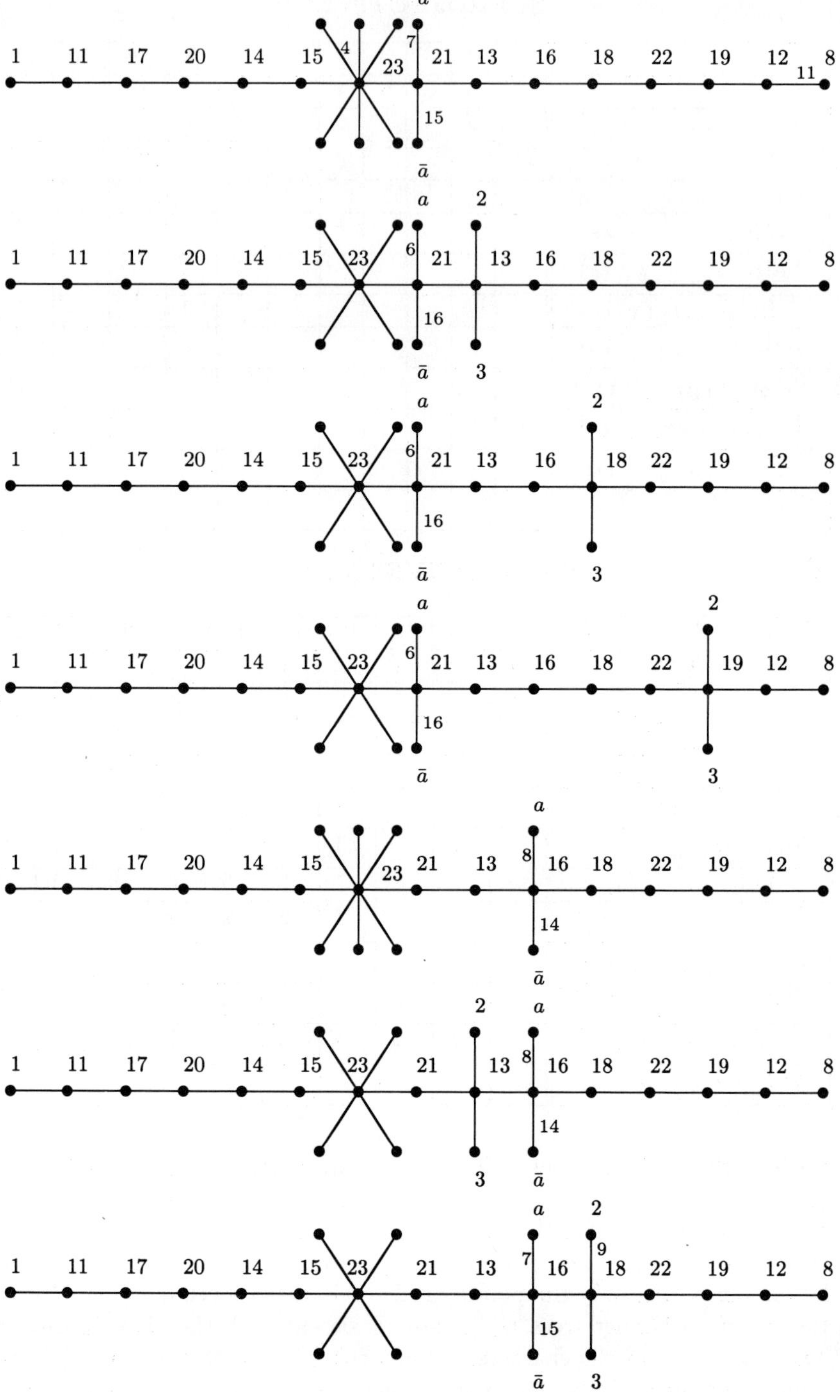
a
1 11 17 20 14 15 4 7 23 21 13 16 18 22 19 12 11 8
15
ā
a 2
1 11 17 20 14 15 23 6 21 13 16 18 22 19 12 8
16
ā 3
a 2
1 11 17 20 14 15 23 6 21 13 16 18 22 19 12 8
16
ā 3
a 2
1 11 17 20 14 15 23 6 21 13 16 18 22 19 12 8
16
ā 3
a
1 11 17 20 14 15 23 21 13 8 16 18 22 19 12 8
14
ā
2 a
1 11 17 20 14 15 23 21 13 8 16 18 22 19 12 8
14
3 ā
a 2
1 11 17 20 14 15 23 21 13 7 16 9 18 22 19 12 8
15
ā 3

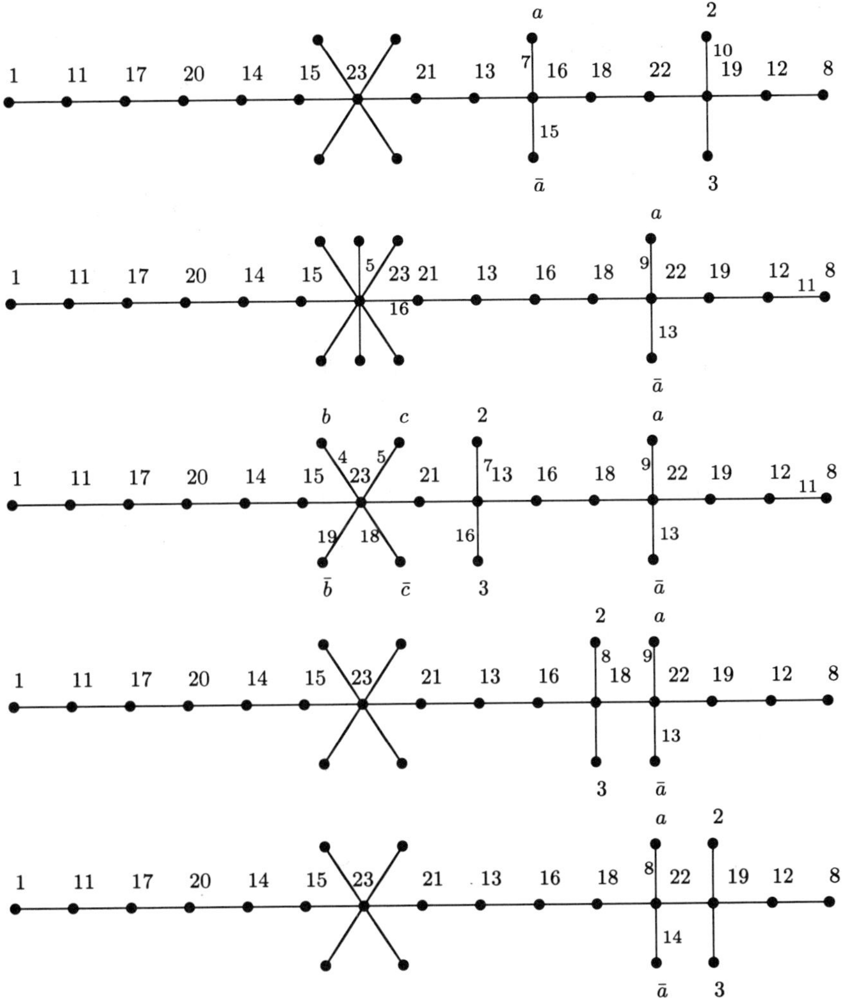

The following notation has been used to denote some nodes: $\{a, \bar{a}\} = \{6, 7\}$ and $\{b, \bar{b}, c, \bar{c}\} = \{4, 5, 9, 10\}$. We have labelled some edges of the trees according to Green correspondence. The number on an edge indicates the socle of the Green correspondent of the simple module on that edge. It also shows the socles of various indecomposable modules around that edge.

We use the following tensor products:

$$\begin{aligned} \chi_2 \otimes \chi_2 &\approx \chi_7 \\ \chi_2 \otimes \chi_{14} &\approx \chi_4 + \chi_{25} + \chi_{52} \end{aligned}$$

(Here and in the following tables $\approx$ denotes the tensor product restricted to the block.) In terms of nodes these tensor products are as follows:

$$\begin{aligned} 2 \otimes 2 &\approx 7 \\ 2 \otimes 8 &\approx 4 + 13 + 16 \end{aligned}$$

In every possible tree, node 2 is a leaf, i.e. it is an irreducible Brauer character. Let $22Ax$ be its Green correspondent. Then by Lemma 4.3.2 $2 \otimes 2$ has Green correspondent $22Ax \otimes 22Ax = 1A(2x-1)$. Here, $2x-1$ has to be read modulo 22. Since 22 is even, $2x-1$ is always odd. Now $2 \otimes 2 = 7$, hence there must be an edge coming from node 7 with odd label $2x-1$. This immediately rules out the second to the fifth and the last tree. Trees number seven, eight and eleven can be ruled out by the same argument, since in these cases $2x-1$ does not label the edge coming from node 7.

In the first tree, x must be equal to 4 (we have assumed without loss of generality that node 2 sits on the upper side of the tree). Now the irreducible module corresponding to node 8 has Green correspondent $1A11$. Hence $2 \otimes 8$ has Green correspondent $22A4 \otimes 1A11 = 22A15$. The Brauer character of this is node 21 which does not occur in $2 \otimes 8$. So this tree is impossible.

We rule out the ninth tree with a similar argument. Here, the Green correspondent of node 2 is $22A5$. Hence $2 \otimes 8$ has Green correspondent $22A5 \otimes 1A5 = 22A16$. This again must have Brauer character 21, which does not occur in $2 \otimes 8$.

The tenth tree must therefore be correct. We have only to determine its planar embedding. Since the irreducible module corresponding to node 2 has Green correspondent $22A7$, we conclude from $2 \otimes 2 = 7$ and $22A7 \otimes 22A7 = 1A13$ that $\bar{a}$ equals 7. Also, from $2 \otimes 8 = 4+13+16$ and $22A7 \otimes 1A11 = 22A18$ we see that $\bar{c}$ equals 4. Unfortunately we cannot decide whether b equals 9 or 10.

Group: J4 Prime: 29 Block: 1

Nr.	CAS-Nr.	Degree	CC	N&C
1	1	1	r	×
2	2	1333	3	∘
3	3	1333	2	∘
4	6	887778	7	×
5	7	887778	6	×
6	9	1187145	10	×
7	10	1187145	9	×
8	12	3403149	13	∘
9	13	3403149	12	∘
10	19	35411145	r	∘
11	20	35411145	r	∘
12	21	95288172	r	×
13	27	393877506	r	∘
14	32	786127419	r	×
15	33	786127419	r	×
16	34	786127419	r	×
17	36	885257856	r	∘
18	37	885257856	r	∘
19	38	1016407168	r	×
20	39	1016407168	r	×
21	41	1089007680	r	∘
22	49	1509863773	r	×
23	53	1981808640	r	∘
24	54	1981808640	r	∘
25	55	1981808640	r	∘
26	56	2001151845	r	×
27	57	2001151845	r	×
28	58	2001151845	r	×
29	61	2727495848	r	∘

We have used the following notation for the nodes in the tree below: $\{e', e''\} = \{15, 16\}$, $\{f, f'\} = \{17, 18\}$, $\{g', g''\} = \{24, 25\}$ and $\{h', h''\} = \{27, 28\}$.

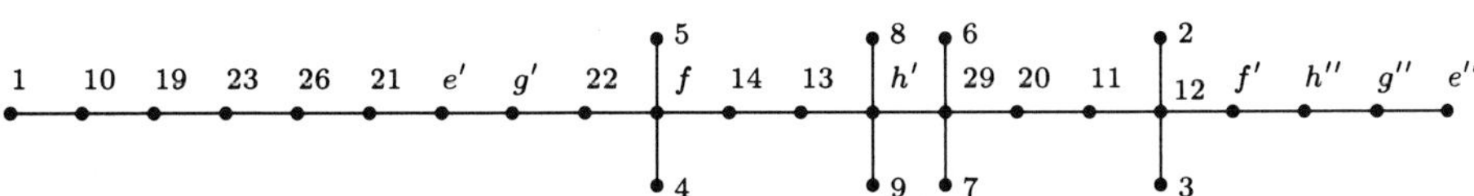

PROJECTIVES:

Nr.:	1	2	3	4	5	6	7	8	9	10	11	12	13	14	15	16
CC:	r	3	2	5	4	7	6	9	8	r	r	r	r	r	r	r
N&C:	×	∘	∘	×	×	×	×	∘	∘	∘	∘	×	∘	×	×	×
$\mathrm{Ind}_{2^{10}:L_5(2)}(\chi_1)$	1									2	2	2		1		
$\mathrm{Ind}_{2^{11}:M_{24}}(\chi_1)$	1									1	1	1				
$\mathrm{Ind}_{2^{11}:M_{24}}(\chi_4)$			1									1				
$\mathrm{Ind}_{2^{11}:M_{24}}(\chi_{13})$				1									1	1	1	1
$\mathrm{Ind}_{2_+^{1+12}\cdot(3M_{22}:2)}(\chi_6)$					1				1					1	1	1
χ_4^{2-}					1				1							
$\mathrm{Ind}_{2^{11}:M_{24}}(\chi_{40})$						1				3	3	9	26	54	54	54
$\mathrm{Ind}_{2^{11}:M_{24}}(\chi_{17})$							1		1					1	1	1
$\mathrm{Ind}_{2^{11}:M_{24}}(\chi_7)$										2	2	2		1	1	1
χ_4^{2+}										1	1		1	1	1	1
$\mathrm{Ind}_{2^{3+12}\cdot(S_5\times L_3(2))}(\chi_2)$													4	2	1	1
$\mathrm{Ind}_{2_+^{1+12}\cdot(3M_{22}:2)}(\chi_{11})$													1	2	2	2
$\mathrm{Ind}_{2^{11}:M_{24}}(\chi_8)$													1			
$\mathrm{Ind}_{2^{11}:M_{24}}(\chi_{11})$														1	2	2
$\mathrm{Ind}_{2^{11}:M_{24}}(\chi_5)$																

PROJECTIVES (continued):

Nr.:	17	18	19	20	21	22	23	24	25	26	27	28	29
CC:	r	r	r	r	r	r	r	r	r	r	r	r	r
N&C:	∘	∘	×	×	∘	×	∘	∘	∘	×	×	×	∘
$\mathrm{Ind}_{2^{10}:L_5(2)}(\chi_1)$	1	1	1	1									
$\mathrm{Ind}_{2^{11}:M_{24}}(\chi_1)$													
$\mathrm{Ind}_{2^{11}:M_{24}}(\chi_4)$													
$\mathrm{Ind}_{2^{11}:M_{24}}(\chi_{13})$	2	2	1	1	2	3	4	4	4	5	5	5	5
$\mathrm{Ind}_{2_+^{1+12}\cdot(3M_{22}:2)}(\chi_6)$	3	3	1	1	3	4	4	4	4	6	6	6	6
χ_4^{2-}	1	1			1	1	1	1	1	2	2	2	1
$\mathrm{Ind}_{2^{11}:M_{24}}(\chi_{40})$	60	60	71	71	73	102	135	135	135	134	134	134	188
$\mathrm{Ind}_{2^{11}:M_{24}}(\chi_{17})$	2	2	1	1	2	4	4	4	4	5	5	5	6
$\mathrm{Ind}_{2^{11}:M_{24}}(\chi_7)$	1	1	3	3			1	1	1				2
χ_4^{2+}			2	2			1	1	1				1
$\mathrm{Ind}_{2^{3+12}\cdot(S_5\times L_3(2))}(\chi_2)$			2	2	1	2	3	3	3	2	2	2	2
$\mathrm{Ind}_{2_+^{1+12}\cdot(3M_{22}:2)}(\chi_{11})$	3	3	2	2	3	4	5	5	5	6	6	6	7
$\mathrm{Ind}_{2^{11}:M_{24}}(\chi_8)$						1	1	1	1	1	1	1	
$\mathrm{Ind}_{2^{11}:M_{24}}(\chi_{11})$	2	2	2	2	3	3	3	3	3	3	3	3	5
$\mathrm{Ind}_{2^{11}:M_{24}}(\chi_5)$						1	1	1	1	1	1	1	1

We have six orbits of algebraically conjugate characters corresponding to the following set of nodes: $\mathcal{O}_1 = \{10, 11\}$, $\mathcal{O}_2 = \{15, 16\}$, $\mathcal{O}_3 = \{17, 18\}$, $\mathcal{O}_4 = \{19, 20\}$, $\mathcal{O}_5 = \{23, 24, 25\}$, $\mathcal{O}_6 = \{26, 27, 28\}$. Let S be the subgroup of the symmetric group on 29 points fixing $\mathcal{O}_1, \ldots, \mathcal{O}_6$ setwise and fixing every element not in one of these sets. Then S acts on all labelled trees which are consistent with all the projectives given above since the characters in an orbit have equal multiplicity in every such projective. The set of all consistent trees falls into six orbits under the action of S. We have printed only one tree from each orbit, using the notation $\{b, \bar{b}\} = \{4, 5\}$, $\{c, \bar{c}\} = \{6, 7\}$ and $\{d, \bar{d}\} = \{8, 9\}$.

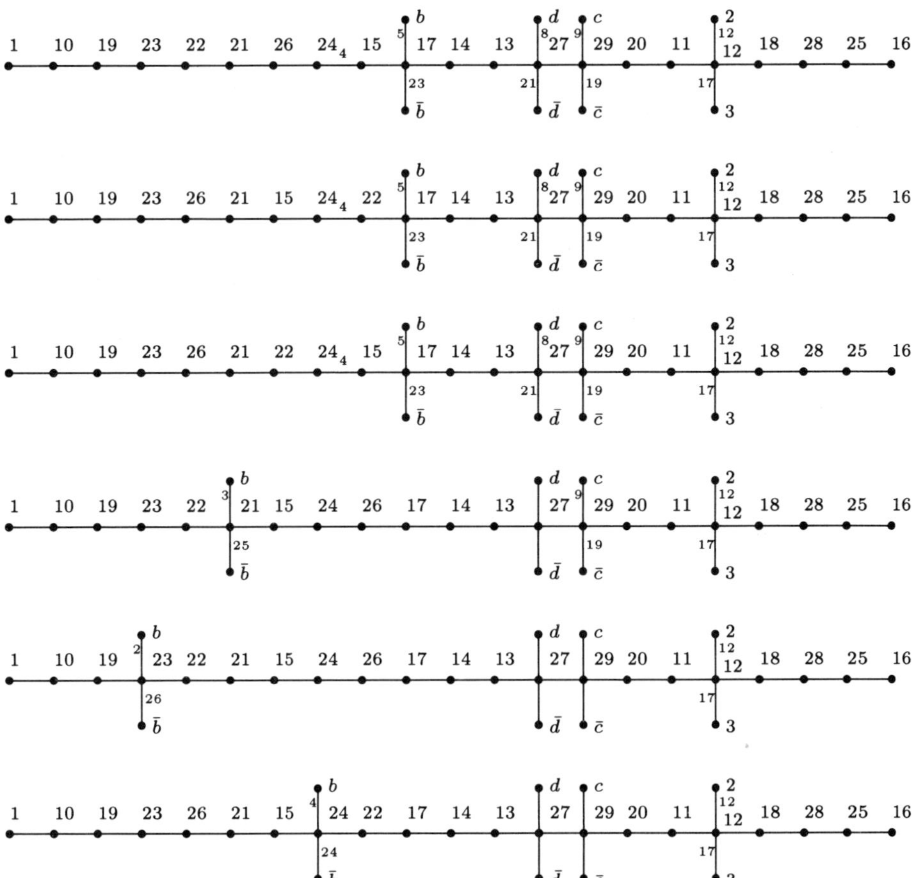

We have labelled certain edges of these trees with numbers representing the socles of the Green correspondents of the simple modules belonging to the edge.

We use the following tensor products of J4:

$$\begin{aligned}\chi_2 \otimes \chi_2 &\approx \chi_7\\ \chi_2 \otimes \chi_{12} &\approx \chi_{10}\\ \chi_2 \otimes \chi_{13} &\approx \chi_{49}\end{aligned}$$

Here the symbol $\approx$ means the tensor product restricted to the block. In terms of nodes of the tree these tensor products read as follows:

$$\begin{aligned}2 \otimes 2 &\approx 5\\ 2 \otimes 8 &\approx 7\\ 2 \otimes 9 &\approx 22\end{aligned}$$

Now in each of the trees we have without loss of generality that the module corresponding to node 2 has Green correspondent $28A12$. By Lemma 4.3.2 we have that $2 \otimes 2$ has an indecomposable direct summand with Green correspondent $1A23$. Since the only indecomposable summand of $2 \otimes 2$ in the block has Brauer character χ_7 which labels node 5, we must have the label 23 at the edge corresponding to node 5. This excludes the last three trees, and shows that b is 4, $\bar{b}$ is 5.

The module corresponding to d has Green correspondent $28A8$. Hence $2 \otimes d$ has an indecomposable direct summand with Green correspondent $1A19$. This is the label on edge $\bar{c}$. It follows that c is 6 and d is 8 (if d were 9 then 8 would have Green correspondent $28A21$ and $2 \otimes 8$ had $1A4$ as correspondent, contradicting their tensor product given above).

Finally, the tensor product $2 \otimes 9$ has now Green correspondent $1A4$. Since only character 22 occurs in $2 \otimes 9$, we conclude from the labelling that the second tree is correct.

Group: J4 Prime: 31 Block: 1

Nr.	CAS-Nr.	Degree	CC	N&C
1	1	1	r	×
2	11	1776888	r	∘
3	36	885257856	r	×
4	37	885257856	r	×
5	42	1182518964	r	×
6	53	1981808640	r	∘
7	54	1981808640	r	∘
8	55	1981808640	r	∘
9	56	2001151845	r	∘
9	57	2001151845	r	∘
9	58	2001151845	r	∘
10	59	2267824128	r	×
11	61	2727495848	r	×

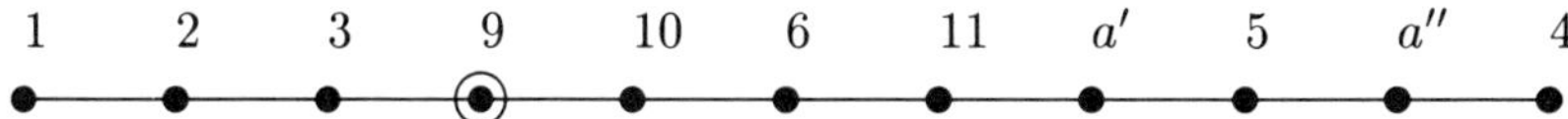

$\{a', a''\} = \{7, 8\}$

PROJECTIVES:

Nr.:	1	2	3	4	5	6	7	8	9	10	11
CC:	r	r	r	r	r	r	r	r	r	r	r
N&C:	×	∘	×	×	×	∘	∘	∘	∘	×	×
$\mathrm{Ind}_{2^{11}:\mathrm{M}_{24}}(\chi_1)$	1	1									
$\mathrm{Ind}_{2^{11}:\mathrm{M}_{24}}(\chi_7)$		1	1	1		1	1	1			2
$\mathrm{Ind}_{2^{11}:\mathrm{M}_{24}}(\chi_{13})$			2	2	2	4	4	4	5	6	5
$\mathrm{Ind}_{2_+^{1+12}\cdot(3\mathrm{M}_{22}:2)}(\chi_8)$					4	4	4	4	3	6	5
$\mathrm{Ind}_{2^{11}:\mathrm{M}_{24}}(\chi_8)$					2	1	1	1	1	2	

We have two orbits of algebraic conjugate characters, generating disjoint fields. The first has length 2 and contains the characters χ_{36} and χ_{37} (nodes 3 and 4). We assume without loss of generality that χ_{36} comes first on the Brauer tree. The second orbit consists of the three characters χ_{53}, χ_{54} and χ_{55} (nodes 6, 7 and 8). We assume without loss of generality that χ_{53} comes first on the tree.

Group: J4 Prime: 37 Block: 1

Nr.	CAS-Nr.	Degree	CC	N&C
1	1	1	r	×
2	2	1333	3	×
3	3	1333	2	×
4	8	889111	r	×
5	15	32307363	16	∘
6	16	32307363	15	∘
7	19	35411145	r	∘
8	20	35411145	r	∘
9	23	259775040	r	×
10	24	259775040	r	×
11	43	1183406741	r	∘
12	44	1183406741	r	∘
13	53	1981808640	r	×
13	54	1981808640	r	×
13	55	1981808640	r	×

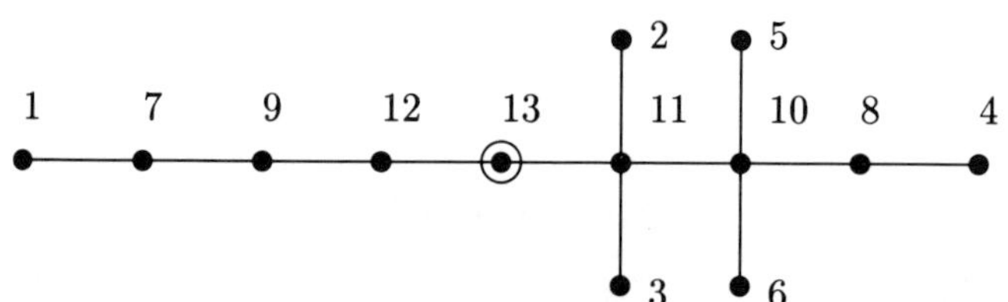

PROJECTIVES:

Nr.:	1	2	3	4	5	6	7	8	9	10	11	12	13
CC:	r	3	2	r	6	5	r	r	r	r	r	r	r
N&C:	×	×	×	×	∘	∘	∘	∘	×	×	∘	∘	×
$\mathrm{Ind}_{2^{11}:\mathrm{M}_{24}}(\chi_1)$	1			1			1	1					
$\mathrm{Ind}_{2^{11}:\mathrm{M}_{24}}(\chi_4)$			1								1		
χ_4^{2+}				1		1	1	1	1	1	1		1
$\mathrm{Ind}_{2^{11}:\mathrm{M}_{24}}(\chi_7)$				1			2	2	2	2	2		1
$\mathrm{Ind}_{2_+^{1+12}\cdot(3\mathrm{M}_{22}:2)}(\chi_8)$						1			1	1	2	3	4
$\mathrm{Ind}_{2_+^{1+12}\cdot(3\mathrm{M}_{22}:2)}(\chi_4)$									1	1	3	1	2
$\chi_6 \otimes \chi_4$											1	1	2

We have two orbits of algebraic conjugate characters whose irrationalities generate disjoint fields. Each orbit has length 2. Hence we can assume without loss of generality that of $\{\chi_{19}, \chi_{20}\}$ χ_{19} comes first on the tree, and that of $\{\chi_{23}, \chi_{24}\}$ χ_{23} comes first. Then we have two trees consistent with the given projectives:

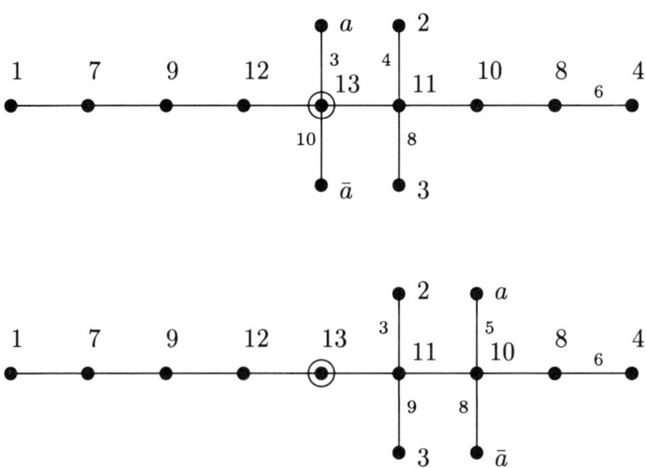

Here $\{a, \bar{a}\} = \{5, 6\}$. Again we have labelled some of the edges with numbers representing the socles of the Green correspondents of the corresponding simple modules. We use the following two tensor products:

$$\begin{aligned} \chi_2 \otimes \chi_2 &\approx \chi_8 \\ \chi_2 \otimes \chi_{15} &\approx \chi_{16} + \chi_{44} + (\chi_{53} + \chi_{54} + \chi_{55}) \end{aligned}$$

In terms of nodes these tensor products are

$$\begin{aligned} 2 \otimes 2 &\approx 4 \\ 2 \otimes 5 &\approx 6 + 12 + 13 \end{aligned}$$

The Green correspondent of the simple module with Brauer character χ_2 is $1Ax$ with x equal to 3 or 5 according to the two possibilities. By Lemma 4.3.2 the Green correspondent of $2 \otimes 2$ is $1A2x$. However, the only character occurring in $\chi_2 \otimes \chi_2$ is χ_8, which corresponds to node 4 in both possibilities. The simple module at this node has correspondent $1A6$, showing that x must be 3. Hence the second tree is correct.

Now χ_{24}, which is node 10, does not appear at all in $\chi_2 \otimes \chi_{15}$. The only way to subtract an ordinary character from this tensor product to leave a projective is to subtract χ_{16}, which is node 6. This shows that the Green correspondent of $\chi_2 \otimes \chi_{15}$ has Brauer character χ_{16}. It follows that χ_{15} has Green correspondent $36A5$, rather than $36A8$, hence a is 5. This completes the proof of the tree and its planar embedding.

Group: J4 Prime: 43 Block: 1

Nr.	CAS-Nr.	Degree	CC	N&C
1	1	1	r	×
2	4	299367	5	×
3	5	299367	4	×
4	9	1187145	10	×
5	10	1187145	9	×
6	11	1776888	r	○
7	15	32307363	16	×
8	16	32307363	15	×
9	42	1182518964	r	×
10	46	1445942610	r	○
10	47	1445942610	r	○
10	48	1445942610	r	○
11	53	1981808640	r	×
12	54	1981808640	r	×
13	55	1981808640	r	×
14	60	2692972480	r	○
15	62	3054840657	r	○

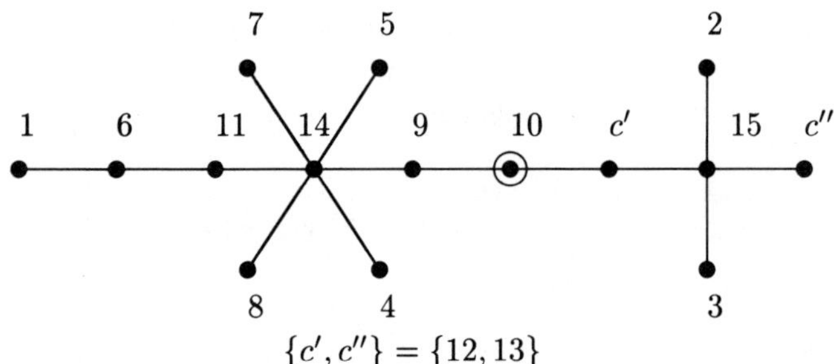

$\{c', c''\} = \{12, 13\}$

PROJECTIVES:

Nr.:	1	2	3	4	5	6	7	8	9	10	11	12	13	14	15
CC:	r	3	2	5	4	r	8	7	r	r	r	r	r	r	r
N&C:	×	×	×	×	×	○	×	×	×	○	×	×	×	○	○
$\mathrm{Ind}_{2^{11}:\mathrm{M}_{24}}(\chi_1)$	1					1									
$\mathrm{Ind}_{2^{11}:\mathrm{M}_{24}}(\chi_4)$			1												1
$\chi_{12} \otimes \chi_2$					1									1	
$\mathrm{Ind}_{2^{11}:\mathrm{M}_{24}}(\chi_7)$						1				1	1	1	1		1
$\chi_{15} \otimes \chi_2$								1	1	1	1	1	1	2	2
$\mathrm{Ind}_{2^{11}:\mathrm{M}_{24}}(\chi_8)$									2	1	1	1	1	2	2
$\chi_{19} \otimes \chi_2$											1	1	1	1	2

There is one orbit of algebraically conjugate characters. It has length 3 and contains the characters χ_{53}, χ_{54} and χ_{55} (nodes 11, 12 and 13). Since every character in this orbit occurs with equal multiplicity in each of the above set of projectives, the symmetric group on 3 letters acts on all Brauer trees which are consistent with this set. There are six orbits under this action and we give a representative of each orbit:

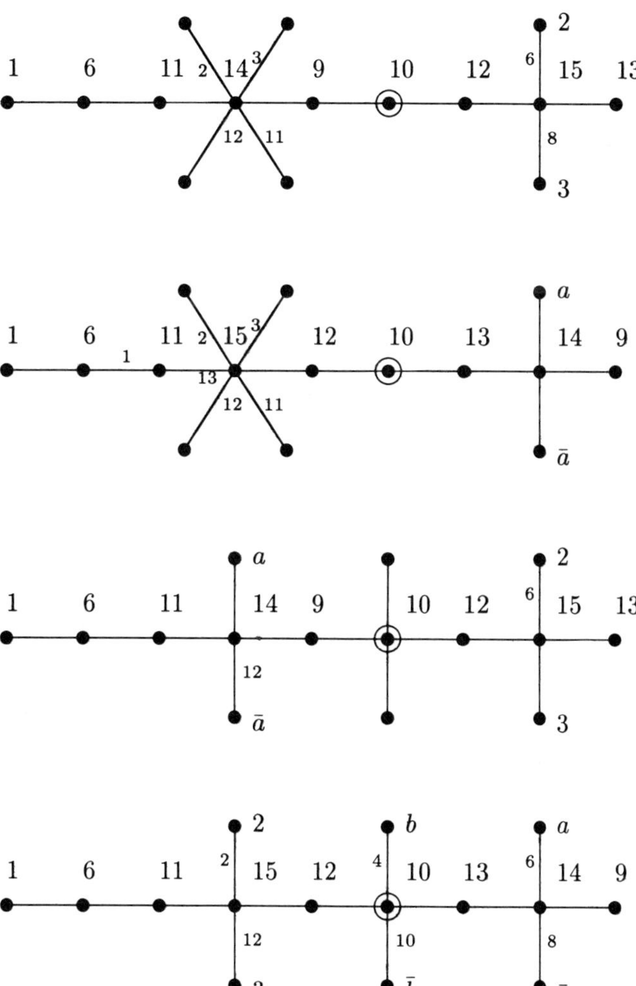

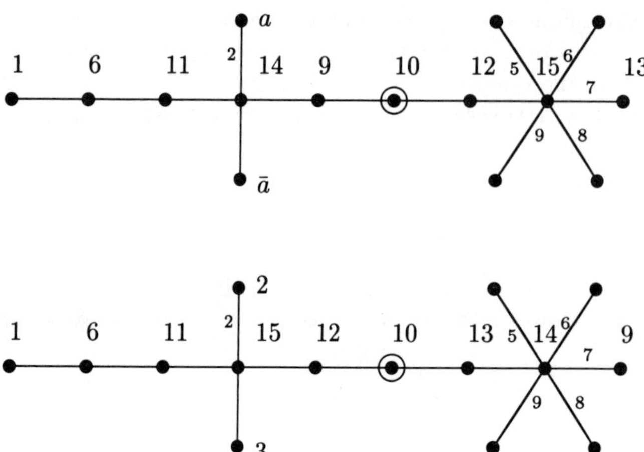

We have used the notation $\{a, \bar{a}\} = \{4, 5\}$ and $\{b, \bar{b}\} = \{7, 8\}$. The nodes which are not labelled in the following trees are not relevant in the proof. Again we have labelled some of the edges of the trees according to Green correspondence. We use the following symmetric square and tensor product:

$$\begin{array}{rcl} \chi_4^{2+} & \approx & \chi_{16} + \chi_{42} + 2\chi_{46} + 2\chi_{47} + 2\chi_{48} + \chi_{53} + \chi_{54} + \chi_{55} + \chi_{60} + \chi_{62} \\ \chi_4 \otimes \chi_{10} & \approx & 4\chi_{42} + 6\chi_{46} + 6\chi_{47} + 6\chi_{48} + 8\chi_{53} + 8\chi_{54} + 8\chi_{55} + 10\chi_{60} + 11\chi_{62} \end{array}$$

In terms of nodes these characters are as follows:

$$\begin{array}{rcl} 2^{2+} & \approx & 8 + 9 + 10^2 + 11 + 12 + 13 + 14 + 15 \\ 2 \otimes 5 & \approx & 9^4 + 10^6 + 11^8 + 12^8 + 13^8 + 14^{10} + 15^{11} \end{array}$$

(Here the exponents on the nodes denote the multiplicities of the corresponding characters.)

Let $1Ax$ be the Green correspondent of the irreducible module with Brauer character 2. Then $1A2x$ is the correspondent of 2^{2+}. Here, $2x$ has to be read modulo 14. It follows that $2x$ is even. Now consider in the last two trees the edge with label 7. From the above, the node on the right-hand side of this edge cannot be the character of the Green correspondent of 2^{2+}. Hence the two nodes of edge 7 can be subtracted from 2^{2+} to leave a projective character plus the character of the Green correspondent. It follows that in these cases the character of the Green correspondent must sit on edge 8 (corresponding to χ_{16}). In the last tree x equals 2, but $2x = 4$ does not label the edge corresponding to 8. In the second to the last tree $x \in \{5, 6, 8, 9\}$ hence $2x \in \{10, 12, 2, 4\}$. Again $2x$ does not label the edge corresponding to node 8.

Now consider in the second tree the edge labelled with 1 and 13. It follows as above that node 11 cannot be the character of the Green correspondent of 2^{2+}. Hence $11 + 15$ can be subtracted from 2^{2+} to leave a projective character plus the character of the Green correspondent. It follows that 8 must be the character of the Green correpondent. We have $x \in \{2, 3, 11, 12\}$, hence $2x \in \{4, 6, 8, 10\}$. This contradiction shows that the second tree is impossible.

The third tree is easy to rule out. It would imply that the character of the Green correspondent of 2^{2+} is 5 or 6. Neither of these, however, occurs in 2^{2+}.

In the fourth tree, the edge corresponding to node 5 is labelled with 6 or 8. Thus $2 \otimes 5$ has Green correspondent $1A2 \otimes 1A6 = 1A8$ or $1A2 \otimes 1A8 = 1A10$. Both labels correspond to leaves, whose characters do not occur at all in $2 \otimes 5$.

This leaves the first tree as the only possibility. By the arguments used to rule the second tree, we have that 8 is the Green correspondent of 2^{2+}. Since the edge corresponding to node 2 has label 6, node 8 must be at the edge with label 12. Now suppose node 5 were at edge 11. Then $2 \otimes 5$ would have Green correspondent $1A6 \otimes 1A11 = 1A3$. But node 4 does not occur at all in $2 \otimes 5$. Hence node 5 must lie on edge 3. This completes the proof of the planar embedding. Since we can assume without loss of generality that of the three algebraic conjugate characters χ_{53} comes first on the tree, we have only the two possible planar embedded trees given above.

6.24 The Triple Cover of the Fischer Group F24′

Group: 3F24′ Prime: 3 Block: 3

Nr.	CAS-Nr.	Degree	CC	N&C
1	94	178514751987	r	×
2	231	178514751987	232	∘
2	232	178514751987	231	∘

1 2

Group: 3F24′ Prime: 5 Block: 4

Nr.	CAS-Nr.	Degree	CC	N&C
1	11	35873145	r	×
2	28	3178094920	r	×
3	32	5775278080	r	∘
4	43	14507059905	r	∘
5	44	17068369920	r	×

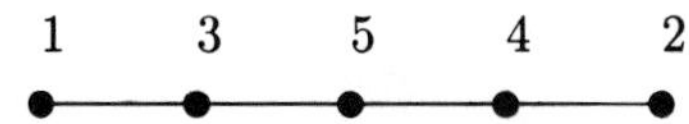

PROJECTIVES:

Nr.	1	2	3	4	5
CC	r	r	r	r	r
N&C	×	×	∘	∘	×
$\chi_3 \otimes \chi_{24}$	2		4	1	3
$\chi_2 \otimes \chi_{24}$		3		4	1
$\chi_2 \otimes \chi_{42}$			4	1	5

Group: 3F24′ Prime: 5 Block: 5

Nr.	CAS-Nr.	Degree	CC	N&C
1	18	159402880	r	×
2	34	7150713570	r	∘
3	36	9100908180	r	×
4	56	44493328880	r	×
5	58	46602926370	r	∘

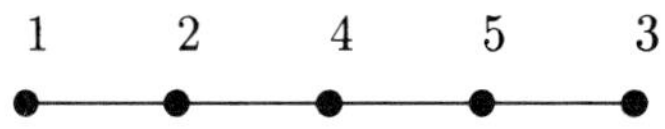

1 — 2 — 4 — 5 — 3

PROJECTIVES:

Nr.	1	2	3	4	5
CC	r	r	r	r	r
N&C	×	∘	×	×	∘
$\chi_3 \otimes \chi_{14}$	4	4			
$\chi_2 \otimes \chi_{60}$	1	6	3	16	14

Group: 3F24′ Prime: 5 Block: 6

Nr.	CAS-Nr.	Degree	CC	N&C
1	48	25027497495	r	×
2	99	200219979960	r	×
2	100	200219979960	r	×
3	105	225247477455	r	∘

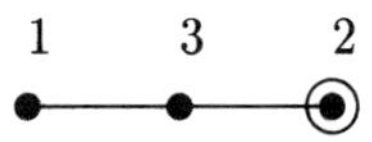

1 — 3 — 2 (circled)

Group: 3F24′ Prime: 5 Block: 11

Nr.	CAS-Nr.	Degree	CC	AC	N&C
1	119	19034730	120	ar	×
2	179	12258366120	180	ar	∘
3	201	78346948680	202	ar	×
4	207	84057367680	208	ar	×
5	221	150164984970	222	ar	∘

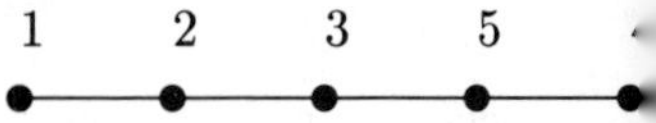

PROJECTIVES:

Nr.	1	2	3	4	5
AC	ar	ar	ar	ar	ar
N&C	×	∘	×	×	∘
$\chi_{109} \otimes \chi_{14}$	1	1			
$\chi_{3} \otimes \chi_{131}$		1	1		
$\chi_{2} \otimes \chi_{143}$			1	1	2

Group: 3F24′ Prime: 5 Block: 12

Complex conjugate to Block 11.

Group: 3F24′ Prime: 5 Block: 13

Nr.	CAS-Nr.	Degree	CC	AC	N&C
1	139	1152161010	141	140	×
1	140	1152161010	142	139	×
2	159	3264456195	160	ar	×
3	163	4416617205	164	ar	∘

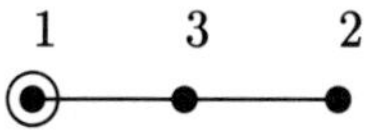

Group: 3F24′ Prime: 5 Block: 14

Complex conjugate to Block 13.

Group: 3F24′ Prime: 5 Block: 15

Nr.	CAS-Nr.	Degree	CC	AC	N&C
1	153	1818548820	154	ar	×
2	155	2400239520	156	ar	×
3	177	10726070355	178	ar	∘
4	181	14507059905	182	ar	∘
5	185	21014341920	186	ar	×

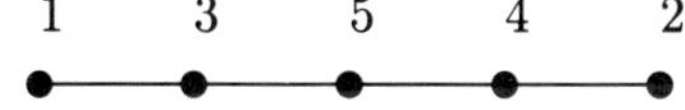

PROJECTIVES:

Nr.	1	2	3	4	5
AC	ar	ar	ar	ar	ar
N&C	×	×	∘	∘	×
$\chi_2 \otimes \chi_{193}$	3	6	8	15	14
$\chi_2 \otimes \chi_{194}$	2	5	4	12	9
$\chi_{109} \otimes \chi_{24}$	1		1		
$\chi_2 \otimes \chi_{143}$			1		1

Group: 3F24′ Prime: 5 Block: 16

Complex conjugate to Block 15.

Group: 3F24′ Prime: 7 Block: 2

Nr.	CAS-Nr.	Degree	CC	N&C
1	3	57477	r	×
2	21	635618984	r	×
3	44	17068369920	r	∘
4	66	67331776512	r	∘
5	85	150201655296	r	×
6	88	156321775827	r	×
7	104	222758961152	r	∘

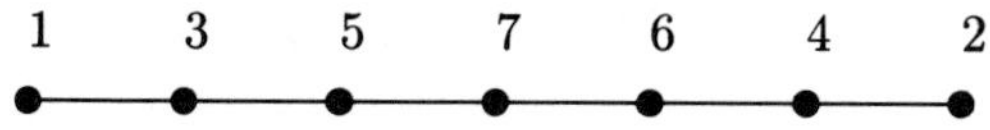

1 — 3 — 5 — 7 — 6 — 4 — 2

PROJECTIVES:

Nr.	1	2	3	4	5	6	7
CC	r	r	r	r	r	r	r
N&C	×	×	∘	∘	×	×	∘
$\chi_2 \otimes \chi_{10}$	1		1				
$\chi_2 \otimes \chi_{41}$		2	2	12	8	13	9
$\chi_2 \otimes \chi_{23}$		2	2	5	2	3	
$\chi_{109} \otimes \chi_{184}$		2	1	3	1	1	
$\chi_2 \otimes \chi_{48}$		1	6	16	26	27	32
$\chi_2 \otimes \chi_{33}$		1	5	10	11	12	9
$\chi_2 \otimes \chi_{24}$		1	1	2	2	1	1
$\chi_2 \otimes \chi_{19}$			1	1	1	1	
$\chi_3 \otimes \chi_{10}$			1		1		
$\chi_2 \otimes \chi_{22}$				1		1	

Group: 3F24′ Prime: 7 Block: 3

Nr.	CAS-Nr.	Degree	CC	N&C
1	5	555611	r	×
2	28	3178094920	r	×
3	38	10169903744	r	∘
4	68	74887473024	r	∘
5	82	142169187069	r	×
6	84	145650089984	r	×
7	103	205940550816	r	∘

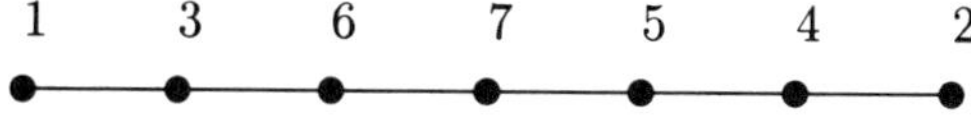

PROJECTIVES:

Nr.	1	2	3	4	5	6	7
CC	r	r	r	r	r	r	r
N&C	×	×	∘	∘	×	×	∘
$\chi_2 \otimes \chi_{26}$	1	5	2	8	3	4	3
$\chi_3 \otimes \chi_{24}$	1		2	11	15	6	9
$\chi_2 \otimes \chi_{24}$		3	1	4	1	4	3
$\chi_2 \otimes \chi_{36}$			1	6	8	9	10
$\chi_2 \otimes \chi_{22}$			1	1	2	1	1
$\chi_2 \otimes \chi_{23}$						1	1

Group: 3F24′ Prime: 7 Block: 4

Nr.	CAS-Nr.	Degree	CC	N&C
1	8	1666833	r	×
2	11	35873145	r	∘
3	16	79452373	r	×
4	17	112168056	r	×
5	45	17161712568	r	∘
6	79	139317477376	r	∘
7	89	156321775827	r	×

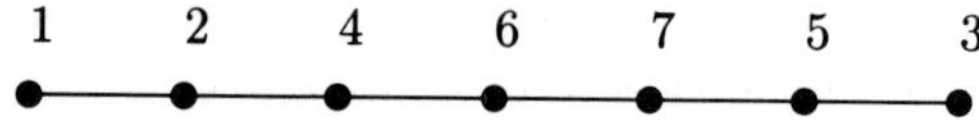

PROJECTIVES:

Nr.	1	2	3	4	5	6	7
CC	r	r	r	r	r	r	r
N&C	×	∘	×	×	∘	∘	×
$\chi_2 \otimes \chi_{10}$	1	2		1			
$\chi_2 \otimes \chi_{26}$	1	1	2		2	1	1
$\chi_2 \otimes \chi_{55}$				1	4	37	40
$\chi_2 \otimes \chi_{24}$					1		1
$\chi_2 \otimes \chi_{22}$						1	1

Group: 3F24′ Prime: 7 Block: 5

Nr.	CAS-Nr.	Degree	CC	N&C
1	13	48893768	r	×
2	18	159402880	r	×
3	27	2346900864	r	∘
4	34	7150713570	r	×
5	71	77108871168	r	∘
6	76	118588933386	r	∘
7	96	190685695200	r	×

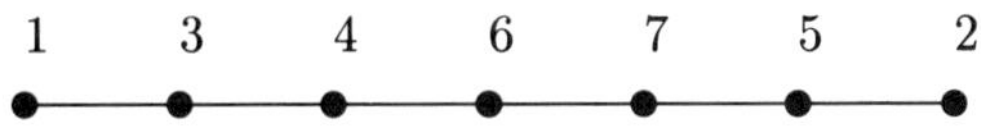

1 — 3 — 4 — 6 — 7 — 5 — 2

PROJECTIVES:

Nr.	1	2	3	4	5	6	7
CC	r	r	r	r	r	r	r
N&C	×	×	∘	×	∘	∘	×
$\chi_2 \otimes \chi_{26}$	3	3	5	2	5		2
$\chi_2 \otimes \chi_{10}$	1	1	1		1		
$\chi_{109} \otimes \chi_{130}$	1		1				
$\chi_3 \otimes \chi_{24}$		8	7	11	15	5	8
$\chi_{109} \otimes \chi_{160}$		2	2	2	2		
$\chi_2 \otimes \chi_{29}$			1	1	5	2	7
$\chi_2 \otimes \chi_{49}$				2	17	28	43
$\chi_2 \otimes \chi_{22}$					1	1	2

Group: 3F24′ Prime: 7 Block: 8

Nr.	CAS-Nr.	Degree	CC	AC	N&C
1	117	6724809	118	ar	×
2	125	195019461	127	ar	×
3	135	330032934	136	ar	○
4	161	4290428142	162	ar	×
5	177	10726070355	178	ar	○
6	216	135605256192	218	ar	○
7	219	142169187069	220	ar	×

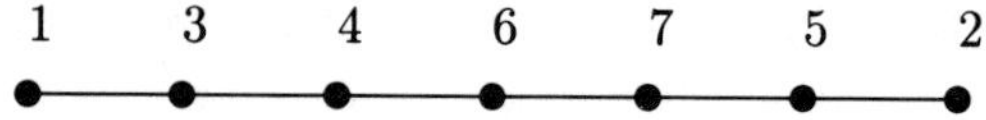

PROJECTIVES:

Nr.	1	2	3	4	5	6	7
AC	ar	ar	ar	ar	ar	ar	ar
N&C	×	×	○	×	○	○	×
$\chi_2 \otimes \chi_{129}$	2	1	3	1	1	1	1
$\chi_2 \otimes \chi_{183}$		4	2	3	8	20	23
$\chi_2 \otimes \chi_{223}$			1	4	15	144	156
$\chi_2 \otimes \chi_{229}$				5	4	143	142
$\chi_2 \otimes \chi_{143}$					1	1	2

Group: 3F24′ Prime: 7 Block: 9

Complex conjugate to Block 8.

Group: 3F24′ Prime: 7 Block: 10

Nr.	CAS-Nr.	Degree	CC	AC	N&C
1	126	195019461	128	ar	×
2	155	2400239520	156	ar	×
3	189	21122107776	190	ar	∘
4	215	135605256192	217	ar	∘
5	227	154455413112	228	ar	×
6	245	274587401088	246	ar	×
7	247	274910709213	248	ar	∘

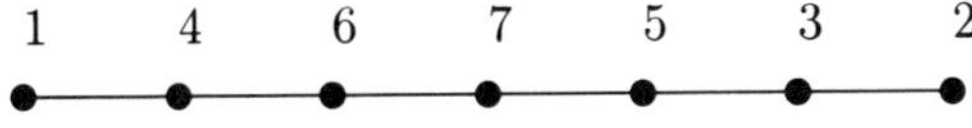

PROJECTIVES:

Nr.	1	2	3	4	5	6	7
CC	r	r	r	r	r	r	r
N&C	×	×	∘	∘	×	×	∘
$\chi_2 \otimes \chi_{151}$	2	2	4	6	3	4	1
$\chi_2 \otimes \chi_{129}$	1	3	3	1			
$\chi_{109} \otimes \chi_{24}$	1			1			
$\chi_2 \otimes \chi_{159}$		4	5	6	2	9	4
$\chi_2 \otimes \chi_{143}$				2	1	3	2
$\chi_2 \otimes \chi_{147}$				1	2	2	3
$\chi_2 \otimes \chi_{139}$				1	1	2	2

Group: 3F24′ Prime: 7 Block: 11

Complex conjugate to Block 10.

Group: 3F24′ Prime: 11 Block: 1

Nr.	CAS-Nr.	Degree	CC	N&C
1	1	1	r	×
2	5	555611	r	×
3	18	159402880	r	∘
4	19	281380736	r	∘
5	25	1337276304	r	×
6	33	6471756928	r	∘
7	36	9100908180	r	×
8	53	37337059200	r	×
9	54	38641860608	r	∘
10	68	74887473024	r	×
11	71	77108871168	r	∘

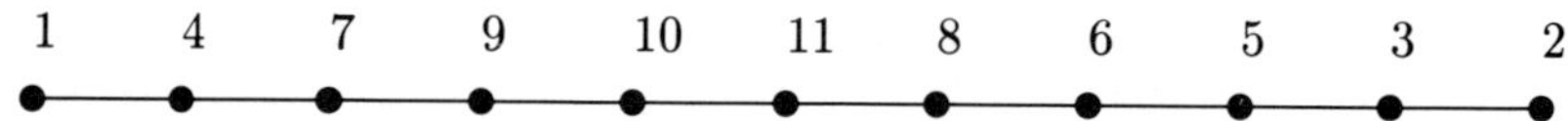

PROJECTIVES:

Nr.	1	2	3	4	5	6	7	8	9	10	11
CC	r	r	r	r	r	r	r	r	r	r	r
N&C	×	×	∘	∘	×	∘	×	×	∘	×	∘
$\chi_4 \otimes \chi_4$	1			1	1	1					
$\chi_2 \otimes \chi_{13}$		2	3		1				1	2	1
$\chi_2 \otimes \chi_{10}$			1		1		1	1	1		1
$\chi_2 \otimes \chi_{42}$				2	1	6	7	16	7	4	13
$\chi_2 \otimes \chi_{23}$				1	1	2	3	3	2		2
$\chi_2 \otimes \chi_{11}$					1	1	1	1	1		1
$\chi_2 \otimes \chi_{17}$						1	1	1	1		
$\chi_2 \otimes \chi_{12}$						1		1			
$\chi_2 \otimes \chi_{39}$								2	1	4	5
$\chi_2 \otimes \chi_{14}$									1	1	
$\chi_2 \otimes \chi_{22}$										1	1

Group: 3F24′ Prime: 11 Block: 2

Nr.	CAS-Nr.	Degree	CC	N&C
1	2	8671	r	×
2	8	1666833	r	×
3	43	14507059905	r	∘
4	44	17068369920	r	∘
5	59	54234085491	r	×
6	73	102385217025	r	×
7	82	142169187069	r	∘
8	84	145650089984	r	×
9	88	156321775827	r	∘
10	89	156321775827	r	∘
11	95	184117100544	r	×

1 — 4 — 5 — 7 — 11 — 9 — 6 — 10 — 8 — 3 — 2

PROJECTIVES:

Nr.	1	2	3	4	5	6	7	8	9	10	11
CC	r	r	r	r	r	r	r	r	r	r	r
N&C	×	×	∘	∘	×	×	∘	×	∘	∘	×
$\chi_2 \otimes \chi_9$	1	1	1	1							
$\chi_2 \otimes \chi_{20}$		1	2	1	1			1	1		1
$\chi_3 \otimes \chi_{11}$		1	1	1	1				1		1
$\chi_2 \otimes \chi_{13}$		1	1								
$\chi_2 \otimes \chi_{28}$			4		2		7	7	2	3	7
$\chi_2 \otimes \chi_{38}$			3	3	7	3	15	15	10	12	18
$\chi_2 \otimes \chi_{42}$			1	5	10	16	10	9	24	10	15
$\chi_2 \otimes \chi_{32}$			1	2	2	5	3	6	8	5	6
$\chi_2 \otimes \chi_{23}$			1	2	2	2		1	3		1
$\chi_2 \otimes \chi_{39}$			1	1	2	7	9	10	11	11	14
$\chi_2 \otimes \chi_{21}$			1		1		1	1	1		1
$\chi_2 \otimes \chi_{14}$			1					1			
$\chi_{109} \otimes \chi_{198}$					1	4	5	2	4	4	6
$\chi_2 \otimes \chi_{22}$						1	2	1	1	1	2
$\chi_{109} \otimes \chi_{173}$						1		1	1	1	

Group: 3F24′ Prime: 11 Block: 3

Nr.	CAS-Nr.	Degree	CC	N&C
1	3	57477	r	×
2	24	1264015025	r	∘
3	52	36858678129	r	∘
4	86	151397207325	r	×
4	87	151397207325	r	×
5	104	222758961152	r	×
6	108	336033532800	r	∘

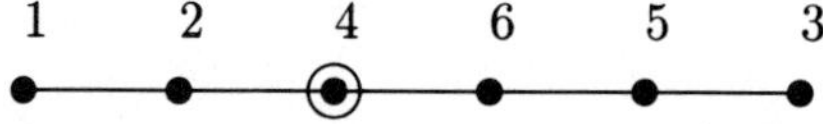

PROJECTIVES:

Nr.	1	2	3	4	5	6
CC	r	r	r	r	r	r
N&C	×	∘	∘	×	×	∘
$\chi_3 \otimes \chi_{14}$	1	3	1	2	1	
$\chi_2 \otimes \chi_{20}$	1	2		1		
$\chi_2 \otimes \chi_{21}$		1		2		1

Group: 3F24′ Prime: 11 Block: 4

Nr.	CAS-Nr.	Degree	CC	AC	N&C
1	109	783	110	ar	×
2	191	21842179632	192	ar	∘
3	223	151397207325	225	ar	×
3	224	151397207325	226	ar	×
4	229	160255122300	230	ar	∘
5	233	185161334400	234	ar	∘
6	239	215861428224	240	ar	×

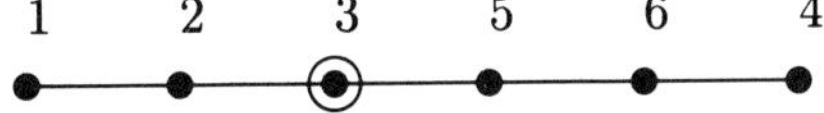

PROJECTIVES:

Nr.	1	2	3	4	5	6
AC	ar	ar	ar	ar	ar	ar
N&C	×	∘	×	∘	∘	×
$\chi_3 \otimes \chi_{119}$	1	2	1			
$\chi_2 \otimes \chi_{137}$		2	3		1	

Group: 3F24′ Prime: 11 Block: 5

Complex conjugate to Block 4.

Group: 3F24′ Prime: 11 Block: 6

Nr.	CAS-Nr.	Degree	CC	AC	N&C
1	111	64584	112	ar	×
2	117	6724809	118	ar	×
3	155	2400239520	156	ar	∘
4	167	8250621525	169	168	×
5	168	8250621525	170	167	×
6	181	14507059905	182	ar	∘
7	203	80256172032	205	ar	×
8	204	80256172032	206	ar	×
9	219	142169187069	220	ar	∘
10	241	256966819200	242	ar	×
11	247	274910709213	248	ar	∘

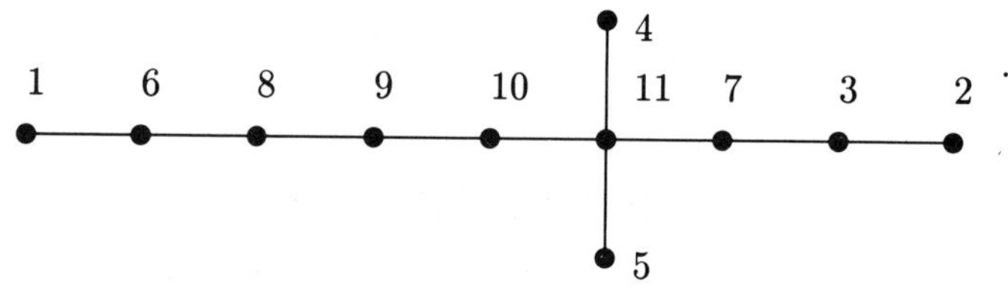

PROJECTIVES:

Nr.	1	2	3	4	5	6	7	8	9	10	11
AC	ar	ar	ar	5	4	ar	ar	ar	ar	ar	ar
N&C	×	×	∘	×	×	∘	×	×	∘	×	∘
$\chi_2 \otimes \chi_{119}$	1	2	2			1					
$\chi_2 \otimes \chi_{137}$		1	3			3	2	5	2		
$\chi_2 \otimes \chi_{121}$		1	2			1	1	1			
$\chi_2 \otimes \chi_{175}$				1	1	2	6	7	7	14	20
$\chi_2 \otimes \chi_{171}$				1	1		5	2	7	18	20
$\chi_2 \otimes \chi_{147}$							1	1	1	2	3
$\chi_{110} \otimes \chi_{173}$							1			1	2
$\chi_2 \otimes \chi_{139}$									1	3	2

The projectives above determine the tree up to the location of the conjugate pair of nodes χ_{167} and χ_{168}. We have the following two possible trees:

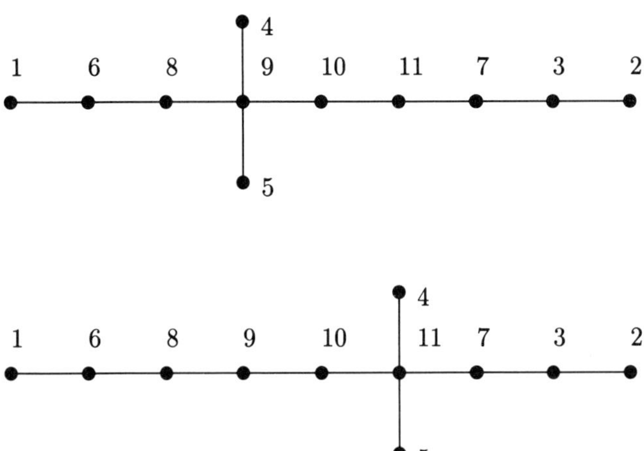

Let $1B0$ denote the Green correspondent of the irreducible module corresponding to the leaf of the Brauer tree of the second block with character χ_2. Since χ_2 is of type cross, $1B0$ is indeed an irreducible module in the normalizer of the Sylow 11-subgroup. Similarly, let $1G0$ denote the irreducible module which is the Green correspondent of a leaf of the stem in Block 8. Since the Brauer correspondent of Block 8 contains only 1-dimensional irreducible Brauer characters, as is easily seen from the character table of 3F24′, the tensor product $1B0 \otimes 1G0 = 1E0$ is an irreducible module. We shall see in a second that $1E0$ lies in the Brauer correspondent of Block 6.

The tree of Block 8 is completely known from the projectives we have given there. The conjugate nodes 5 and 6 are joined to node 11. Hence the Green correspondents of the edges joining them are $10G3$ and $10G8$ resp. Nodes 5 and 6 of Block 8 correspond to the ordinary characters χ_{131} and χ_{132}. Thus the Green correspondents of $\chi_2 \otimes \chi_{131}$ and $\chi_2 \otimes \chi_{132}$ are $1B0 \otimes 10G3 = 10E3$ and $1B0 \otimes 10G8 = 10E8$. We have $\chi_2 \otimes \chi_{131} \approx \chi_{247} \approx \chi_2 \otimes \chi_{132}$. Since χ_{247} is node 11 in Block 6, we indeed have that $1E0$ lies in the Brauer correspondent of Block 6. Furthermore, there must be two edges incident to node 11 whose labels differ by 5. This can only be the case in the second of the above trees.

Group: 3F24′ Prime: 11 Block: 7

Complex conjugate to Block 6.

Group: 3F24′ Prime: 11 Block: 8

Nr.	CAS-Nr.	Degree	CC	AC	N&C
1	113	306153	115	ar	×
2	114	306153	116	ar	×
3	125	195019461	127	ar	∘
4	126	195019461	128	ar	∘
5	131	216154575	133	132	∘
6	132	216154575	134	131	∘
7	193	34128405675	195	ar	×
8	194	34128405675	196	ar	×
9	215	135605256192	217	ar	∘
10	216	135605256192	218	ar	∘
11	235	203775436800	236	ar	×

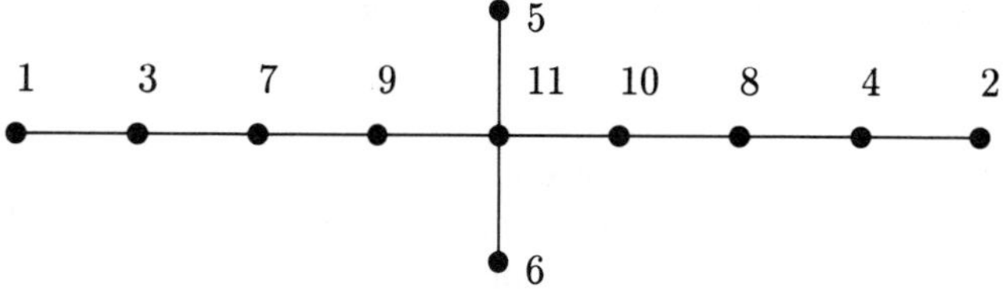

PROJECTIVES:

Nr.	1	2	3	4	5	6	7	8	9	10	11
AC	ar	ar	ar	ar	6	5	ar	ar	ar	ar	ar
N&C	×	×	∘	∘	∘	∘	×	×	∘	∘	×
$\chi_3 \otimes \chi_{119}$	2	2	4	4			3	3	1	1	
$\chi_2 \otimes \chi_{123}$	1	1	1	2			1	1	1		
$\chi_2 \otimes \chi_{137}$	1		3	2			6	4	5	2	1
$\chi_{109} \otimes \chi_{16}$	1		2				1				
$\chi_{109} \otimes \chi_9$		1		1							
$\chi_{109} \otimes \chi_{14}$				1				1			
$\chi_{110} \otimes \chi_{173}$					1						1
$\chi_2 \otimes \chi_{171}$						1	1	2	5	8	11
$\chi_{131} \otimes \chi_4$						1		1	4	5	9
$\chi_2 \otimes \chi_{135}$							1		1	1	1
$\chi_2 \otimes \chi_{139}$									1	1	2

Group: 3F24′ Prime: 11 Block: 9

Complex conjugate to Block 8.

Group: 3F24′ Prime: 13 Block: 1

Nr.	CAS-Nr.	Degree	CC	N&C
1	1	1	r	×
2	4	249458	r	×
3	6	1603525	7	×
4	7	1603525	6	×
5	8	1666833	r	∘
6	13	48893768	r	×
7	26	1540153692	r	∘
8	29	3208653525	r	∘
9	35	8529641472	r	×
10	53	37337059200	r	∘
11	59	54234085491	r	×
12	76	118588933386	r	×
13	79	139317477376	r	∘

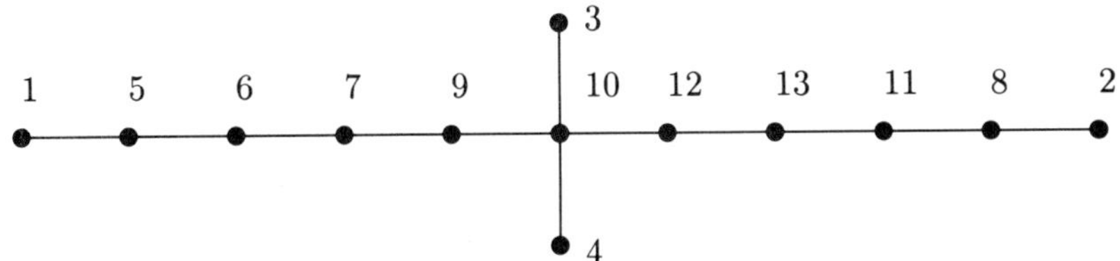

PROJECTIVES:

Nr.	1	2	3	4	5	6	7	8	9	10	11	12	13
CC	r	r	4	3	r	r	r	r	r	r	r	r	r
N&C	×	×	×	×	∘	×	∘	∘	×	∘	×	×	∘
$\chi_2 \otimes \chi_2$	1				1								
$\chi_4 \otimes \chi_{19}$		1	1	1				6	3	13	9	9	5
$\chi_{42} \otimes \chi_2$			1	1				4	4	16	10	14	10
$\chi_5 \otimes \chi_2$					1	2	1						
$\chi_{10} \otimes \chi_2$					1	1	1		2	1			
$\chi_8 \otimes \chi_2$					1	1	1		1				
$\chi_{19} \otimes \chi_2$										1		1	
$\chi_{22} \otimes \chi_2$												1	1

Group: 3F24′ Prime: 13 Block: 2

Nr.	CAS-Nr.	Degree	CC	N&C
1	3	57477	r	×
2	11	35873145	r	∘
3	22	1069551175	r	×
4	44	17068369920	r	×
5	89	156321775827	r	∘
6	97	197813862400	r	∘
6	98	197813862400	r	∘
7	108	336033532800	r	×

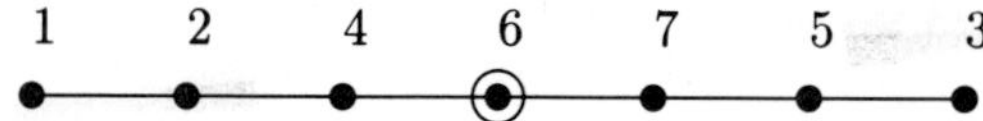

PROJECTIVES:

Nr.	1	2	3	4	5	6	7
CC	r	r	r	r	r	r	r
N&C	×	∘	×	×	∘	∘	×
$\chi_2 \otimes \chi_2$	1	1					
$\chi_9 \otimes \chi_2$		1		1			
$\chi_{24} \otimes \chi_2$				1	1	2	2
$\chi_{19} \otimes \chi_2$				1		1	
$\chi_{28} \otimes \chi_2$					3	3	6

Group: 3F24′ Prime: 13 Block: 3

Nr.	CAS-Nr.	Degree	CC	N&C
1	5	555611	r	×
2	27	2346900864	r	∘
3	69	77007684600	r	×
3	70	77007684600	r	×
4	71	77108871168	r	×
5	84	145650089984	r	∘
6	85	150201655296	r	×
7	88	156321775827	r	∘

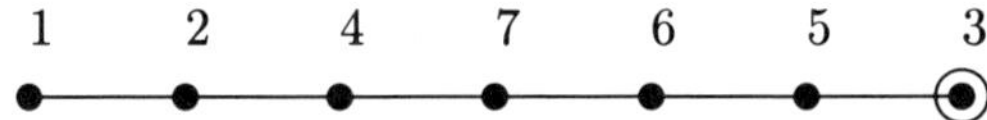

PROJECTIVES:

Nr.	1	2	3	4	5	6	7
CC	r	r	r	r	r	r	r
N&C	×	∘	×	×	∘	×	∘
$\chi_8 \otimes \chi_2$	1	1					
$\chi_{20} \otimes \chi_2$		2		3	1	1	1
$\chi_{10} \otimes \chi_2$		1		1			
$\chi_{16} \otimes \chi_2$			1		1		
$\chi_{23} \otimes \chi_2$				2	1	2	3
$\chi_{14} \otimes \chi_2$					1	1	
$\chi_{19} \otimes \chi_2$						1	1

Group: 3F24′ Prime: 13 Block: 4

Nr.	CAS-Nr.	Degree	CC	N&C
1	14	74837400	15	×
2	15	74837400	14	×
3	37	9441555200	r	◦
4	39	10776585600	r	×
5	40	10776585600	r	×
6	46	18481844304	r	×
7	47	18481844304	r	×
8	64	65393917952	r	◦
9	65	65393917952	r	◦
10	91	164572397352	r	◦
11	92	164572397352	r	◦
12	101	205353825600	r	×
13	102	205353825600	r	×

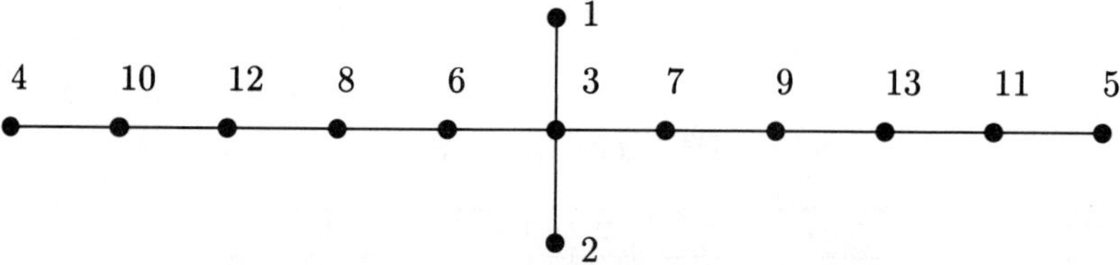

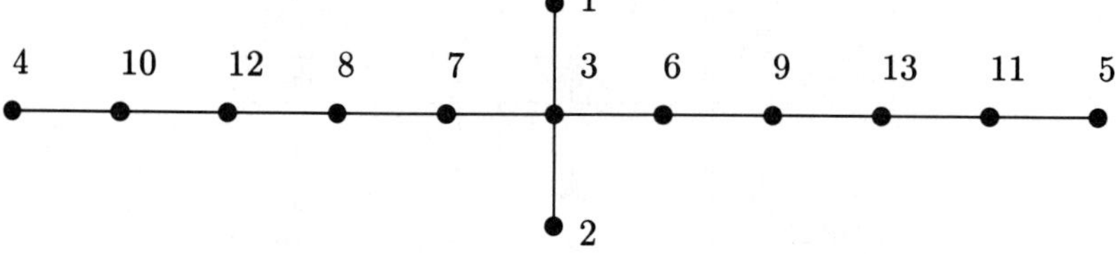

PROJECTIVES:

Nr.	1	2	3	4	5	6	7	8	9	10	11	12	13
CC	2	1	r	r	r	r	r	r	r	r	r	r	r
N&C	×	×	∘	×	×	×	×	∘	∘	∘	∘	×	×
$\chi_{13} \otimes \chi_2$	1	1	2										
$\chi_4 \otimes \chi_{16}$			8	1	1	6	6	2	2	2	2	1	1
$\chi_{21} \otimes \chi_2$			2			1	1						
$\chi_3 \otimes \chi_{19}$				1	1					1	1		
$\chi_{22} \otimes \chi_2$						1	1	1	1	1	1	1	1
$\chi_{24} \otimes \chi_2$								1	1	1	1	2	2

We have two pairs of algebraically conjugate characters. The first is $\{\chi_{91}, \chi_{92}\}$ (nodes $\{10, 11\}$), the second $\{\chi_{101}, \chi_{102}\}$ (nodes $\{12, 13\}$). The characters in the two orbits generate disjoint fields. Hence we can assume without loss of generality that χ_{91} (node 10) comes left of χ_{92} (node 11) and that χ_{101} (node 12) comes left of χ_{102} (node 13) on the Brauer tree.

There are also 3 pairs of rational valued characters conjugate under the outer automorphism. These are $\{\chi_{39}, \chi_{40}\}$ (nodes $\{4, 5\}$), $\{\chi_{46}, \chi_{47}\}$ (nodes $\{6, 7\}$) and finally $\{\chi_{64}, \chi_{65}\}$ (nodes $\{8, 9\}$). The two characters χ_{39} and χ_{40} agree on all conjugacy classes except the three pairs (12I,12J), (24C,24D) and (36A,36B) (ATLAS notation for the conjugacy classes of the simple group). The two characters χ_{46} and χ_{47} agree on all classes except (12I,12J) and (36A,36B). Finally the two characters χ_{64} and χ_{65} differ only on the pair of classes (27B,27C). Furthermore, all other irreducible characters of the simple group agree on these last two classes. By fixing notation we can therefore assume that χ_{39} (node 4) comes left of χ_{40} (node 5) and that χ_{64} (node 8) comes left of χ_{65} (node 9) on the Brauer tree. With these conventions we have only the two trees given above consistent with the projectives.

Group: 3F24′ Prime: 13 Block: 5

Nr.	CAS-Nr.	Degree	CC	AC	N&C
1	109	783	110	ar	×
2	113	306153	115	ar	×
3	129	203843871	130	ar	∘
4	143	1255560075	145	144	×
5	144	1255560075	146	143	×
6	153	1818548820	154	ar	∘
7	155	2400239520	156	ar	×
8	167	8250621525	169	168	×
9	168	8250621525	170	167	×
10	203	80256172032	205	ar	∘
11	215	135605256192	217	ar	×
12	233	185161334400	234	ar	×
13	243	259900935525	244	ar	∘

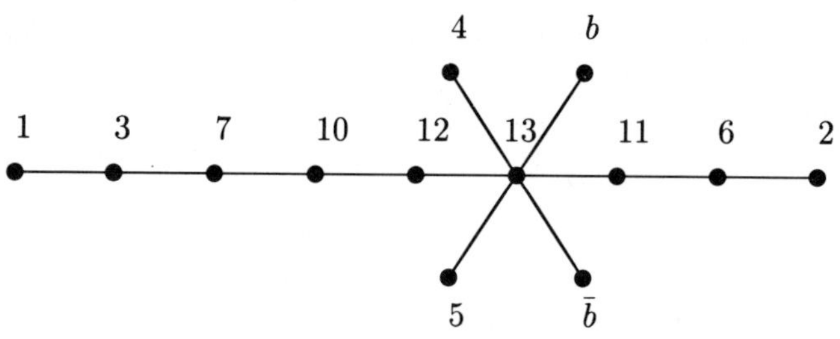

$\{b, \bar{b}\} = \{8, 9\}$

PROJECTIVES:

Nr.	1	2	3	4	5	6	7	8	9	10	11	12	13
AC	ar	ar	ar	5	4	ar	ar	9	8	ar	ar	ar	ar
N&C	×	×	∘	×	×	∘	×	×	×	∘	×	×	∘
$\chi_3 \otimes \chi_{119}$	1	2	1			3					1		
$\chi_{111} \otimes \chi_2$	1		1										
$\chi_{151} \otimes \chi_2$		1	1			4	2			2	6	1	3
$\chi_{123} \otimes \chi_2$		1	1			2	1				1		
$\chi_{129} \otimes \chi_2$			3			1	3			1	1	1	
$\chi_{177} \otimes \chi_2$			1	1	1	1	2			10	16	18	26
$\chi_{161} \otimes \chi_2$			1				3			7	3	7	5
$\chi_{121} \otimes \chi_2$			1				2			1			
$\chi_{167} \otimes \chi_2$								1	1	4	5	9	12
$\chi_{139} \otimes \chi_2$											1	1	2

The following three types of trees are consistent with the above set of projectives:

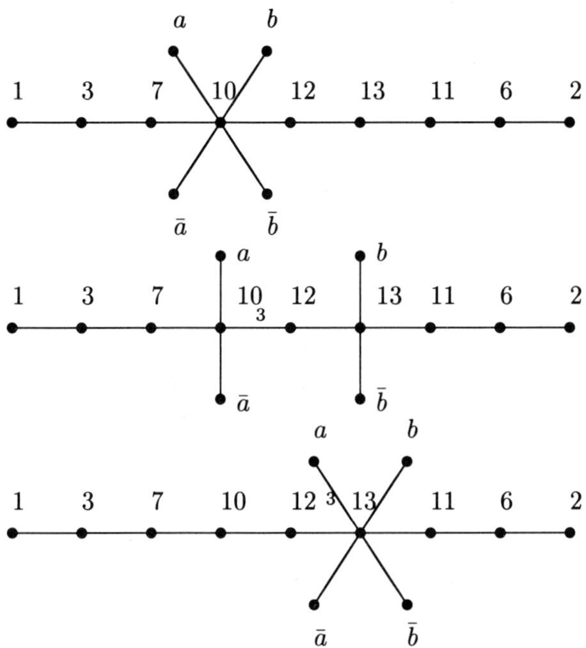

We have used the notation $\{a, \bar{a}, b, \bar{b}\} = \{4, 5, 8, 9\}$. Now consider $\chi_{109} \otimes \chi_{143} \approx_F \chi_{145} + 2\chi_{146} + \chi_{244}$ where $\approx_F$ denotes restriction to Block 6, which is complex conjugate to the present block. There are irreducible trivial source modules with Brauer characters χ_{109} and χ_{143}. Hence every indecomposable direct summand of their tensor product is a trivial source module. Since there are no exceptional characters in Block 6, the Brauer character of every non-projective indecomposable direct summand is an ordinary irreducible of type cross. Thus there must be a projective involving χ_{244} and one of χ_{145} or χ_{146}. It follows that nodes 4 and 5 are joined to node 13, so the first tree cannot be correct.

To prove that the third tree actually is correct, consider the tensor product $\chi_6 \otimes \chi_{109} \approx_E \chi_{143}$. Now χ_6 is a leaf in the principal block whith Green correspondent $1A3$. Let $1E0$ denote the Green correspondent of the leaf at node 1_E with Brauer character χ_{109}. From the above tensor product we must have an edge with label 3 incident to node 4_E. This proves that the third tree is correct and furthermore that a equals 4.

In Block 4 there are characters with the same irrationalities as χ_{167} and χ_{168}. We do not know the embedding consistent with that block.

Group: 3F24′ Prime: 13 Block: 6

Complex conjugate to Block 5.

Group: 3F24′ Prime: 13 Block: 7

Nr.	CAS-Nr.	Degree	CC	AC	N&C
1	114	306153	116	ar	×
2	121	25356672	122	ar	×
3	135	330032934	136	ar	○
4	147	1349587008	149	148	×
5	148	1349587008	150	147	×
6	165	4620461076	166	ar	○
7	189	21122107776	190	ar	×
8	204	80256172032	206	ar	○
9	211	115511526900	212	ar	×
10	216	135605256192	218	ar	×
11	237	210555861318	238	ar	○
12	247	274910709213	248	ar	○
13	249	295709508864	250	ar	×

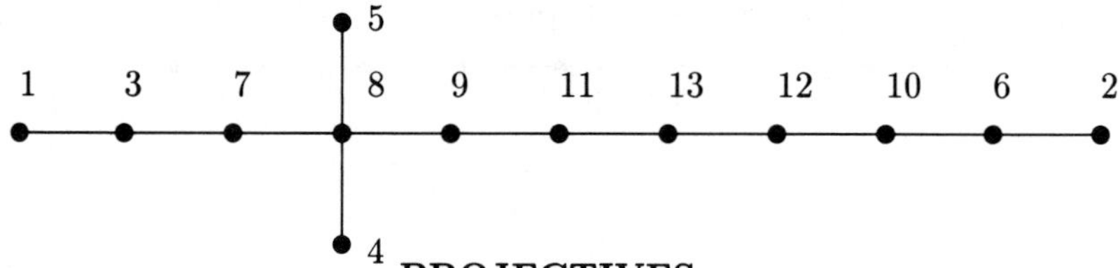

5

1 3 7 8 9 11 13 12 10 6 2

4

PROJECTIVES:

Nr.	1	2	3	4	5	6	7	8	9	10	11	12	13
AC	ar	ar	ar	5	4	ar	ar	ar	ar	ar	ar	ar	ar
N&C	×	×	○	×	×	○	×	○	×	×	○	○	×
$\chi_3 \otimes \chi_{119}$	2	4	3			5	1			1			
$\chi_3 \otimes \chi_{125}$	1	4	2			8	2	2	2	5	2	1	1
$\chi_{129} \otimes \chi_2$	1	1	3			2	3	1		1			
$\chi_{117} \otimes \chi_2$	1	1	2			1	1						
$\chi_{111} \otimes \chi_2$	1		1										
$\chi_{135} \otimes \chi_2$			2			1	3	1	1	1	1		
$\chi_{147} \otimes \chi_2$					1			1		1	2	3	4
$\chi_{125} \otimes \chi_2$							1	2	1				
$\chi_{139} \otimes \chi_2$									1	1	2	2	2
$\chi_{109} \otimes \chi_{31}$										1		1	

The consistency with Block 4 follows from $\chi_{110} \otimes \chi_{114} \approx_D \chi_{15}$ ($1_F \otimes 1_G \approx_D 2_D$) and $\chi_{110} \otimes \chi_{147} \approx_D \chi_{102}$ ($1_F \otimes 4_G \approx_D 13_D$).

Group: 3F24′ Prime: 13 Block: 8

Complex conjugate to Block 7.

Group: 3F24′ Prime: 17 Block: 1

Nr.	CAS-Nr.	Degree	CC	N&C
1	1	1	r	×
2	2	8671	r	×
3	10	32715683	r	∘
4	17	112168056	r	∘
5	50	29444114700	r	×
6	71	77108871168	r	×
7	76	118588933386	r	∘
8	86	151397207325	r	∘
9	87	151397207325	r	∘
10	88	156321775827	r	×
11	89	156321775827	r	×
12	93	169598100672	r	×
13	94	178514751987	r	∘
14	95	184117100544	r	×
15	97	197813862400	r	∘
16	98	197813862400	r	∘
17	104	222758961152	r	×

1 — 4 — 10 — 15 — 14 — 8 — 5 — 7 — 17 — 13 — 12 — 16 — 11 — 9 — 6 — 3 — 2

PROJECTIVES:

Nr.	1	2	3	4	5	6	7	8	9	10	11	12	13	14	15	16	17
CC	r	r	r	r	r	r	r	r	r	r	r	r	r	r	r	r	r
N&C	×	×	∘	∘	×	×	∘	∘	∘	×	×	×	∘	×	∘	∘	×
$\chi_3 \otimes \chi_3$	1			1													
$\chi_2 \otimes \chi_{11}$		1	2			1											
$\chi_2 \otimes \chi_{20}$			2	1		3		1	1	1				1			
$\chi_2 \otimes \chi_{26}$			1			5		4	4	1	1			4	1	1	
$\chi_{109} \otimes \chi_{190}$			1			4		3	3	1	2		1	4	2	2	1
$\chi_2 \otimes \chi_{24}$			1			2	1	1	1	1	1	1		2	2	2	1
$\chi_{18} \otimes \chi_3$				2	2	5	1	5	5	2				4			
$\chi_2 \otimes \chi_{32}$					3	3	8	5	5	8	5	9	5	6	10	10	9
$\chi_{109} \otimes \chi_{202}$					2	3	5	7	7	8	8	10	10	8	10	10	10
$\chi_2 \otimes \chi_{23}$					2	2	1	2	2	3		4	1	1	3	3	
$\chi_2 \otimes \chi_{22}$					2	1	1	2	2	1	1	2		2	2	2	
$\chi_2 \otimes \chi_{21}$					1	2		2	2	1		1		1	1	1	
$\chi_2 \otimes \chi_{19}$					1		1			1		1			1	1	
$\chi_{109} \otimes \chi_{173}$							1			1	1		2		1	1	3

We have two orbits of algebraically conjugate characters. The characters of the first orbit are χ_{86} and χ_{87} (nodes 8 and 9). They generate a field disjoint from the character field of χ_{97} and χ_{98} (nodes 15 and 16), which constitute the second orbit. Hence we can assume without loss of generality that node 8 is left of node 9 and that node 15 is left of node 16 on the Brauer tree. Then only the one tree given above is consistent with our projectives.

Group: 3F24′ Prime: 17 Block: 2

Nr.	CAS-Nr.	Degree	CC	AC	N&C
1	109	783	110	ar	×
2	111	64584	112	ar	×
3	121	25356672	122	ar	∘
4	129	203843871	130	ar	∘
5	139	1152161010	141	140	×
6	140	1152161010	142	139	×
7	147	1349587008	149	148	×
8	148	1349587008	150	147	×
9	163	4416617205	164	ar	×
10	167	8250621525	169	168	∘
11	168	8250621525	170	167	∘
12	187	21096751104	188	ar	×
13	223	151397207325	225	ar	∘
14	224	151397207325	226	ar	∘
15	231	178514751987	232	ar	∘
16	237	210555861318	238	ar	×
17	241	256966819200	242	ar	×

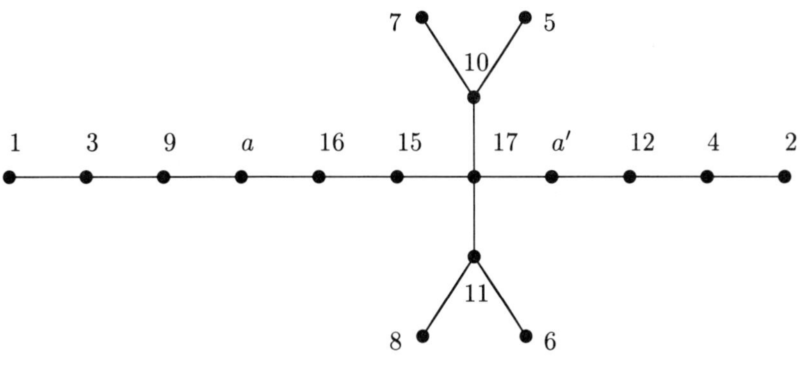

$$\{a, a'\} = \{13, 14\}.$$

PROJECTIVES:

Nr.	1	2	3	4	5	6	7	8	9	10	11	12	13	14	15	16	17
AC	ar	ar	ar	ar	6	5	8	7	ar	11	10	ar	ar	ar	ar	ar	ar
N&C	×	×	∘	∘	×	×	×	×	×	∘	∘	×	∘	∘	∘	×	×
$\chi_2 \otimes \chi_{117}$	1	1	1	2								1					
$\chi_{109} \otimes \chi_3$	1		1														
$\chi_2 \otimes \chi_{135}$		1		3								3	1	1		1	
$\chi_{125} \otimes \chi_3$			4						8			3	6	6		2	3
$\chi_2 \otimes \chi_{137}$			2	2					3			5	3	3		2	
$\chi_2 \otimes \chi_{201}$					1	1			2	5	5	8	81	81	96	112	144
$\chi_2 \otimes \chi_{175}$							1	1	1	1	1	1	8	8	11	11	14
$\chi_2 \otimes \chi_{171}$										1	1		7	7	13	11	18
$\chi_2 \otimes \chi_{157}$												2	3	3	1	4	1
$\chi_2 \otimes \chi_{143}$													2	2	1	2	3
$\chi_{110} \otimes \chi_{173}$															2	1	1

The projectives above determine the stem of the tree up to the location of the two algebraically conjugate characters χ_{223} and χ_{224} (nodes 13 and 14). The irrationalities of these characters are dependent on the irrationalities of χ_{86} and χ_{87} in the principal block. Since we have made a choice there, we cannot fix an ordering of nodes 13 and 14. Thus there are the following two possible stems:

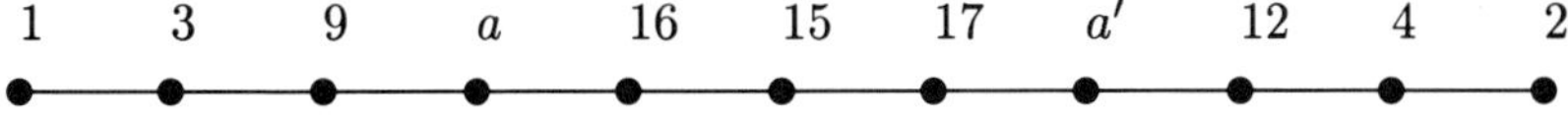

where $\{a, a'\} = \{13, 14\}$. We also see that node 10 is joined to one of 16 or 17.

We shall use the following tensor products:

$$\begin{array}{lll} \chi_{109} \otimes \chi_{141} & \approx_A & \chi_{93} \\ \chi_{109} \otimes \chi_{149} & \approx_A & \chi_{104} \\ \chi_{109} \otimes \chi_{169} & \approx_A & 2\chi_{76} + \chi_{93} + \chi_{94} + \chi_{95} + \chi_{97} + \chi_{98} + 2\chi_{104} \end{array}$$

Here, of course $\approx_A$ denotes the tensor product restricted to the principal block. In terms of nodes these tensor products are as follows:

$$\begin{array}{lll} 1_B \otimes 5_C & \approx & 12_A \\ 1_B \otimes 7_C & \approx & 17_A \\ 1_B \otimes 10_C & \approx & 7_A^2 + 12_A + 13_A + 14_A + 15_A + 16_A + 17_A^2 \end{array}$$

Here, as usual, the exponents denote the multiplicities of the characters. The star indicates complex conjugation, i.e., duality.

let $1B0$ be the Green correspondent of the leaf at node 1_B. The second tensor product implies that node 5_C must be incident to an edge with label 5 or 11, since the edges incident to node 12_A in the principal block have these two labels. Similarly, the third tensor product implies that node 7_B must be incident to an edge with label 4 or 12, since this is true for node 17_A in the principal block.

We assume without loss of generality that node 10_B is located on the upper half of the tree, i.e. the labels on the edges incident to it are smaller than 8. Suppose now that node 10_B is joined to node 16_B. Then there is an edge with label 3 meeting it. In other words, in the dual block there is a uniserial indecomposable module with Brauer character χ_{169} and Green correspondent $16C14$. The Green correspondent of the tensor product of this module with the irreducible node 1_B then is $1B0 \otimes 16C14 = 16A14$. From the tree of the principal block, the node of type naught incident to edge 14 is 8. However, this node does not occur at all in $\chi_{109} \otimes \chi_{169}$.

It follows that node 10_B is joined to node 17_B. From the above remarks one now easily concludes that each of the nodes 5_B and 7_B is joined to one of 10_B or 11_B. One also can derive the planar embedding from the above tensor products. The two nodes in any of the sets $\{5_B, 6_B\}$, $\{7_B, 8_B\}$ and $\{10_B, 11_B\}$ can be interchanged independently of the others, since the character fields of any two of these sets intersect only in the third cyclotomic field. This final observation completes the proof of the Brauer tree.

Group: 3F24′ Prime: 17 Block: 3

Complex conjugate to Block 2.

Group: 3F24′ Prime: 23 Block: 1

Nr.	CAS-Nr.	Degree	CC	N&C
1	1	1	r	×
2	6	1603525	7	∘
2	7	1603525	6	∘
3	11	35873145	r	∘
4	19	281380736	r	×
5	32	5775278080	r	∘
6	35	8529641472	r	×
7	72	77379702400	r	∘
8	79	139317477376	r	×
9	94	178514751987	r	∘
10	97	197813862400	r	×
11	98	197813862400	r	×
12	107	282049015248	r	∘

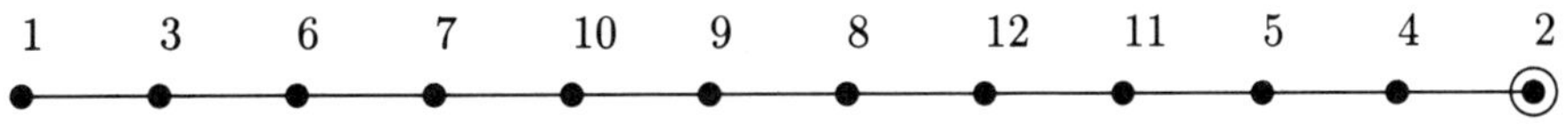

1 3 6 7 10 9 8 12 11 5 4 2

PROJECTIVES:

Nr.	1	2	3	4	5	6	7	8	9	10	11	12
CC	r	r	r	r	r	r	r	r	r	r	r	r
N&C	×	∘	∘	×	∘	×	∘	×	∘	×	×	∘
$\chi_2 \otimes \chi_2$	1		1									
$\chi_{17} \otimes \chi_3$			1		1	1			1	1	1	
$\chi_{25} \otimes \chi_2$			1			4	4	1	1	2	2	3
$\chi_{11} \otimes \chi_2$			1			2	1					
$\chi_{23} \otimes \chi_2$				1	2	2	4		1	3	3	2
$\chi_{30} \otimes \chi_2$						3	8	2	2	6	6	7
$\chi_{22} \otimes \chi_2$							2	1		2	2	3
$\chi_{19} \otimes \chi_2$							1			1	1	1

There are two algebraically conjugate characters, namely χ_{97} and χ_{98} (nodes 10 and 11). We assume without loss of generality that node 10 comes first on the Brauer tree. Then only the tree given above is consistent with all the projectives. Notice that the two edges incident to node 12 have the labels 4 and 8 in the Green correspondence labelling.

Group: 3F24′ Prime: 23 Block: 2

Nr.	CAS-Nr.	Degree	CC	AC	N&C
1	109	783	110	ar	×
2	119	19034730	120	ar	∘
3	139	1152161010	141	140	∘
4	140	1152161010	142	139	∘
5	143	1255560075	145	144	∘
5	144	1255560075	146	143	∘
6	159	3264456195	160	ar	×
7	171	8993909925	173	172	∘
8	172	8993909925	174	171	∘
9	197	41785039296	198	ar	∘
10	201	78346948680	202	ar	×
11	229	160255122300	230	ar	×
12	231	178514751987	232	ar	∘

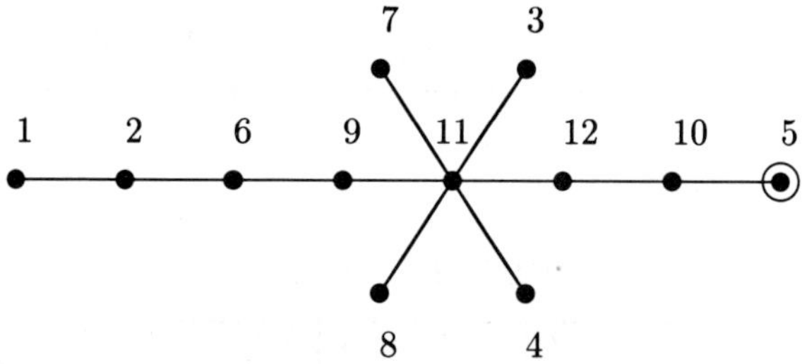

PROJECTIVES:

Nr.	1	2	3	4	5	6	7	8	9	10	11	12
AC	ar	ar	4	3	ar	ar	8	7	ar	ar	ar	ar
N&C	×	∘	∘	∘	∘	×	∘	∘	∘	×	×	∘
$\chi_{117} \otimes \chi_2$	1	2				1						
$\chi_{137} \otimes \chi_2$		2				4			2			
$\chi_{201} \otimes \chi_2$			1	1	1	1	4	4	22	43	85	96
$\chi_{109} \otimes \chi_{22}$					1					1		
$\chi_{131} \otimes \chi_2$								1			1	
$\chi_{147} \otimes \chi_2$									1		3	2
$\chi_{139} \otimes \chi_2$										1	1	2

We have the following two trees consistent with the above set of projectives:

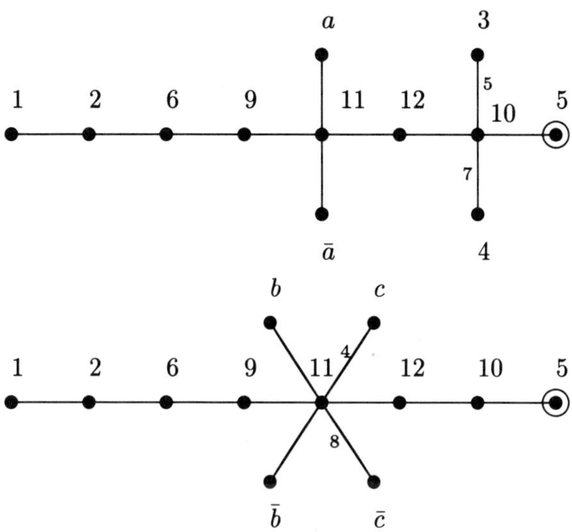

Here, $\{a, \bar{a}\} = \{7, 8\}$ and $\{b, \bar{b}, c, \bar{c}\} = \{3, 4, 7, 8\}$.

Let $1B0$ denote the Green correspondent of the leaf at node 1, i.e., of the irreducible module with Brauer character χ_{109}. Then we can label the edges of the tree as usual according to Green correspondence. In the first case, the edges incident to the nodes 3 and 4 have the labels 5 and 7 resp. This means that the Green correspondents of the corresponding simple modules are $22B5 = 1B0 \otimes 22A5$ and $22B7 = 1B0 \otimes 22A7$. Its dual modules are $22C7$ and $22C5$ resp. Now $1B0 \otimes 22C5 = 22A5$ and $1B0 \otimes 22C7 = 22A7$. However, $\chi_{109} \otimes \chi_{141} \approx \chi_{107}$. In terms of nodes, $1 \otimes 3^* \approx 12$. The labels on the edges incident to node 12 in the principal block are 4 and 12. This shows that the second tree is correct. Furthermore, c must equal 3. Since the irrationalities of the characters involved around node 11 are independent up to a third root of unity, we can choose the order of the nodes 7 and 8 arbitrarily.

Group: $3F24'$ Prime: 23 Block: 3

Complex conjugate to Block 2.

Group: 3F24′ Prime: 29 Block: 1

Nr.	CAS-Nr.	Degree	CC	N&C
1	1	1	r	×
2	3	57477	r	∘
3	6	1603525	7	∘
4	7	1603525	6	∘
5	18	159402880	r	×
6	37	9441555200	r	∘
7	58	46602926370	r	×
8	66	67331776512	r	∘
9	76	118588933386	r	×
10	86	151397207325	r	×
11	87	151397207325	r	×
12	90	160313753600	r	∘
13	91	164572397352	r	×
13	92	164572397352	r	×
14	97	197813862400	r	∘
15	98	197813862400	r	∘

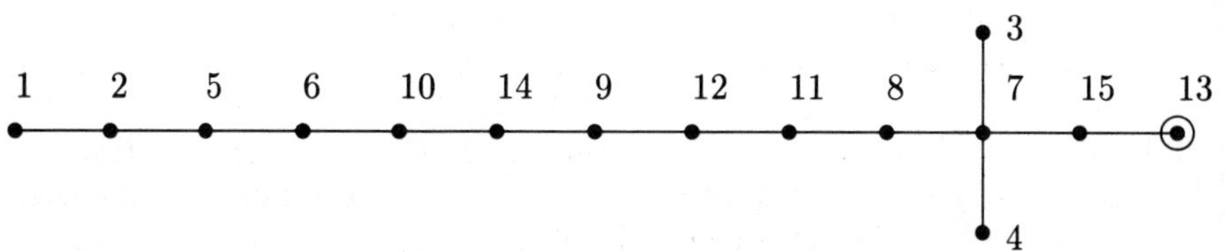

PROJECTIVES:

Nr.	1	2	3	4	5	6	7	8	9	10	11	12	13	14	15
CC	r	r	4	3	r	r	r	r	r	r	r	r	r	r	r
N&C	×	∘	∘	∘	×	∘	×	∘	×	×	×	∘	×	∘	∘
$\chi_2 \otimes \chi_2$	1	1													
$\chi_3 \otimes \chi_{14}$		1			4	5			1	2	2	3			
$\chi_8 \otimes \chi_2$		1			1										
$\chi_{145} \otimes \chi_{109}$				1			1								
$\chi_{26} \otimes \chi_2$					3	6		2		4	4	2	1	1	1
$\chi_{21} \otimes \chi_2$					1	2	2	3		2	2			1	1
$\chi_{39} \otimes \chi_2$							4	4	11	10	10	13	12	15	15
$\chi_{19} \otimes \chi_2$							2	1	1					1	1

We have two orbits of algebraically conjugate characters, each of length two. The first orbit contains the characters χ_{86} and χ_{87} (nodes 10 and 11), the second the characters χ_{97} and χ_{98} (nodes 14 and 15). The characters in the two orbits generate disjoint fields. Hence we can assume without loss of generality that node 10 comes left of node 11 and that node 14 comes left of node 15 on the Brauer tree.

Group: 3F24′ Prime: 29 Block: 2

Nr.	CAS-Nr.	Degree	CC	AC	N&C
1	111	64584	112	ar	×
2	117	6724809	118	ar	∘
3	131	216154575	133	132	×
4	132	216154575	134	131	×
5	147	1349587008	149	148	×
6	148	1349587008	150	147	×
7	151	1553430879	152	ar	×
8	167	8250621525	169	168	×
8	168	8250621525	170	167	×
9	209	103562058600	210	ar	∘
10	215	135605256192	217	ar	∘
11	216	135605256192	218	ar	∘
12	223	151397207325	225	ar	×
13	224	151397207325	226	ar	×
14	239	215861428224	240	ar	∘
15	247	274910709213	248	ar	×

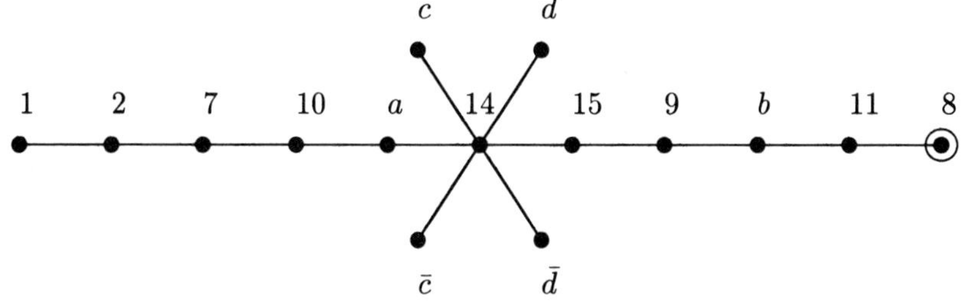

Here, $\{a, b\} = \{12, 13\}$ and $\{c, \bar{c}, d, \bar{d}\} = \{3, 4, 5, 6\}$.

PROJECTIVES:

Nr.	1	2	3	4	5	6	7	8	9	10	11	12	13	14	15
AC	ar	ar	4	3	6	5	ar	ar	ar	ar	ar	ar	ar	ar	ar
N&C	×	∘	×	×	×	×	×	×	∘	∘	∘	×	×	∘	×
$\chi_{129} \otimes \chi_2$	1	2					1			1	1	1	1		
$\chi_{109} \otimes \chi_2$	1	1													
$\chi_{137} \otimes \chi_2$		1					3		1	5	2	3	3		
$\chi_{123} \otimes \chi_2$		1					2			1					
$\chi_{52} \otimes \chi_{109}$			1	1						2	3	3	3	10	7
$\chi_{173} \otimes \chi_{110}$			1											3	2
$\chi_{147} \otimes \chi_2$						1				1	1	1	1	4	3
$\chi_{149} \otimes \chi_{110}$						1								1	
$\chi_{76} \otimes \chi_{109}$								2	6	7	10	8	8	20	25
$\chi_{169} \otimes \chi_{110}$								1			1			2	2
$\chi_{143} \otimes \chi_2$									3	2	1	2	2		2
$\chi_{139} \otimes \chi_2$									1	1	1	1	1	1	2

Group: 3F24′ Prime: 29 Block: 3

Complex conjugate to Block 2.

6.25 The Double Cover of the Baby Monster

Group: 2BM Prime: 3 Block: 5

Nr.	CAS-Nr.	Degree	CC	N&C
1	73	140763279350304	r	×
2	85	422966584586250	r	×
3	86	563729863936554	r	∘

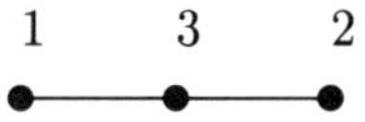

1 — 3 — 2

Group: 2BM Prime: 3 Block: 6

Nr.	CAS-Nr.	Degree	CC	N&C
1	114	1928549534519790	r	×
2	124	2991219686193960	r	×
3	144	4919769220713750	r	∘

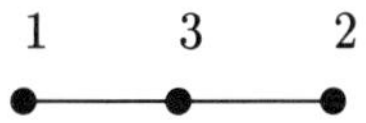

1 — 3 — 2

Group: 2BM Prime: 5 Block: 9

Nr.	CAS-Nr.	Degree	CC	N&C
1	204	60863961600000	r	×
2	205	77683916800000	206	×
2	206	77683916800000	205	×
3	209	138547878400000	r	∘

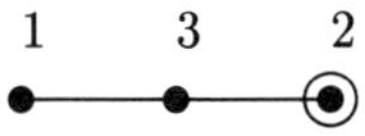

1 — 3 — 2

Group: 2BM Prime: 5 Block: 3

Nr.	CAS-Nr.	Degree	CC	N&C
1	20	80426400000	r	×
2	75	181844090400000	r	×
3	121	2788834713600000	r	∘
4	181	13940902029600000	r	∘
5	184	16547812226400000	r	×

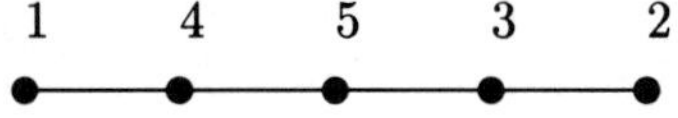

PROJECTIVES:

Nr.	1	2	3	4	5
CC	r	r	r	r	r
N&C	×	×	∘	∘	×
$\chi_2 \otimes \chi_{77}$		1	3		2
$\chi_2 \otimes \chi_{99}$			1	5	6

Group: 2BM Prime: 5 Block: 5

Nr.	CAS-Nr.	Degree	CC	N&C
1	31	641238778125	r	×
2	69	94981065075000	r	×
3	149	5485156508081250	r	∘
4	150	5992026614971875	r	∘
5	175	11381560819200000	r	×

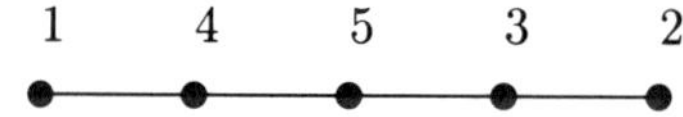

PROJECTIVES:

Nr.	1	2	3	4	5
CC	r	r	r	r	r
N&C	×	×	∘	∘	×
$\chi_2 \otimes \chi_{68}$		1	1		
$\chi_3 \otimes \chi_{47}$			1		1
$\chi_2 \otimes \chi_{52}$				1	1

Group: 2BM Prime: 5 Block: 6

Nr.	CAS-Nr.	Degree	CC	N&C
1	38	4836998684375	r	×
2	42	6697382793750	r	×
3	103	1042755084800000	r	∘
4	116	2437847336925000	r	∘
5	126	3469068040246875	r	×

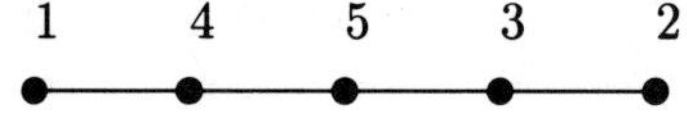

PROJECTIVES:

Nr.	1	2	3	4	5
CC	r	r	r	r	r
N&C	×	×	∘	∘	×
$\chi_2 \otimes \chi_{89}$	1			1	
$\chi_3 \otimes \chi_{52}$			1		1
$\chi_2 \otimes \chi_{83}$				1	1

Group: 2BM   Prime: 5   Block: 7

Nr.	CAS-Nr.	Degree	CC	N&C
1	50	10529575087500	r	×
2	60	22348485900000	r	×
3	79	260522896303125	r	∘
4	87	581759023584375	r	∘
5	98	809403858900000	r	×

PROJECTIVES:

Nr.	1	2	3	4	5
CC	r	r	r	r	r
N&C	×	×	∘	∘	×
$\chi_2 \otimes \chi_{77}$	1			2	1
$\chi_2 \otimes \chi_{104}$			1		1

Group: 2BM Prime: 7 Block: 3

Nr.	CAS-Nr.	Degree	CC	N&C
1	5	9458750	r	×
2	47	9958526703125	r	∘
3	56	19246163186250	r	×
4	75	181844090400000	r	∘
5	151	6414993199558125	r	×
6	157	7698465274500000	r	×
7	181	13940902029600000	r	∘

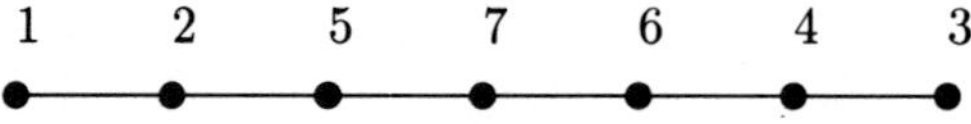

PROJECTIVES:

Nr.	1	2	3	4	5	6	7
CC	r	r	r	r	r	r	r
N&C	×	∘	×	∘	×	×	∘
$\chi_2 \otimes \chi_{80}$		1			2		1
$\chi_2 \otimes \chi_{49}$			1	1			
$\chi_3 \otimes \chi_{39}$				1		2	1

Group: 2BM Prime: 7 Block: 5

Nr.	CAS-Nr.	Degree	CC	N&C
1	8	347643114	r	×
2	30	635966233056	r	×
3	68	89626740328125	r	○
4	125	3068619025500000	r	○
5	155	7575407965125000	r	×
6	171	9596829826538955	r	×
7	182	14014628339712000	r	○

PROJECTIVES:

Nr.	1	2	3	4	5	6	7
CC	r	r	r	r	r	r	r
N&C	×	×	○	○	×	×	○
$\chi_2 \otimes \chi_{49}$		1	1				
$\chi_3 \otimes \chi_{39}$			1		1		
$\chi_2 \otimes \chi_{48}$				1		1	
$\chi_2 \otimes \chi_{58}$					1		1
$\chi_2 \otimes \chi_{62}$						1	1

Group: 2BM Prime: 7 Block: 6

Nr.	CAS-Nr.	Degree	CC	N&C
1	14	4622913750	r	×
2	23	108348770530	r	×
3	102	968283412561920	r	○
4	149	5485156508081250	r	×
5	153	6934321313920000	r	○
6	158	7744013981383680	r	○
7	174	10161349228100070	r	×

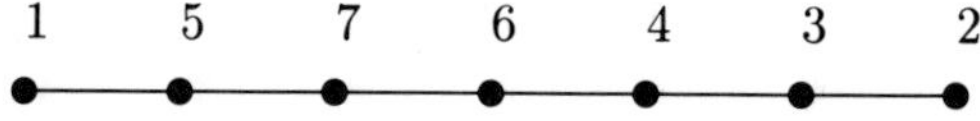

PROJECTIVES:

Nr.	1	2	3	4	5	6	7
CC	r	r	r	r	r	r	r
N&C	×	×	○	×	○	○	×
$\chi_2 \otimes \chi_{54}$		1	1				
$\chi_2 \otimes \chi_{71}$			1	2		1	
$\chi_3 \otimes \chi_{35}$					1	1	2

Group: 2BM Prime: 7 Block: 7

Nr.	CAS-Nr.	Degree	CC	N&C
1	16	12501781215	r	×
2	34	2624476798125	r	×
3	70	118479380574555	r	∘
4	116	2437847336925000	r	×
5	154	6979512601119240	r	∘
6	162	8388837260960625	r	∘
7	180	13046344927150080	r	×

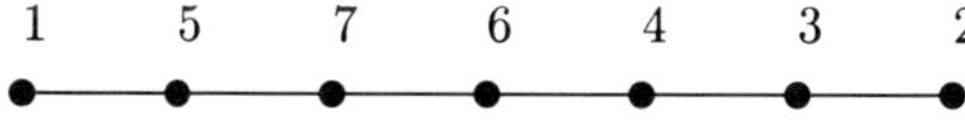

PROJECTIVES:

Nr.	1	2	3	4	5	6	7
CC	r	r	r	r	r	r	r
N&C	×	×	∘	×	∘	∘	×
$\chi_2 \otimes \chi_{48}$	1	1	1		1		
$\chi_2 \otimes \chi_{44}$		1	1				
$\chi_3 \otimes \chi_{61}$			2	3	1	1	1
$\chi_3 \otimes \chi_{39}$				1		1	
$\chi_2 \otimes \chi_{58}$					1		1
$\chi_2 \otimes \chi_{62}$						1	1

Group: 2BM Prime: 7 Block: 8

Nr.	CAS-Nr.	Degree	CC	N&C
1	18	27416186875	r	×
2	139	4570963756734375	r	∘
3	159	7853379563452500	r	∘
4	178	12424315904000000	r	×
4	179	12424315904000000	r	×

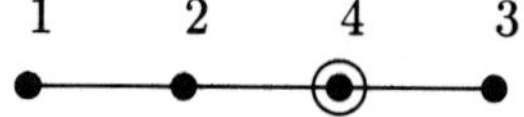

PROJECTIVES:

Nr.	1	2	3	4
CC	r	r	r	r
N&C	×	∘	∘	×
$\chi_2 \otimes \chi_{41}$	1	1		

Group: 2BM Prime: 7 Block: 9

Nr.	CAS-Nr.	Degree	CC	N&C
1	28	536105794455	r	×
2	29	551549171250	r	×
3	51	10853720317440	r	∘
4	57	19946023665750	r	∘
5	109	1458928219297500	r	×
6	110	1523654585578125	r	×
7	123	2952870715858140	r	∘

PROJECTIVES:

Nr.	1	2	3	4	5	6	7
CC	r	r	r	r	r	r	r
N&C	×	×	∘	∘	×	×	∘
$\chi_2 \otimes \chi_{54}$	1			1			
$\chi_2 \otimes \chi_{50}$		1	1				
$\chi_3 \otimes \chi_{44}$			1		1		
$\chi_3 \otimes \chi_{62}$				1		1	
$\chi_2 \otimes \chi_{95}$					1		1
$\chi_2 \otimes \chi_{37}$						1	1

Group: 2BM Prime: 7 Block: 11

Nr.	CAS-Nr.	Degree	CC	N&C
1	189	36657653760	r	×
2	192	100406462464	r	×
3	202	40345142190080	r	∘
4	208	110949141022720	r	∘
5	222	1793872427483136	r	×
6	225	2995626807613440	r	×
7	234	4638342016000000	r	∘

PROJECTIVES:

Nr.	1	2	3	4	5	6	7
CC	r	r	r	r	r	r	r
N&C	×	×	∘	∘	×	×	∘
$\chi_{185} \otimes \chi_{41}$	2	3	3	4		3	1
$\chi_{2} \otimes \chi_{203}$	1			1		1	1
$\chi_{185} \otimes \chi_{48}$		1	1				
$\chi_{2} \otimes \chi_{219}$			1	1	1	4	3
$\chi_{185} \otimes \chi_{19}$				1		1	
$\chi_{2} \otimes \chi_{214}$					1	1	2

Group: 2BM Prime: 7 Block: 12

Nr.	CAS-Nr.	Degree	CC	N&C
1	193	772566552576	r	×
2	197	17069098618880	r	∘
3	205	77683916800000	206	∘
3	206	77683916800000	205	∘
4	207	93980448866304	r	×

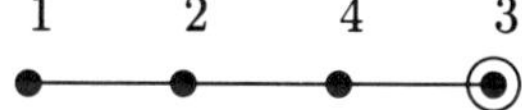

PROJECTIVES:

Nr.	1	2	3	4
CC	r	r	r	r
N&C	×	∘	∘	×
$\chi_2 \otimes \chi_{216}$	1	1		

Group: 2BM Prime: 11 Block: 1

Nr.	CAS-Nr.	Degree	CC	N&C
1	1	1	r	×
2	17	13508418144	r	×
3	54	12927978301875	r	∘
4	62	35301999416320	r	∘
5	72	123485536419840	r	×
6	85	422966584586250	r	×
7	92	665029816320000	r	∘
8	98	809403858900000	r	×
9	101	844994666880000	r	∘
10	121	2788834713600000	r	∘
11	124	2991219686193960	r	×

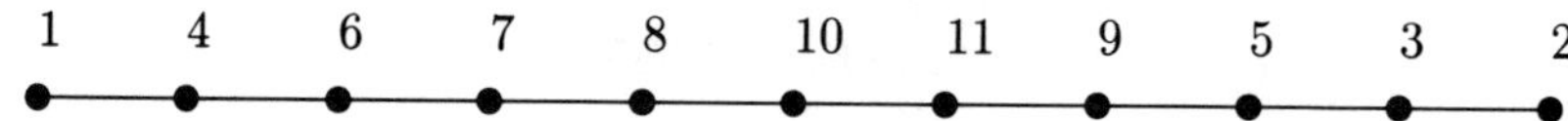

PROJECTIVES:

Nr.	1	2	3	4	5	6	7	8	9	10	11
CC	r	r	r	r	r	r	r	r	r	r	r
N&C	×	×	∘	∘	×	×	∘	×	∘	∘	×
$\chi_2 \otimes \chi_{13}$		1	1								
$\chi_2 \otimes \chi_{67}$			2		3				1		
$\chi_2 \otimes \chi_{148}$				1		2	1	1	3	9	11
$\chi_2 \otimes \chi_{132}$					1				4	3	6
$\chi_2 \otimes \chi_{70}$						3	5	4		6	4
$\chi_3 \otimes \chi_{23}$						1	1	1		1	
$\chi_2 \otimes \chi_{49}$							1	2		1	
$\chi_3 \otimes \chi_{29}$							1	1		1	1
$\chi_3 \otimes \chi_{47}$									1		1
$\chi_2 \otimes \chi_{55}$										1	1

Group: 2BM Prime: 11 Block: 2

Nr.	CAS-Nr.	Degree	CC	N&C
1	2	4371	r	×
2	5	9458750	r	×
3	60	22348485900000	r	∘
4	100	835517991997440	r	×
5	136	4250407956480000	r	∘
6	162	8388837260960625	r	∘
7	166	9211433539600384	r	×
8	173	10134024677130240	r	∘
9	177	11854109736960000	r	×
10	180	13046344927150080	r	∘
11	181	13940902029600000	r	×

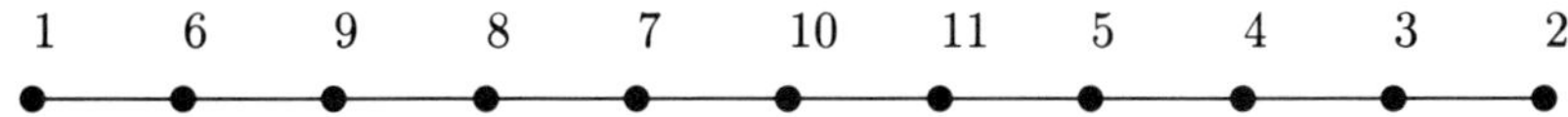

PROJECTIVES:

Nr.	1	2	3	4	5	6	7	8	9	10	11
CC	r	r	r	r	r	r	r	r	r	r	r
N&C	×	×	∘	×	∘	∘	×	∘	×	∘	×
$\chi_2 \otimes \chi_{18}$		1	1								
$\chi_3 \otimes \chi_{61}$			4	10	8	1	2	4	3	1	3
$\chi_2 \otimes \chi_{42}$			1	1							
$\chi_2 \otimes \chi_{49}$				1	1						
$\chi_3 \otimes \chi_{45}$					3	1	1	1	1	2	5
$\chi_2 \otimes \chi_{81}$					1	4	3	3	4	1	2
$\chi_2 \otimes \chi_{80}$						3	5	6	5	2	1
$\chi_3 \otimes \chi_{47}$						2	3	4	4	2	1
$\chi_3 \otimes \chi_{42}$						2	2	1	2	1	
$\chi_2 \otimes \chi_{66}$						1	1	1	1		
$\chi_2 \otimes \chi_{65}$							1	1	1	1	

Group: 2BM Prime: 11 Block: 3

Nr.	CAS-Nr.	Degree	CC	N&C
1	3	96255	r	×
2	7	63532485	r	×
3	34	2624476798125	r	∘
4	57	19946023665750	r	∘
5	69	94981065075000	r	×
6	91	662947625486250	r	×
7	109	1458928219297500	r	∘
8	122	2908795117921875	r	∘
9	140	4619851069640625	r	×
10	150	5992026614971875	r	×
11	154	6979512601119240	r	∘

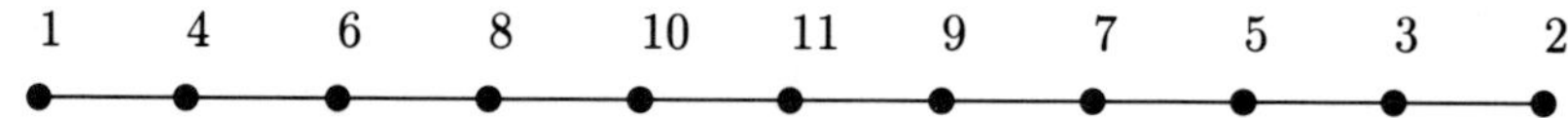

PROJECTIVES:

Nr.	1	2	3	4	5	6	7	8	9	10	11
CC	r	r	r	r	r	r	r	r	r	r	r
N&C	×	×	∘	∘	×	×	∘	∘	×	×	∘
$\chi_3 \otimes \chi_{13}$	1			1							
$\chi_2 \otimes \chi_{11}$		1	1								
$\chi_2 \otimes \chi_{28}$			1		1						
$\chi_3 \otimes \chi_{28}$				1		1					
$\chi_2 \otimes \chi_{40}$					1		1				
$\chi_2 \otimes \chi_{47}$						1		1			
$\chi_2 \otimes \chi_{66}$							1		1		
$\chi_2 \otimes \chi_{52}$								1		1	
$\chi_2 \otimes \chi_{65}$									1		1
$\chi_3 \otimes \chi_{35}$										1	1

Group: 2BM Prime: 11 Block: 4

Nr.	CAS-Nr.	Degree	CC	N&C
1	4	1139374	r	×
2	8	347643114	r	×
3	41	6619124890560	r	∘
4	58	21346507000000	59	∘
5	59	21346507000000	58	∘
6	73	140763279350304	r	∘
7	77	211069033500000	r	×
8	89	646398915093750	r	×
9	113	1807462370115584	r	∘
10	114	1928549534519790	r	∘
11	125	3068619025500000	r	×

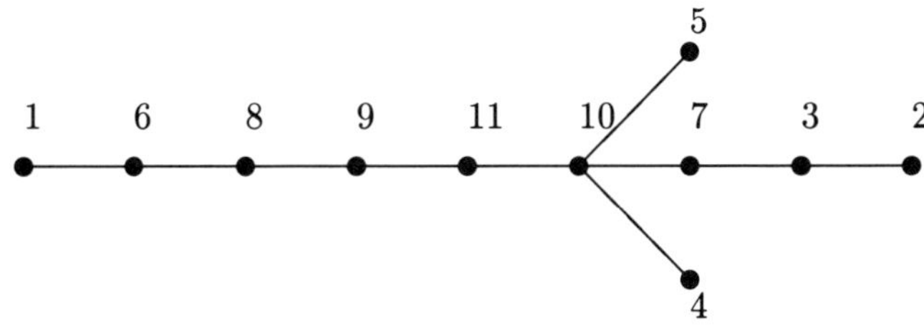

PROJECTIVES:

Nr.	1	2	3	4	5	6	7	8	9	10	11
CC	r	r	r	5	4	r	r	r	r	r	r
N&C	×	×	∘	∘	∘	∘	×	×	∘	∘	×
$\chi_3 \otimes \chi_{18}$	1					1					
$\chi_2 \otimes \chi_{18}$		1	1								
$\chi_2 \otimes \chi_{29}$			1				1				
$\chi_2 \otimes \chi_{133}$				1	1	1		3	6	5	11
$\chi_2 \otimes \chi_{31}$						1		1			
$\chi_2 \otimes \chi_{103}$							1			1	
$\chi_3 \otimes \chi_{19}$								1	1		
$\chi_2 \otimes \chi_{66}$									1		1
$\chi_2 \otimes \chi_{104}$										1	1

There are only two possible embedded Brauer trees which are consistent with the above projectives (without loss of generality we can assume that node 5 lies in the upper half of the tree):

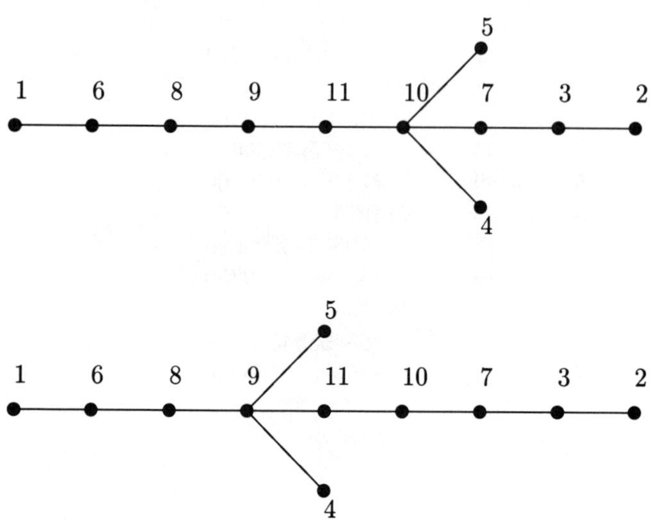

Let us assume the following: node 2 (χ_8) has Green correspondent $1D0$, the Green correspondent of the character χ_3 in Block 3 is $1C0$ and node 1 (χ_7) in Block 4 has Green correspondent $1E0$. Under these assumptions the tensor product

$$\chi_3 \otimes \chi_4 \approx \chi_7$$

tells us that $1C0 \otimes 1D0 \approx 1E0$. We rule out the second tree by the following argument. The tensor product

$$\chi_3 \otimes \chi_{58} \approx \overset{6}{\chi}_{145} + \overset{5}{\chi}_{151}$$

must contain the Green correspondent of $10Ex$ if $10Cx$ is the Green correspondent of χ_{58}. Because χ_{145} has the Green correspondents $10E3$ and $10E8$, we conclude that the second tree is wrong.

Group: 2BM Prime: 11 Block: 5

Nr.	CAS-Nr.	Degree	CC	N&C
1	6	9550635	r	×
2	9	356054375	r	×
3	33	1917448516791	r	∘
4	64	38658236225625	r	∘
5	75	181844090400000	r	×
6	78	226187908369605	r	×
7	79	260522896303125	r	∘
8	82	349055414150625	r	×
9	111	1679281227336960	r	∘
10	145	5191701169700864	r	∘
11	151	6414993199558125	r	×

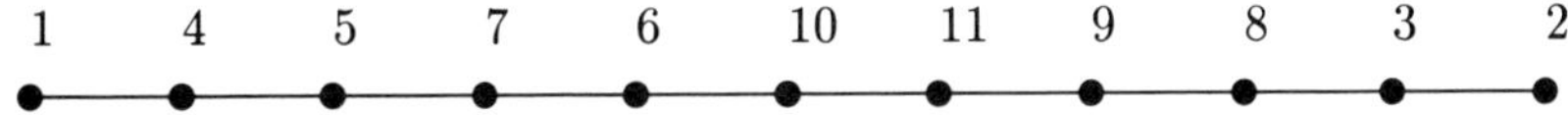

PROJECTIVES:

Nr.	1	2	3	4	5	6	7	8	9	10	11
CC	r	r	r	r	r	r	r	r	r	r	r
N&C	×	×	∘	∘	×	×	∘	×	∘	∘	×
$\chi_2 \otimes \chi_{16}$	1			1							
$\chi_2 \otimes \chi_{13}$		1	1								
$\chi_2 \otimes \chi_{38}$			1					1			
$\chi_2 \otimes \chi_{42}$				1	1						
$\chi_2 \otimes \chi_{55}$					1	1	2				
$\chi_2 \otimes \chi_{133}$						2			1	36	35
$\chi_2 \otimes \chi_{104}$						1	1			4	4
$\chi_2 \otimes \chi_{37}$								1	1		
$\chi_2 \otimes \chi_{45}$									1		1

Group: 2BM Prime: 11 Block: 6

Nr.	CAS-Nr.	Degree	CC	N&C
1	10	1407126890	r	×
2	14	4622913750	r	×
3	24	156871952640	r	∘
4	44	8165038927500	r	×
5	50	10529575087500	r	∘
6	86	563729863936554	r	∘
7	102	968283412561920	r	×
8	144	4919769220713750	r	∘
9	165	9211433539600384	r	×
10	176	11854109736960000	r	×
11	184	16547812226400000	r	∘

1 — 5 — 7 — 8 — 9 — 11 — 10 — 6 — 4 — 3 — 2

PROJECTIVES:

Nr.	1	2	3	4	5	6	7	8	9	10	11
CC	r	r	r	r	r	r	r	r	r	r	r
N&C	×	×	∘	×	∘	∘	×	∘	×	×	∘
$\chi_2 \otimes \chi_{11}$	1				1						
$\chi_3 \otimes \chi_{23}$		3	4	1	1		1				
$\chi_2 \otimes \chi_{39}$				1		1	1	1			
$\chi_2 \otimes \chi_{51}$					1		1				
$\chi_2 \otimes \chi_{80}$						1	1	2	1	1	
$\chi_2 \otimes \chi_{84}$						1		4	6	8	9
$\chi_2 \otimes \chi_{68}$							1	1			
$\chi_3 \otimes \chi_{52}$								1	2	3	4
$\chi_3 \otimes \chi_{35}$									1	1	2
$\chi_4 \otimes \chi_{19}$				1		1		1	1		

Group: 2BM Prime: 11 Block: 7

Nr.	CAS-Nr.	Degree	CC	N&C
1	12	4221380670	r	×
2	20	80426400000	r	×
3	27	309720864375	r	∘
4	32	1095935366250	r	∘
5	43	7331420799495	r	×
6	63	38348970335820	r	×
7	87	581759023584375	r	∘
8	93	665029816320000	r	∘
9	95	714865488180480	r	∘
10	96	718889622405120	r	×
11	105	1198405322994375	r	×

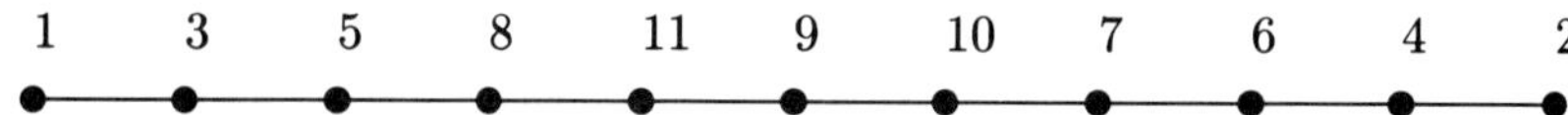

PROJECTIVES:

Nr.	1	2	3	4	5	6	7	8	9	10	11
CC	r	r	r	r	r	r	r	r	r	r	r
N&C	×	×	∘	∘	×	×	∘	∘	∘	×	×
$\chi_2 \otimes \chi_{23}$	1	1	2	2	1	1					
$\chi_2 \otimes \chi_{11}$	1		1								
$\chi_2 \otimes \chi_{31}$		2	1	2	2			1			
$\chi_2 \otimes \chi_{13}$		1		1							
$\chi_2 \otimes \chi_{61}$			1	1	2	2	1	2			1
$\chi_2 \otimes \chi_{40}$				1		1		1			1
$\chi_2 \otimes \chi_{80}$					1			4			3
$\chi_3 \otimes \chi_{39}$						1	1				
$\chi_2 \otimes \chi_{42}$							1			1	
$\chi_2 \otimes \chi_{133}$									2	1	1

Group: 2BM Prime: 11 Block: 8

Nr.	CAS-Nr.	Degree	CC	N&C
1	185	96256	r	×
2	188	8844386304	r	×
3	193	772566552576	r	∘
4	201	31123395403776	r	∘
5	205	77683916800000	206	∘
6	206	77683916800000	205	∘
7	210	168430710030336	211	×
8	211	168430710030336	210	×
9	219	791851002292224	r	×
10	231	4250243698421760	r	×
11	237	5191701169700864	r	∘

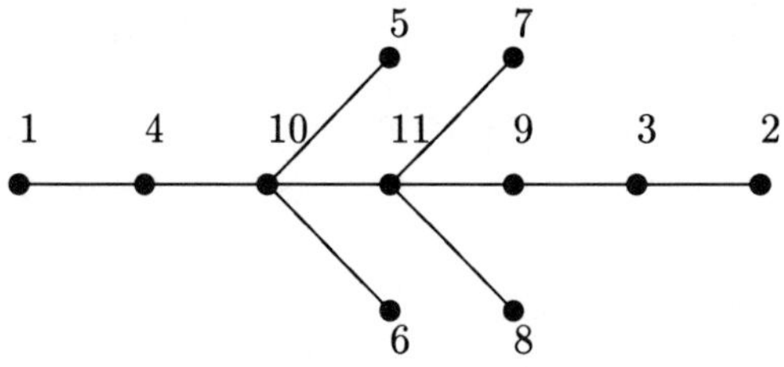

PROJECTIVES:

Nr.	1	2	3	4	5	6	7	8	9	10	11
CC	r	r	r	r	6	5	8	7	r	r	r
N&C	×	×	∘	∘	∘	∘	×	×	×	×	∘
$\chi_{185} \otimes \chi_{11}$	1	2	2	1							
$\chi_{185} \otimes \chi_{29}$		3	4	1					1	2	1
$\chi_{3} \otimes \chi_{192}$		3	4						1		
$\chi_{185} \otimes \chi_{55}$				1					1	5	5
$\chi_{2} \otimes \chi_{223}$					1	1	1	1	1	12	13
$\chi_{2} \otimes \chi_{228}$					1	1			5	32	35
$\chi_{2} \otimes \chi_{232}$							1	1	3	16	21

Without loss of generality we can assume that the nodes 5 and 7 lie on the upper half of the embedded tree.

Group: 2BM Prime: 11 Block: 9

Nr.	CAS-Nr.	Degree	CC	N&C
1	186	10506240	r	×
2	209	138547878400000	r	∘
3	212	235755961319424	213	∘
3	213	235755961319424	212	∘
4	227	3824419701473280	r	×
5	243	21065543276560384	r	×
6	245	24515659148820480	r	∘

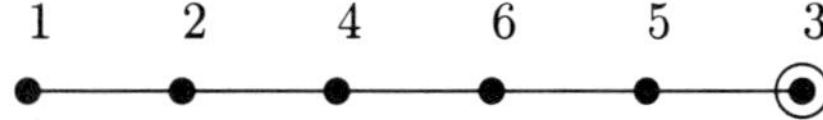

PROJECTIVES:

Nr.	1	2	3	4	5	6
CC	r	r	r	r	r	r
N&C	×	∘	∘	×	×	∘
$\chi_3 \otimes \chi_{192}$	1	1				
$\chi_2 \otimes \chi_{196}$		1		1		
$\chi_2 \otimes \chi_{229}$			2	13	94	105
$\chi_2 \otimes \chi_{207}$				1	1	2

Group: 2BM Prime: 11 Block: 10

Nr.	CAS-Nr.	Degree	CC	N&C
1	187	410132480	r	×
2	195	4322693806080	r	∘
3	220	828829551513600	r	×
4	235	5191647233310720	236	∘
4	236	5191647233310720	235	∘
5	241	12604975138529280	r	∘
6	242	16972115104000000	r	×

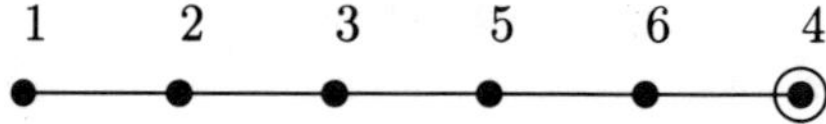

PROJECTIVES:

Nr.	1	2	3	4	5	6
CC	r	r	r	r	r	r
N&C	×	∘	×	∘	∘	×
$\chi_2 \otimes \chi_{192}$	1	1				
$\chi_{185} \otimes \chi_{15}$		1	1			
$\chi_2 \otimes \chi_{207}$			1		2	1

Group: 2BM Prime: 11 Block: 11

Nr.	CAS-Nr.	Degree	CC	N&C
1	189	36657653760	r	×
2	190	53936390144	191	∘
2	191	53936390144	190	∘
3	194	864538761216	r	∘
4	198	21400636907520	r	×
5	202	40345142190080	r	×
6	204	60863961600000	r	∘

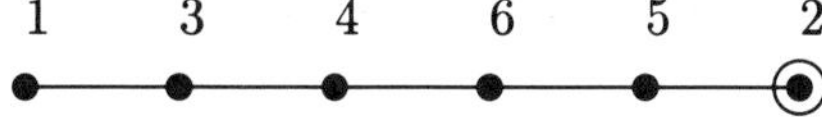

PROJECTIVES:

Nr.	1	2	3	4	5	6
CC	r	r	r	r	r	r
N&C	×	∘	∘	×	×	∘
$\chi_{185} \otimes \chi_{13}$	1		2	1		
$\chi_{185} \otimes \chi_{11}$	1		1		1	1
$\chi_{185} \otimes \chi_{28}$			1	2		1

The above projectives determine the tree uniquely, if we also take into account that the node 2 cannot sit at end of the tree followed by node 1, because it has a larger degree than node 1.

Group: 2BM Prime: 13 Block: 1

Nr.	CAS-Nr.	Degree	CC	N&C
1	1	1	r	×
2	9	356054375	r	×
3	18	27416186875	r	∘
4	37	4375623425250	r	∘
5	50	10529575087500	r	×
6	76	210919336847010	r	∘
7	79	260522896303125	r	×
8	92	665029816320000	r	×
9	97	775438738408125	r	∘
10	99	820429392234375	r	×
11	115	2123804897280000	r	∘
12	159	7853379563452500	r	∘
13	165	9211433539600384	r	×

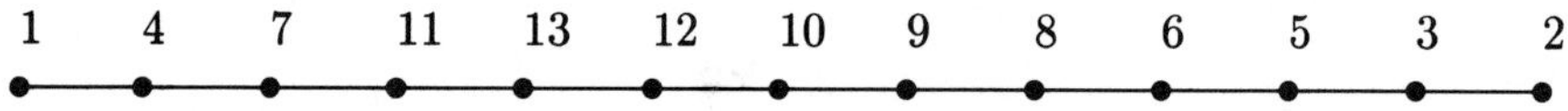

1 — 4 — 7 — 11 — 13 — 12 — 10 — 9 — 8 — 6 — 5 — 3 — 2

PROJECTIVES:

Nr.	1	2	3	4	5	6	7	8	9	10	11	12	13
CC	r	r	r	r	r	r	r	r	r	r	r	r	r
N&C	×	×	∘	∘	×	∘	×	×	∘	×	∘	∘	×
$\chi_2 \otimes \chi_8$		1	1										
$\chi_2 \otimes \chi_{56}$			1		1								
$\chi_2 \otimes \chi_{49}$				1			2	1	1		1		
$\chi_2 \otimes \chi_{55}$				1			2				1		
$\chi_2 \otimes \chi_{27}$					1	1							
$\chi_2 \otimes \chi_{43}$						1		1					
$\chi_2 \otimes \chi_{91}$									1	1		3	3
$\chi_2 \otimes \chi_{80}$										1		2	1
$\chi_2 \otimes \chi_{81}$											1	1	2

Group: 2BM Prime: 13 Block: 2

Nr.	CAS-Nr.	Degree	CC	N&C
1	2	4371	r	×
2	3	96255	r	×
3	28	536105794455	r	∘
4	72	123485536419840	r	∘
5	85	422966584586250	r	×
6	86	563729863936554	r	×
7	104	1093905856312500	r	∘
8	109	1458928219297500	r	×
9	131	4068660047203125	r	∘
10	169	9573585768140625	r	∘
11	170	9573585768140625	r	∘
12	173	10134024677130240	r	×
13	176	11854109736960000	r	×

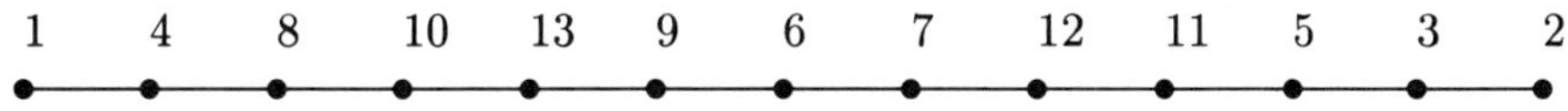

1 — 4 — 8 — 10 — 13 — 9 — 6 — 7 — 12 — 11 — 5 — 3 — 2

PROJECTIVES:

Nr.	1	2	3	4	5	6	7	8	9	10	11	12	13
CC	r	r	r	r	r	r	r	r	r	r	r	r	r
N&C	×	×	∘	∘	×	×	∘	×	∘	∘	∘	×	×
$\chi_3 \otimes \chi_8$		1	1										
$\chi_2 \otimes \chi_{34}$			1		1								
$\chi_2 \otimes \chi_{38}$				1				1					
$\chi_2 \otimes \chi_{73}$					1					1	1		1
$\chi_2 \otimes \chi_{39}$						1	1						
$\chi_2 \otimes \chi_{55}$						1			1				
$\chi_2 \otimes \chi_{65}$							1					1	
$\chi_3 \otimes \chi_{58}$								1		5	5	5	4
$\chi_3 \otimes \chi_{45}$									2	1	1	1	3
$\chi_3 \otimes \chi_{35}$										1	1	1	1

The characters χ_{169},χ_{170} form a pair of algebraic conjugate characters, whose irrationalities are on the $34C/D$-classes. Without loss of generality we can therefore assume that the node 11 belonging to χ_{169} comes first on the real stem. It follows that the tree given above is the only tree which is consistent with the given projectives.

Group: 2BM Prime: 13 Block: 3

Nr.	CAS-Nr.	Degree	CC	N&C
1	4	1139374	r	×
2	5	9458750	r	×
3	20	80426400000	r	○
4	25	252984703125	26	○
5	26	252984703125	25	○
6	30	635966233056	r	○
7	33	1917448516791	r	×
8	71	119483690332160	r	○
9	75	181844090400000	r	×
10	100	835517991997440	r	○
11	105	1198405322994375	r	×
12	119	2642676197359616	r	×
13	125	3068619025500000	r	○

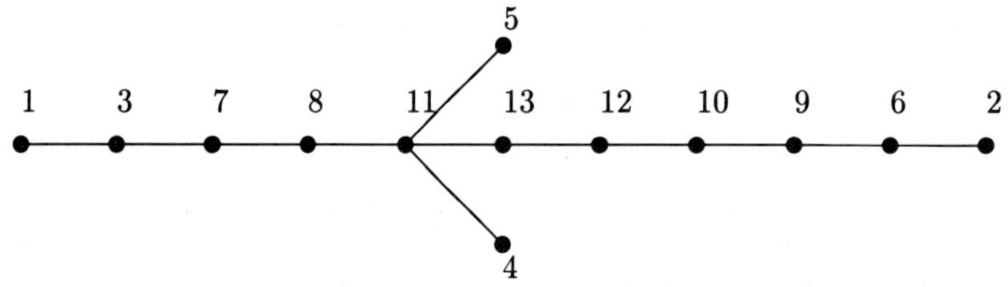

PROJECTIVES:

Nr.	1	2	3	4	5	6	7	8	9	10	11	12	13
CC	r	r	r	5	4	r	r	r	r	r	r	r	r
N&C	×	×	○	○	○	○	×	○	×	○	×	×	○
$\chi_2 \otimes \chi_8$	1		1										
$\chi_2 \otimes \chi_{17}$		1				1							
$\chi_2 \otimes \chi_{24}$			1				2	1					
$\chi_3 \otimes \chi_{122}$				1	1			1	35	110	33	181	136
$\chi_2 \otimes \chi_{27}$						1			1				
$\chi_2 \otimes \chi_{57}$							1	1			1		1
$\chi_2 \otimes \chi_{44}$								1			1		
$\chi_2 \otimes \chi_{51}$									1	1			
$\chi_2 \otimes \chi_{143}$										1	3	5	7
$\chi_2 \otimes \chi_{48}$											1		1
$\chi_3 \otimes \chi_{62}$												1	1

There are three different embedded Brauer trees consistent with the projectives given above (without loss of generality we can assume that the node 5 lies on the upper half of the real stem):

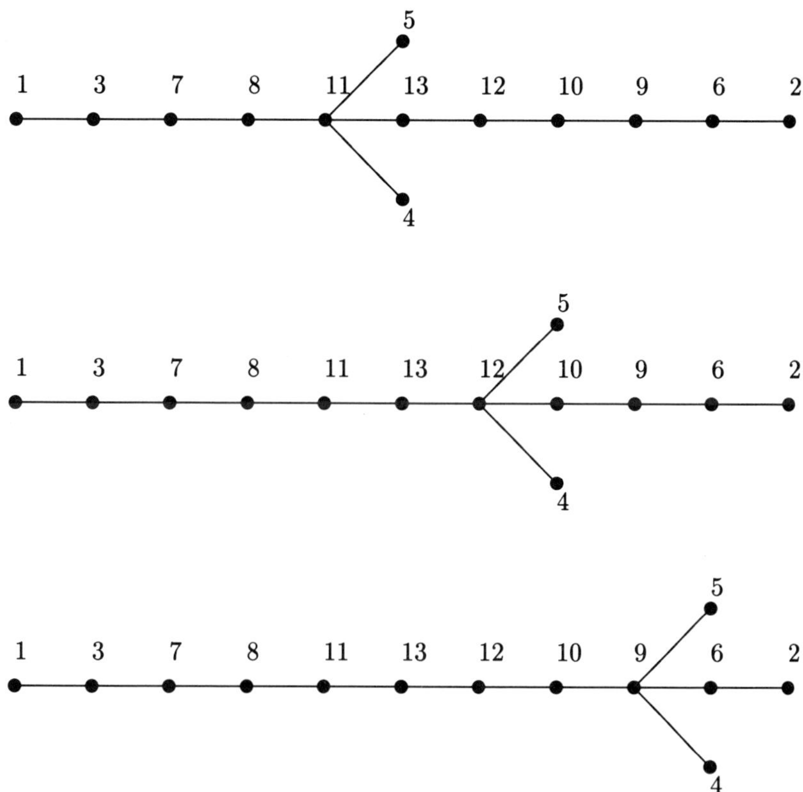

If we define that χ_2 in Block 2 has Green correspondent $1B0$ and that χ_5 has Green correspondent $1C0$, the tensor product

$$\chi_2 \otimes \chi_5 \approx_B \chi_2$$

tells us that $1B0 \otimes 1C0 \approx 1B0$. Looking at the tensor product

$$\chi_2 \otimes \chi_{25} \approx_B \chi_{104}$$

in terms of nodes

$$1_B \otimes 4_C \approx_B 7_B$$

we derive that the node 4_C must have Green correspondent $12C4$ because χ_{104} has Green correspondents $12B4$ and $12B9$. It follows that the first tree is correct.

Group: 2BM Prime: 13 Block: 4

Nr.	CAS-Nr.	Degree	CC	N&C
1	6	9550635	r	×
2	7	63532485	r	×
3	31	641238778125	r	∘
4	32	1095935366250	r	∘
5	52	11890281046875	53	∘
6	53	11890281046875	52	∘
7	64	38658236225625	r	×
8	74	162746401888125	r	×
9	96	718889622405120	r	∘
10	136	4250407956480000	r	×
11	150	5992026614971875	r	∘
12	168	9569488431021750	r	∘
13	177	11854109736960000	r	×

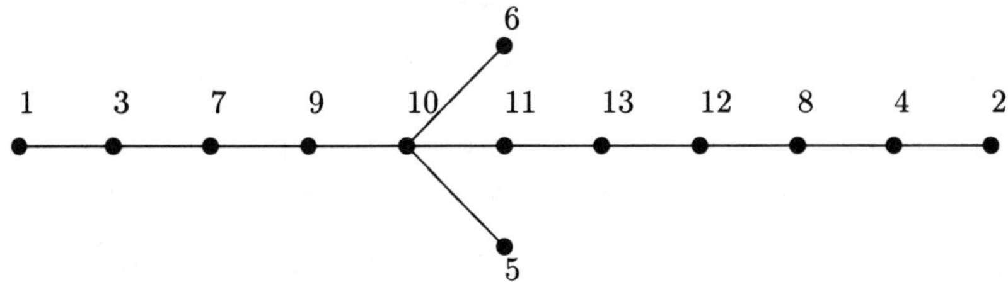

PROJECTIVES:

Nr.	1	2	3	4	5	6	7	8	9	10	11	12	13
CC	r	r	r	r	6	5	r	r	r	r	r	r	r
N&C	×	×	∘	∘	∘	∘	×	×	∘	×	∘	∘	×
$\chi_2 \otimes \chi_{10}$	1		1										
$\chi_2 \otimes \chi_8$		1		1									
$\chi_2 \otimes \chi_{27}$			1				1						
$\chi_2 \otimes \chi_{38}$				1				1					
$\chi_2 \otimes \chi_{122}$					1	1		1		5	7	43	46
$\chi_3 \otimes \chi_{40}$							1		2	1			
$\chi_2 \otimes \chi_{44}$								1				1	
$\chi_2 \otimes \chi_{55}$										1	1		
$\chi_3 \otimes \chi_{35}$											1		1
$\chi_2 \otimes \chi_{65}$												1	1

Let us define the Green correspondent of the node 1_D belonging to χ_6 to be $1D0$. The tensor product

$$\chi_2 \otimes \chi_5 \approx_D \chi_7$$

tells us that $1B0 \otimes 1C0 \approx 1D6$. From the tensor product

$$\chi_2 \otimes \chi_{25} \approx_D \chi_{52}$$

in terms of nodes

$$1_B \otimes 4_C \approx_D 5_D$$

we derive that the node 5_D has Green correspondent $12D10$, because we have already determined that the Green correspondent of χ_{25} is $12C4$. Using this and the above projectives, which already determine the real stem, we are left with the embedded tree given as above.

Group: 2BM Prime: 13 Block: 5

Nr.	CAS-Nr.	Degree	CC	N&C
1	11	3214743741	r	×
2	16	12501781215	r	×
3	29	551549171250	r	∘
4	42	6697382793750	r	∘
5	63	38348970335820	r	×
6	83	364635285437500	r	∘
7	93	665029816320000	r	×
8	123	2952870715858140	r	∘
9	126	3469068040246875	r	×
10	162	8388837260960625	r	×
11	163	8602825703915520	r	∘
12	166	9211433539600384	r	×
13	172	9845152706812500	r	∘

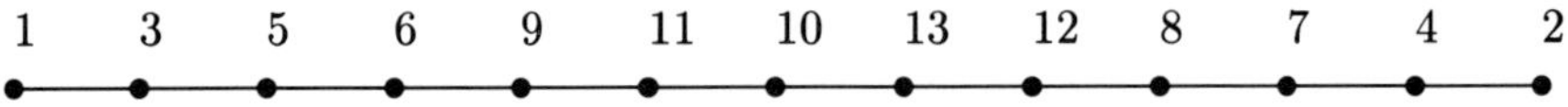

1 — 3 — 5 — 6 — 9 — 11 — 10 — 13 — 12 — 8 — 7 — 4 — 2

PROJECTIVES:

Nr.	1	2	3	4	5	6	7	8	9	10	11	12	13
CC	r	r	r	r	r	r	r	r	r	r	r	r	r
N&C	×	×	∘	∘	×	∘	×	∘	×	×	∘	×	∘
$\chi_{185} \otimes \chi_{201}$	1		1		1	1	1	3	1	7	6	6	6
$\chi_{2} \otimes \chi_{10}$	1		1										
$\chi_{2} \otimes \chi_{14}$		1		1									
$\chi_{2} \otimes \chi_{43}$			1		1								
$\chi_{2} \otimes \chi_{60}$				2	1	1	3	1	1	1	2		
$\chi_{2} \otimes \chi_{40}$					1	2	1	1	1				
$\chi_{2} \otimes \chi_{44}$					1	1							
$\chi_{2} \otimes \chi_{57}$						1	1	1	1	1			1
$\chi_{2} \otimes \chi_{101}$						1		2	3	7	6	6	7
$\chi_{2} \otimes \chi_{78}$							1	2	2	3	4	3	3
$\chi_{2} \otimes \chi_{65}$								1	1		1	1	
$\chi_{2} \otimes \chi_{113}$									1	12	7	12	18
$\chi_{3} \otimes \chi_{35}$										1	1	1	1
$\chi_{3} \otimes \chi_{45}$										1		1	2

Group: 2BM Prime: 13 Block: 6

Nr.	CAS-Nr.	Degree	CC	N&C
1	185	96256	r	×
2	187	410132480	r	×
3	192	100406462464	r	∘
4	197	17069098618880	r	∘
5	205	77683916800000	206	×
6	206	77683916800000	205	×
7	209	138547878400000	r	×
8	214	261841888000000	215	∘
9	215	261841888000000	214	∘
10	219	791851002292224	r	×
11	234	4638342016000000	r	∘
12	242	16972115104000000	r	∘
13	243	21065543276560384	r	×

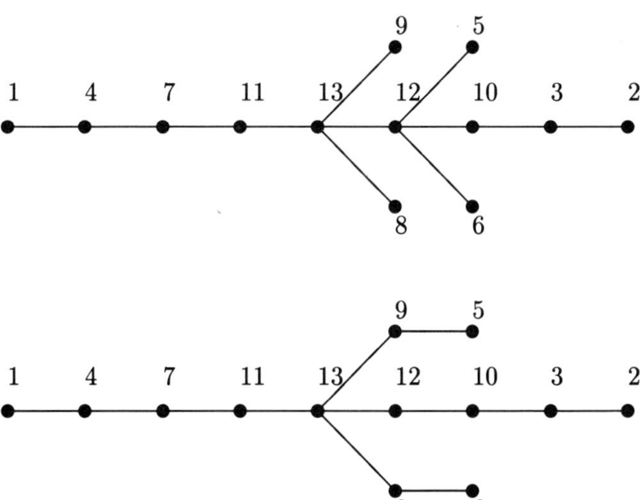

PROJECTIVES:

Nr.	1	2	3	4	5	6	7	8	9	10	11	12	13
CC	r	r	r	r	6	5	r	9	8	r	r	r	r
N&C	×	×	∘	∘	×	×	×	∘	∘	×	∘	∘	×
$\chi_{185} \otimes \chi_8$	1	2	2	1									
$\chi_2 \otimes \chi_{194}$		1	1	1			1						
$\chi_2 \otimes \chi_{188}$		1	1										
$\chi_2 \otimes \chi_{198}$			1	1			1			2		1	
$\chi_{185} \otimes \chi_{39}$				1			1				1	2	3
$\chi_2 \otimes \chi_{238}$					1	1		2	2	6	41	154	191
$\chi_2 \otimes \chi_{222}$							1	1	1	1	11	29	40
$\chi_2 \otimes \chi_{201}$							1				1		
$\chi_3 \otimes \chi_{210}$								1	1	2	18	65	83

Let us assume that χ_{185} has Green correspondent $1F0$. In the proof for the third block we defined that the Green correspondent of χ_5 is $1C0$, and so it follows from this that χ_4 has the Green correspondent $1C6$. The tensor product

$$\chi_4 \otimes \chi_{185} \approx_F \chi_{185} + \chi_{187} + \chi_{192}$$

tells us that $1C6 \otimes 1F0 \approx 1F0$.

The tensor product

$$\chi_{25} \otimes \chi_{185} \approx_F \chi_{215}$$

tells us that χ_{215} has Green correspondent $12F10$ because χ_{25} has Green correspondent $12C4$. This leaves the following two possible embedded Brauer trees.

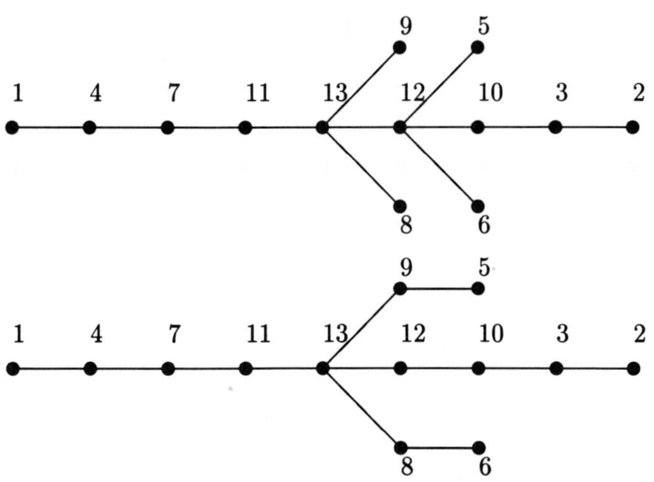

(Without loss of generality we can assume that the node 5_F lies on the upper half of the embedded tree.)

Group: 2BM Prime: 13 Block: 7

Nr.	CAS-Nr.	Degree	CC	N&C
1	186	10506240	r	×
2	196	10177847623680	r	∘
3	216	312199319900160	r	×
4	223	2642676197359616	224	×
4	224	2642676197359616	223	×
5	226	3370903132962816	r	∘
6	244	24089453696000000	r	∘
7	245	24515659148820480	r	×

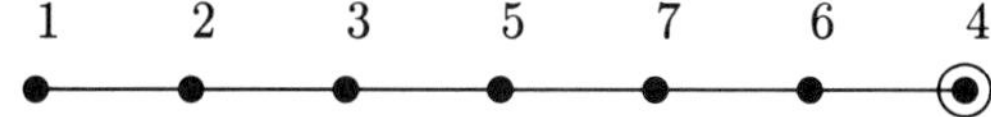

PROJECTIVES:

Nr.	1	2	3	4	5	6	7
CC	r	r	r	r	r	r	r
N&C	×	∘	×	×	∘	∘	×
$\chi_2 \otimes \chi_{188}$	1	1					
$\chi_2 \otimes \chi_{195}$		2	3		1		
$\chi_{185} \otimes \chi_{66}$			1	1	3	9	10
$\chi_2 \otimes \chi_{210}$				1		4	3
$\chi_2 \otimes \chi_{207}$					1	3	4

Group: 2BM Prime: 17 Block: 1

Nr.	CAS-Nr.	Degree	CC	N&C
1	1	1	r	×
2	3	96255	r	×
3	8	347643114	r	∘
4	21	90807234375	22	×
5	22	90807234375	21	×
6	28	536105794455	r	∘
7	29	551549171250	r	×
8	50	10529575087500	r	∘
9	58	21346507000000	59	∘
10	59	21346507000000	58	∘
11	60	22348485900000	r	×
12	73	140763279350304	r	×
13	85	422966584586250	r	∘
14	140	4619851069640625	r	∘
15	155	7575407965125000	r	×
16	166	9211433539600384	r	×
17	176	11854109736960000	r	∘

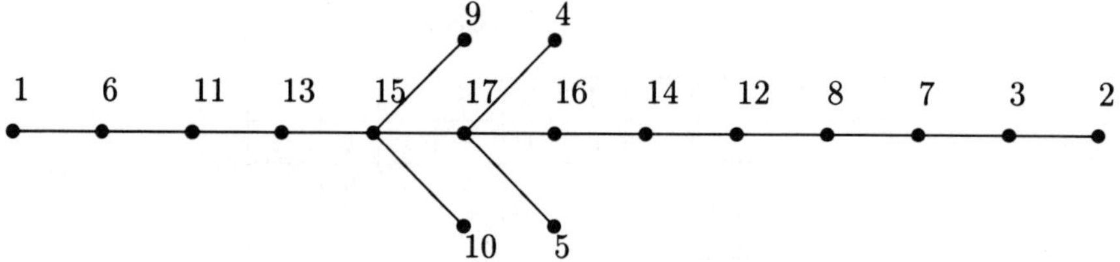

PROJECTIVES:

Nr.	1	2	3	4	5	6	7	8	9	10	11	12	13	14	15	16	17
CC	r	r	r	5	4	r	r	r	10	9	r	r	r	r	r	r	r
N&C	×	×	○	×	×	○	×	○	○	○	×	×	○	○	×	×	○
$\chi_2 \otimes \chi_4$		1	1														
$\chi_2 \otimes \chi_{18}$			1			1	1				1						
$\chi_2 \otimes \chi_{10}$			1				1										
$\chi_2 \otimes \chi_{84}$				1	1										5	1	8
$\chi_2 \otimes \chi_{69}$						1	1	1			2		2		2		1
$\chi_3 \otimes \chi_{37}$						1					1				2		2
$\chi_2 \otimes \chi_{11}$							1	1									
$\chi_2 \otimes \chi_{24}$								1				1					
$\chi_2 \otimes \chi_{133}$									1	1		1	2	23	38	39	51
$\chi_2 \otimes \chi_{33}$											1		1				
$\chi_2 \otimes \chi_{48}$												1		1			
$\chi_2 \otimes \chi_{82}$													1		6	1	6
$\chi_2 \otimes \chi_{65}$														1		1	

The projectives given above prove that there exist only the following two embedded Brauer trees:

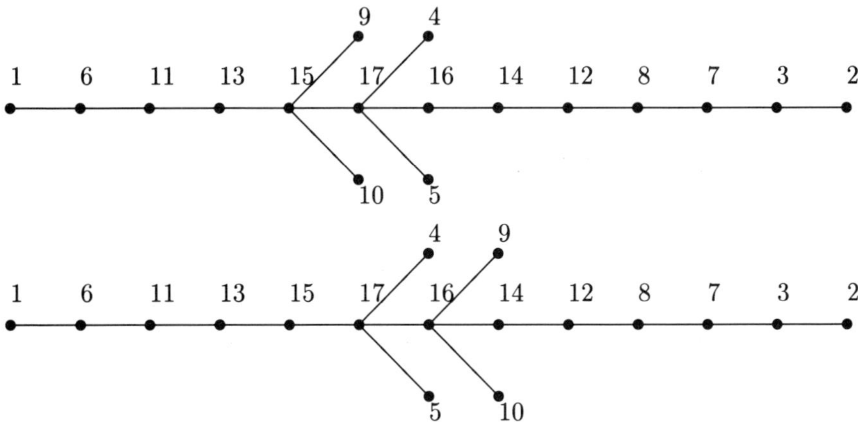

(Without loss of generality we can assume that the nodes 4 and 9 lie on the upper half of the embedded tree.) By Green correspondence the tensor product

$$\chi_3 \otimes \chi_{21} \approx \chi_{22}$$

should contain the Green correspondent of $1A(x+8)$, if $1Ax$ is the Green correspondent of the node 4 (χ_{21}). This contradicts the second tree.

Group: 2BM Prime: 17 Block: 2

Nr.	CAS-Nr.	Degree	CC	N&C
1	2	4371	r	×
2	20	80426400000	r	∘
3	43	7331420799495	r	×
4	47	9958526703125	r	×
5	117	2565217512000000	r	∘
6	119	2642676197359616	r	∘
7	167	9558000449712000	r	×
8	169	9573585768140625	r	×
8	170	9573585768140625	r	×
9	181	13940902029600000	r	∘

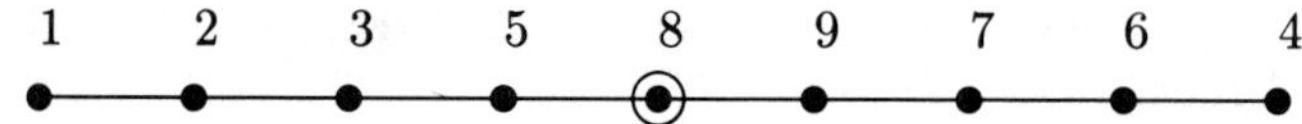

PROJECTIVES:

Nr.	1	2	3	4	5	6	7	8	9
CC	r	r	r	r	r	r	r	r	r
N&C	×	∘	×	×	∘	∘	×	×	∘
$\chi_3 \otimes \chi_4$	1	1							
$\chi_2 \otimes \chi_{23}$		1	1						
$\chi_2 \otimes \chi_{61}$			1		1				
$\chi_2 \otimes \chi_{69}$					1			1	
$\chi_3 \otimes \chi_{57}$						1	2	4	5
$\chi_2 \otimes \chi_{62}$							1		1
$\chi_2 \otimes \chi_{72}$								1	1

Group: 2BM Prime: 17 Block: 3

Nr.	CAS-Nr.	Degree	CC	N&C
1	5	9458750	r	×
2	6	9550635	r	×
3	13	4275362520	r	○
4	14	4622913750	r	○
5	30	635966233056	r	×
6	35	4097337118875	36	×
7	36	4097337118875	35	×
8	40	6145833622500	r	○
9	68	89626740328125	r	×
10	97	775438738408125	r	×
11	137	4331775591000000	r	○
12	138	4331775591000000	r	○
13	151	6414993199558125	r	○
14	152	6692129944500000	r	×
15	165	9211433539600384	r	×
16	174	10161349228100070	r	×
17	177	11854109736960000	r	○

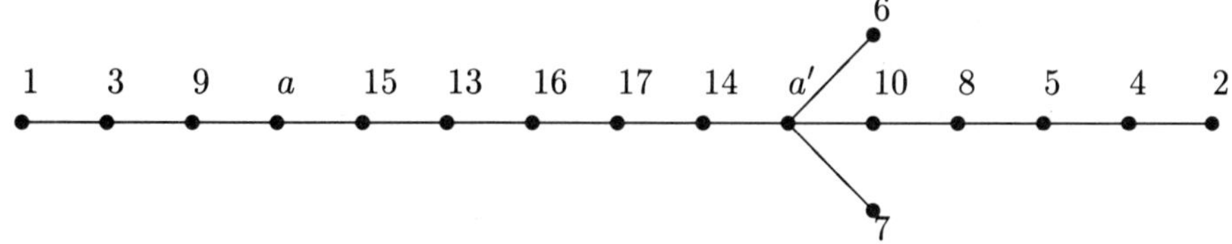

$\{a, a'\} = \{11, 12\}$

PROJECTIVES:

Nr.	1	2	3	4	5	6	7	8	9	10	11	12	13	14	15	16	17
CC	r	r	r	r	r	7	6	r	r	r	r	r	r	r	r	r	r
N&C	×	×	∘	∘	×	×	×	∘	×	×	∘	∘	∘	×	×	×	∘
$\chi_2 \otimes \chi_7$	1	1	1	1													
$\chi_2 \otimes \chi_{17}$	1		1		1			1									
$\chi_2 \otimes \chi_4$		1		1													
$\chi_3 \otimes \chi_{57}$			1		1			2	4	5	8	8	2	4	6	3	2
$\chi_3 \otimes \chi_{15}$			1		1			1	1								
$\chi_2 \otimes \chi_{27}$				1	1												
$\chi_2 \otimes \chi_{49}$					1			2	1	1	1	1		1			
$\chi_2 \otimes \chi_{74}$					1			1	3	1	4	4	1	3	2		
$\chi_2 \otimes \chi_{63}$					1			1	1		1	1		1			
$\chi_2 \otimes \chi_{161}$						1	1			4	33	33	62	53	81	92	104
$\chi_2 \otimes \chi_{105}$								1	3	7	17	17	3	15	17	4	8
$\chi_2 \otimes \chi_{55}$									1		1	1		1			
$\chi_2 \otimes \chi_{91}$										1	1	1	3	2	3	11	12
$\chi_2 \otimes \chi_{76}$											1	1		2	1	1	2
$\chi_2 \otimes \chi_{80}$													2		1	6	5
$\chi_2 \otimes \chi_{75}$														1		2	3

Note that the characters χ_{137} and χ_{138} are a pair of algebraic conjugate characters. There are two possibilities for the real stem consistent with the projectives given above, according to whether the node 11 belonging to χ_{137} or the node 12 belonging to χ_{138} comes first on the real stem. There are 8 possible embedded Brauer trees (without loss of generality we can assume that the node 6 lies on the upper half of the real stem).

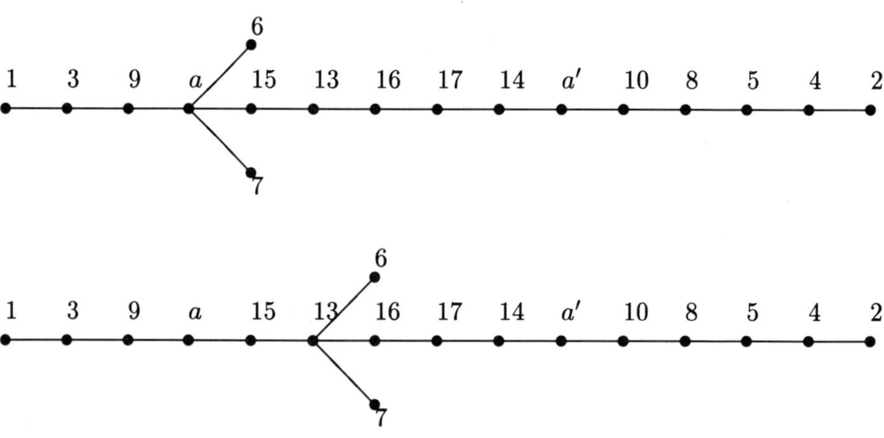

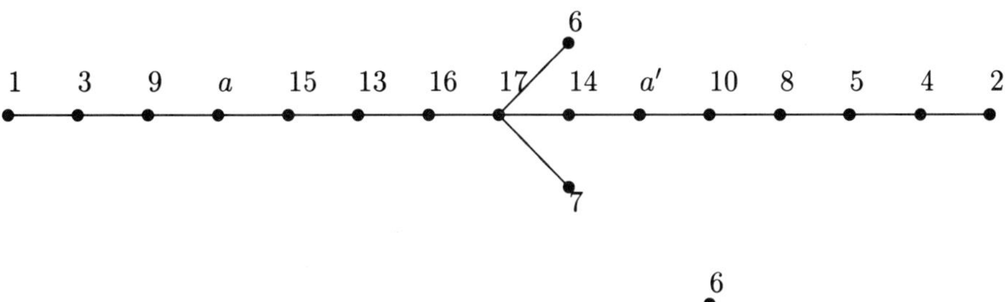

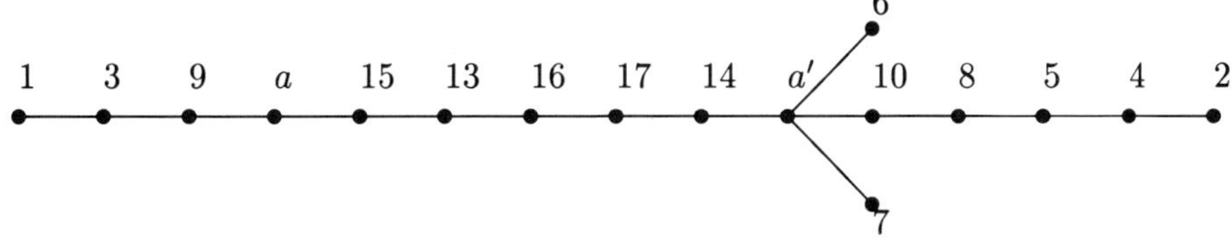

$$\{a, a'\} = \{11, 12\}$$

By Green correspondence the tensor product

$$\chi_3 \otimes \chi_{35} \approx_C \chi_{151} + \chi_{165} + \chi_{174}^2 + \chi_{177}$$

in terms of nodes

$$2_A \otimes 6_C \approx_C 13_C + 15_C + 16_C^2 + 17_C$$

should involve the Green correspondent of $1A8 \otimes 1Cx = 1C(x+8)$, if $1Cx$ is the Green correspondent of the node 6_C belonging to χ_{35}. Because the nodes 10_C, 14_C and 7_C do not occur in the tensor product we can rule out all but the two trees given above.

Group: 2BM Prime: 17 Block: 4

Nr.	CAS-Nr.	Degree	CC	N&C
1	185	96256	r	×
2	186	10506240	r	×
3	188	8844386304	r	∘
4	189	36657653760	r	×
5	192	100406462464	r	∘
6	204	60863961600000	r	×
7	205	77683916800000	206	∘
8	206	77683916800000	205	∘
9	207	93980448866304	r	∘
10	210	168430710030336	211	∘
11	211	168430710030336	210	∘
12	217	447224633856000	218	×
13	218	447224633856000	217	×
14	223	2642676197359616	224	∘
15	224	2642676197359616	223	∘
16	227	3824419701473280	r	∘
17	240	8740741152000000	r	×

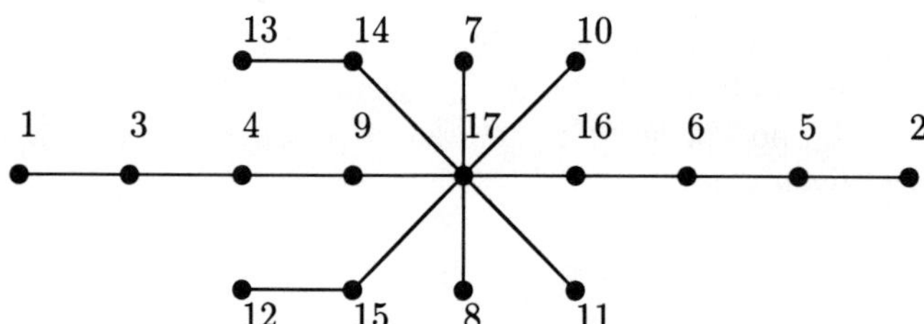

PROJECTIVES:

Nr.	1	2	3	4	5	6	7	8	9	10	11	12	13	14	15	16	17
CC	r	r	r	r	r	r	8	7	r	11	10	13	12	15	14	r	r
N&C	×	×	∘	×	∘	×	∘	∘	∘	∘	∘	×	×	∘	∘	∘	×
$\chi_{185} \otimes \chi_4$	1	1	1		1												
$\chi_{185} \otimes \chi_9$		1	2	2	1												
$\chi_2 \otimes \chi_{193}$			1	1	2	2											
$\chi_{185} \otimes \chi_{17}$			1	1													
$\chi_2 \otimes \chi_{203}$				1		2			1							4	2
$\chi_2 \otimes \chi_{197}$					1	2										2	1
$\chi_2 \otimes \chi_{228}$						1	1	1	2					12	12	33	60
$\chi_2 \otimes \chi_{201}$						1										1	
$\chi_3 \otimes \chi_{214}$							1	1	1	1	1	2	2	16	16	24	57
$\chi_{185} \otimes \chi_{37}$									1							2	3
$\chi_2 \otimes \chi_{232}$										1	1	2	2	12	12	14	36
$\chi_3 \otimes \chi_{199}$												1	1	2	2	2	4
$\chi_2 \otimes \chi_{214}$														1	1	1	3

Without loss of generality we can assume that the nodes 14_D, 10_D and 7_D lie on the upper half of the embedded tree. We also assume that node 1_D (χ_{185}) has the Green correspondent $1D0$. The tensor product

$$\chi_{185} \otimes \chi_{58} \approx_D \chi_{223} + \chi_{224} + \chi_{240}$$

in terms of nodes

$$1_D \otimes 9_A \approx_D 14_D + 15_D + 17_D$$

tells us that node 14_D has the Green correspondent $1D0 \otimes 16A3 = 16D3$. Using this result and the projectives above we are left with the following possibilities for the embedded Brauer tree:

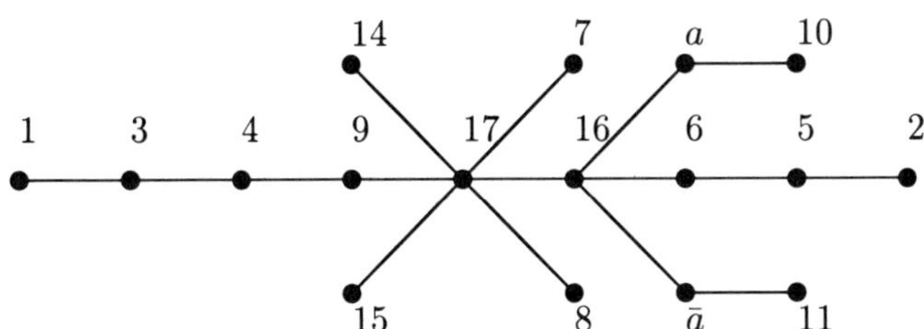

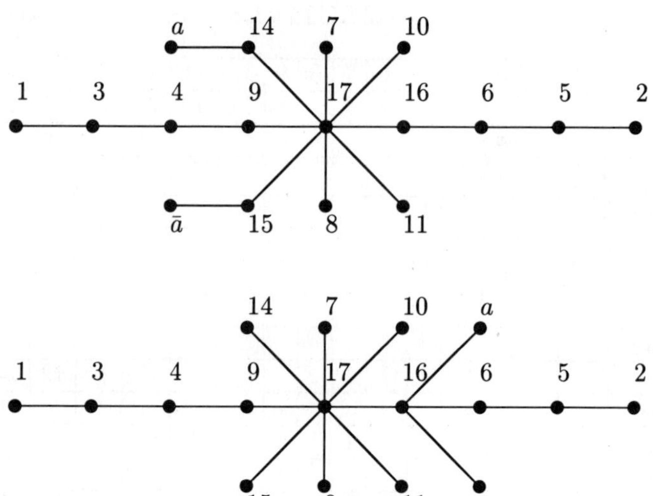

$$\{a, \bar{a}\} = \{12, 13\}$$

We assume that χ_5 has Green correspondent $1C0$. The tensor product

$$\chi_5 \otimes \chi_{185} \approx_D \chi_{186} + \chi_{188} + \chi_{189}$$

in terms of nodes

$$1_C \otimes 1_D \approx_D 2_D + 3_D + 4_D$$

tells us that $1C0 \otimes 1D0 \approx 1D8$ because $2_D + 3_D$ is not projective. We already know that χ_{35}, χ_{36} have the Green correspondents $1C5$, resp. $1C11$. It follows that the tensor products

$$\chi_{35} \otimes \chi_{185} \approx \chi_{217}$$

$$\chi_{36} \otimes \chi_{185} \approx \chi_{218},$$

where χ_{217} is belonging to the node 12_D and χ_{218} is belonging to the node 13_D, should contain $1D13$ resp. $1D3$. We conclude that the node 13_D has Green correspondent $1D13$. This shows that the following tree is correct:

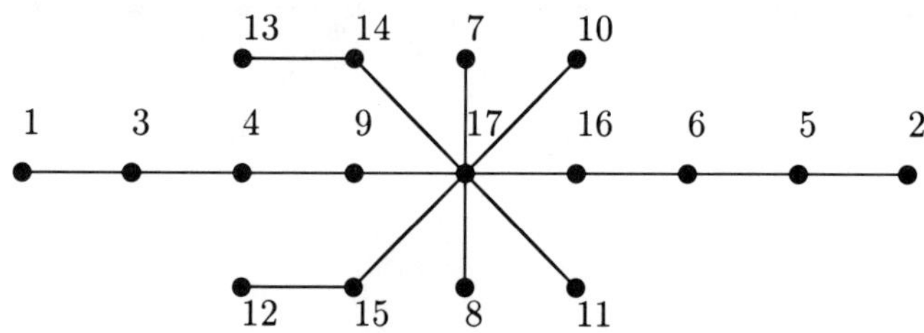

Group: 2BM Prime: 19 Block: 1

Nr.	CAS-Nr.	Degree	CC	N&C
1	1	1	r	×
2	2	4371	r	×
3	9	356054375	r	∘
4	20	80426400000	r	×
5	30	635966233056	r	∘
6	45	8379477898506	46	∘
7	46	8379477898506	45	∘
8	56	19246163186250	r	×
9	72	123485536419840	r	∘
10	96	718889622405120	r	∘
11	103	1042755084800000	r	×
12	111	1679281227336960	r	∘
13	126	3469068040246875	r	×
14	166	9211433539600384	r	∘
15	169	9573585768140625	r	×
16	170	9573585768140625	r	×
17	177	11854109736960000	r	∘
18	181	13940902029600000	r	×
19	182	14014628339712000	r	∘

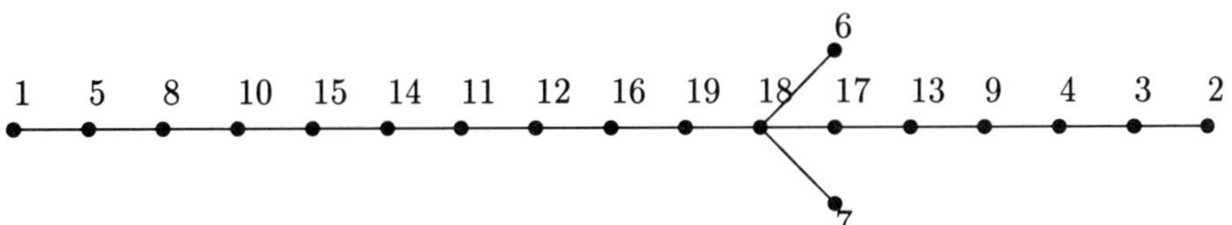

The characters χ_{169}, χ_{170} are algebraically conjugate. Without loss of generality we can assume that node 15 belonging to χ_{169} is the first on the real stem and that node 6 lies on the upper half of the embedded tree. Then the projectives given below determine the embedded tree up to three possibilities.

PROJECTIVES:

Nr.	1	2	3	4	5	6	7	8	9	10	11	12	13	14	15	16	17	18	19
CC	r	r	r	r	r	7	6	r	r	r	r	r	r	r	r	r	r	r	r
N&C	×	×	∘	×	∘	∘	∘	×	∘	∘	×	∘	×	∘	×	×	∘	×	∘
$\chi_2 \otimes \chi_6$		1	1																
$\chi_2 \otimes \chi_8$			1	1															
$\chi_2 \otimes \chi_{28}$				1					1										
$\chi_2 \otimes \chi_{69}$					1			4		4					1	1			1
$\chi_2 \otimes \chi_{63}$					1			1		1		1			1	1			
$\chi_2 \otimes \chi_{17}$					1			1											
$\chi_2 \otimes \chi_{164}$						1	1			9	1	6	17	71	80	80	95	138	133
$\chi_2 \otimes \chi_{151}$						1	1			2	3	6	13	58	59	59	75	109	100
$\chi_2 \otimes \chi_{50}$								1		1									
$\chi_2 \otimes \chi_{40}$									1				1						
$\chi_2 \otimes \chi_{84}$										1				1	2	2	3	5	4
$\chi_2 \otimes \chi_{65}$											2	1	1	1			1		
$\chi_2 \otimes \chi_{39}$											1	1							
$\chi_3 \otimes \chi_{39}$												1		1	1	1	1	1	
$\chi_2 \otimes \chi_{104}$													2	8	8	8	12	17	15
$\chi_3 \otimes \chi_{62}$														3	3	3	4	7	6

The three possible trees are

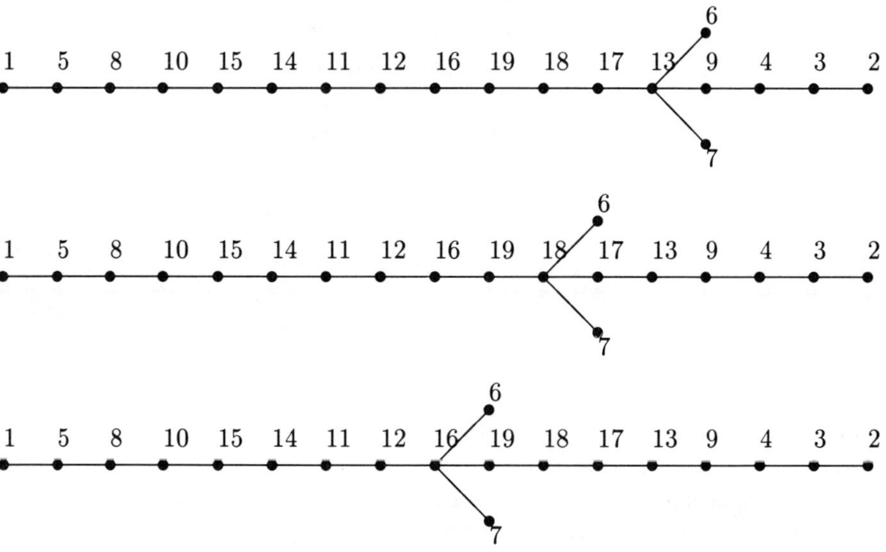

Node 2 (χ_2) has Green correspondent $1A9$. The tensor product

$$\chi_2 \otimes \chi_{45} \approx \chi_{111}$$

tell us that χ_{45} has Green correspondent $18A6$ or $18A13$, because χ_{111} has Green correspondent $18A4$ and $18A15$. This proves that the second tree is correct.

Group: 2BM Prime: 19 Block: 2

Nr.	CAS-Nr.	Degree	CC	N&C
1	3	96255	r	×
2	4	1139374	r	×
3	5	9458750	r	∘
4	18	27416186875	r	∘
5	34	2624476798125	r	×
6	35	4097337118875	36	×
7	36	4097337118875	35	×
8	44	8165038927500	r	×
9	58	21346507000000	59	×
10	59	21346507000000	58	×
11	114	1928549534519790	r	∘
12	144	4919769220713750	r	∘
13	157	7698465274500000	r	×
14	161	8351460096890625	r	×
15	165	9211433539600384	r	∘
16	176	11854109736960000	r	∘
17	178	12424315904000000	r	×
18	179	12424315904000000	r	×
19	180	13046344927150080	r	∘

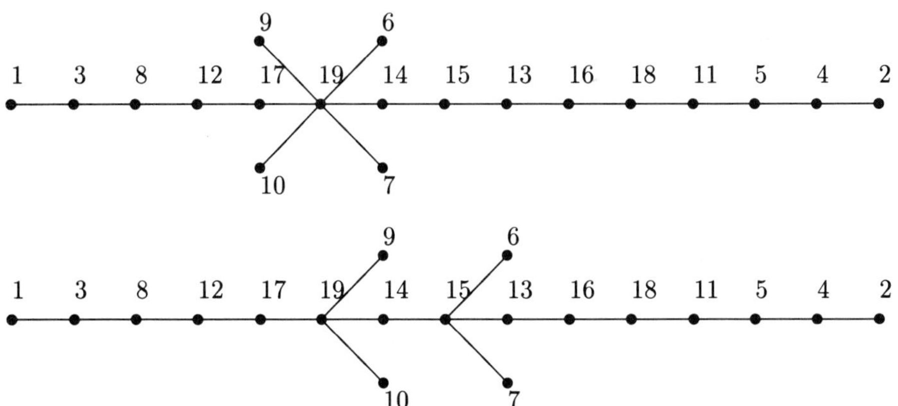

PROJECTIVES:

Nr.	1	2	3	4	5	6	7	8	9	10	11	12	13	14	15	16	17	18	19
CC	r	r	r	r	r	7	6	r	10	9	r	r	r	r	r	r	r	r	r
N&C	×	×	∘	∘	×	×	×	×	×	×	∘	∘	×	×	∘	∘	×	×	∘
$\chi_2 \otimes \chi_7$	1		1																
$\chi_2 \otimes \chi_6$		1		1															
$\chi_2 \otimes \chi_{17}$			1					1											
$\chi_2 \otimes \chi_{11}$				1	1														
$\chi_2 \otimes \chi_{48}$					1						1								
$\chi_{185} \otimes \chi_{218}$							1				16	42	80	96	95	123	123	123	147
$\chi_2 \otimes \chi_{69}$								1			1	2	2		1	1	1	1	
$\chi_2 \otimes \chi_{71}$								1				3	1	1	2	2	2	2	
$\chi_2 \otimes \chi_{39}$								1				1							
$\chi_2 \otimes \chi_{132}$									1	1	9	12	24	38	30	42	47	47	65
$\chi_3 \otimes \chi_{49}$											2	1					2	2	1
$\chi_2 \otimes \chi_{84}$												4	7	1	6	8	6	6	2
$\chi_3 \otimes \chi_{37}$												1	1		1	1	1	1	
$\chi_2 \otimes \chi_{78}$														2	1	2	2	2	3
$\chi_2 \otimes \chi_{62}$														1					1

The characters χ_{178}, χ_{179} are a pair of algebraic conjugate characters with irrationalities on the $56A/B$ classes. Without loss of generality we can assume that the node 17_B belonging to χ_{178} is the first on the real stem and that the nodes 6_B and 9_B lie on the upper half of the embedded tree. Under this assumption the projectives given above prove that the real stem is as follows:

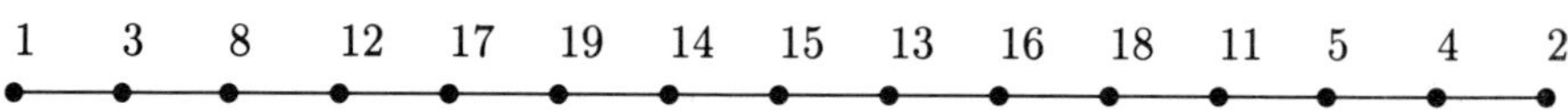

Observe that $\chi_{157} + \chi_{180}$ is not a projective character, because the nodes 13_B and 19_B are not linked on the real stem. In the tensor product

$$\chi_2 \otimes \chi_{58} \approx_B \chi_{59} + \chi_{157} + \chi_{180}$$

in terms of nodes

$$2_A \otimes 9_B \approx_B 10_B + 13_B + 18_B$$

we have to subtract some characters of type ∘ by Green correspondence, leaving a projective character. This tells us that $\chi_{59} + \chi_{180}$ is projective. It also tells us that the node 13_B has the Green correspondent $1B(x+9)$ if $1Bx$ is the Green correspondent of the node 9_B. In particular it follows that nodes 9_B and 10_B are linked to node 19_B. We are left with the two embedded trees given above.

Group: 2BM Prime: 19 Block: 3

Nr.	CAS-Nr.	Degree	CC	N&C
1	185	96256	r	×
2	189	36657653760	r	∘
3	198	21400636907520	r	×
4	210	168430710030336	211	∘
4	211	168430710030336	210	∘
5	221	1566852857856000	r	∘
6	222	1793872427483136	r	×
7	227	3824419701473280	r	×
8	238	8580585368070144	239	×
9	239	8580585368070144	238	×
10	243	21065543276560384	r	∘

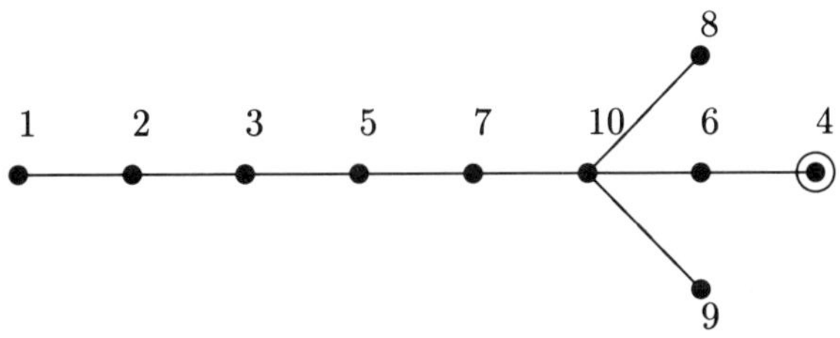

PROJECTIVES:

Nr.	1	2	3	4	5	6	7	8	9	10
CC	r	r	r	r	r	r	r	9	8	r
N&C	×	∘	×	∘	∘	×	×	×	×	∘
$\chi_2 \otimes \chi_{186}$	1	1								
$\chi_2 \otimes \chi_{188}$		1	1							
$\chi_{185} \otimes \chi_{28}$			1		1					
$\chi_2 \otimes \chi_{223}$				1	3	5	9	25	25	60
$\chi_2 \otimes \chi_{201}$					1		1			
$\chi_2 \otimes \chi_{214}$						1	1	2	2	6

The above tree does not give the emdedding. To get the embedding right you might have to interchange the nodes 8 and 9.

Group: 2BM Prime: 23 Block: 1

Nr.	CAS-Nr.	Degree	CC	N&C
1	1	1	r	×
2	13	4275362520	r	∘
3	25	252984703125	26	∘
3	26	252984703125	25	∘
4	30	635966233056	r	×
5	66	47097393520320	r	×
6	79	260522896303125	r	∘
7	98	809403858900000	r	∘
8	137	4331775591000000	r	×
9	138	4331775591000000	r	×
10	145	5191701169700864	r	×
11	148	5257393731060471	r	∘
12	155	7575407965125000	r	∘

1 — 2 — 4 — 7 — 8 — 11 — 10 — 12 — 9 — 6 — 5 — 3 (exceptional)

PROJECTIVES:

Nr.	1	2	3	4	5	6	7	8	9	10	11	12
CC	r	r	r	r	r	r	r	r	r	r	r	r
N&C	×	∘	∘	×	×	∘	∘	×	×	×	∘	∘
$\chi_3 \otimes \chi_3$	1	1										
$\chi_2 \otimes \chi_9$		1		1								
$\chi_2 \otimes \chi_{42}$				1			1					
$\chi_2 \otimes \chi_{55}$					1	2		1	1		1	
$\chi_2 \otimes \chi_{87}$						1	1	3	3	1	2	3
$\chi_2 \otimes \chi_{76}$							1	1	1			1
$\chi_3 \otimes \chi_{47}$								1	1		1	1
$\chi_2 \otimes \chi_{62}$										1	1	
$\chi_2 \otimes \chi_{58}$										1		1

Observing that the characters χ_{137} and χ_{138} are a pair of algebraic conjugate characters, we can assume that the node 8 belonging to χ_{137} is the first on the tree. Under this assumption the tree given above is the only possible tree consistent with the given projectives.

Group: 2BM Prime: 23 Block: 2

Nr.	CAS-Nr.	Degree	CC	N&C
1	2	4371	r	×
2	7	63532485	r	∘
3	27	309720864375	r	×
4	51	10853720317440	r	∘
5	52	11890281046875	53	∘
5	53	11890281046875	52	∘
6	88	595404085985280	r	×
7	96	718889622405120	r	×
8	108	1391162882484375	r	∘
9	109	1458928219297500	r	∘
10	117	2565217512000000	r	×
11	136	4250407956480000	r	×
12	147	5257393731060471	r	∘

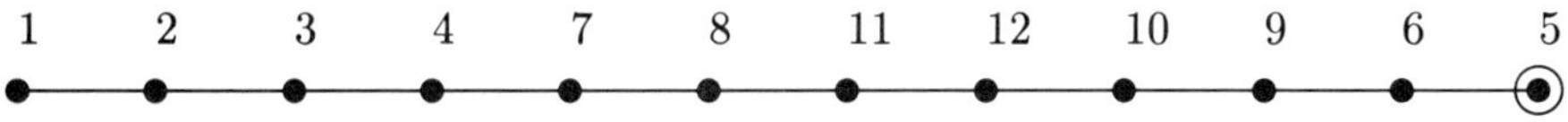

1 — 2 — 3 — 4 — 7 — 8 — 11 — 12 — 10 — 9 — 6 — (5)

PROJECTIVES:

Nr.	1	2	3	4	5	6	7	8	9	10	11	12
CC	r	r	r	r	r	r	r	r	r	r	r	r
N&C	×	∘	×	∘	∘	×	×	∘	∘	×	×	∘
$\chi_2 \otimes \chi_3$	1	1										
$\chi_2 \otimes \chi_8$		1	1									
$\chi_2 \otimes \chi_{12}$			1	1								
$\chi_2 \otimes \chi_{32}$				1			1					
$\chi_2 \otimes \chi_{122}$					1	6			11	13	5	12
$\chi_2 \otimes \chi_{57}$						1			1			
$\chi_2 \otimes \chi_{33}$							1	1				
$\chi_3 \otimes \chi_{45}$								1			3	2
$\chi_2 \otimes \chi_{40}$									1	1		
$\chi_2 \otimes \chi_{86}$										1	1	2

Group: 2BM   Prime: 23   Block: 3

Nr.	CAS-Nr.	Degree	CC	N&C
1	185	96256	r	×
2	187	410132480	r	∘
3	189	36657653760	r	×
4	190	53936390144	191	×
5	191	53936390144	190	×
6	203	60780833777664	r	∘
7	214	261841888000000	215	∘
7	215	261841888000000	214	∘
8	217	447224633856000	218	∘
9	218	447224633856000	217	∘
10	220	828829551513600	r	×
11	231	4250243698421760	r	∘
12	234	4638342016000000	r	×

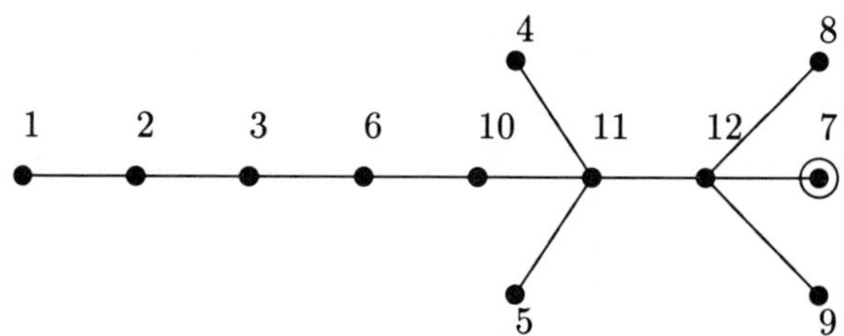

PROJECTIVES:

Nr.	1	2	3	4	5	6	7	8	9	10	11	12
CC	r	r	r	5	4	r	r	9	8	r	r	r
N&C	×	∘	×	×	×	∘	∘	∘	∘	×	∘	×
$\chi_{185} \otimes \chi_3$	1	1										
$\chi_2 \otimes \chi_{192}$		1	1									
$\chi_2 \otimes \chi_{195}$			1			2				1		
$\chi_3 \otimes \chi_{236}$					1		28	64	64	51	435	539
$\chi_2 \otimes \chi_{235}$							1	4	4		13	22
$\chi_{185} \otimes \chi_{48}$										1	1	

The projectives given above leave only four possible embedded Brauer trees (without loss of generality we can assume that node 4_C and 8_C lie on the upper half of the emdedded tree):

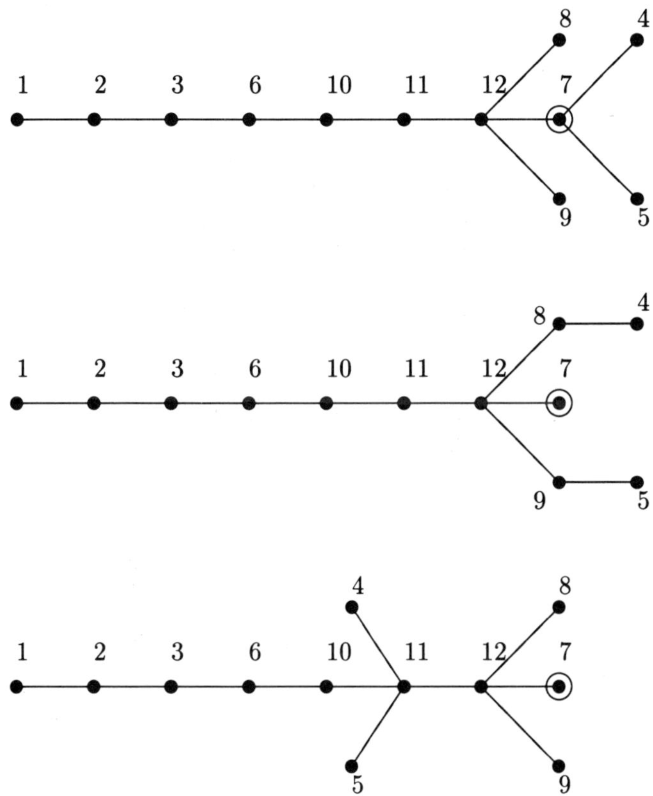

We show that the Green correspondent of the node 4_C is $1C3$, proving that the fourth tree is correct. Assuming that the Green correspondent of the node 1_C (χ_{185}) is $1C0$ the tensor product

$$\chi_{185} \otimes \chi_{185} \approx \chi_1$$

tells us that $1C0 \otimes 1C0 \approx 1A0$. The tensor product

$$\chi_{185} \otimes \chi_{190} \approx \chi_{145}$$

shows, because χ_{145} has Green correspondent $1A3$ and $1A8$, that the Green correspondent of χ_{190} has to be either $1C3$ or $1C8$. This shows that the last tree is correct.

Group: 2BM   Prime: 23   Block: 4

Nr.	CAS-Nr.	Degree	CC	N&C
1	186	10506240	r	×
2	188	8844386304	r	∘
3	196	10177847623680	r	×
4	199	23459577856000	200	×
5	200	23459577856000	199	×
6	210	168430710030336	211	×
7	211	168430710030336	210	×
8	212	235755961319424	213	×
9	213	235755961319424	212	×
10	221	1566852857856000	r	∘
11	232	4490309824000000	233	∘
11	233	4490309824000000	232	∘
12	237	5191701169700864	r	×

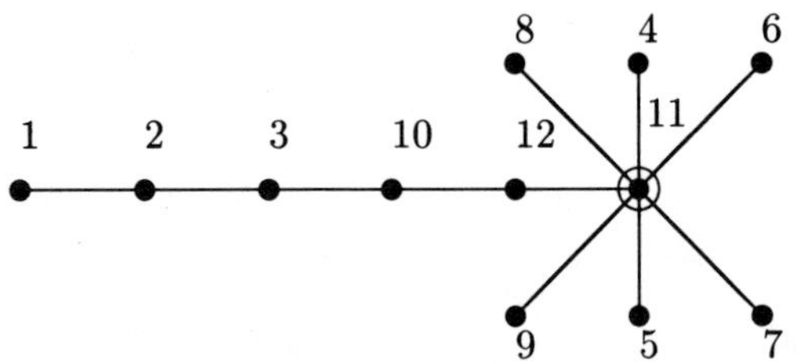

PROJECTIVES:

Nr.	1	2	3	4	5	6	7	8	9	10	11	12
CC	r	r	r	5	4	7	6	9	8	r	r	r
N&C	×	∘	×	×	×	×	×	×	×	∘	∘	×
$\chi_{185} \otimes \chi_3$	1	1										
$\chi_2 \otimes \chi_{192}$		1	1									
$\chi_2 \otimes \chi_{202}$			1							2		1
$\chi_2 \otimes \chi_{222}$				1	1					2	8	8
$\chi_2 \otimes \chi_{223}$						1	1			3	12	13
$\chi_{185} \otimes \chi_{58}$								1	1		4	2

Let us define the Green correspondent of χ_2 to be $1B0$ and the Green correspondent of χ_{186} to be $1D0$. In the proof for the third block we defined that χ_{185} has Green correspondent $1C0$. The tensor product

$$\chi_2 \otimes \chi_{185} \approx_D \chi_{186}$$

tells us

$$1B0 \otimes 1C0 \approx 1D0$$

In the proof for the third block we determined that the Green correspondent of χ_{190} is $1C3$. The tensor product

$$\chi_2 \otimes \chi_{190} \approx \chi_{212}$$

must therefore contain the Green correspondent of $1D3$. So it follows that the Green correspondent of node 8_D (χ_{212}) is $1D3$, proving that there are only the following possibilities for the embedded Brauer tree:

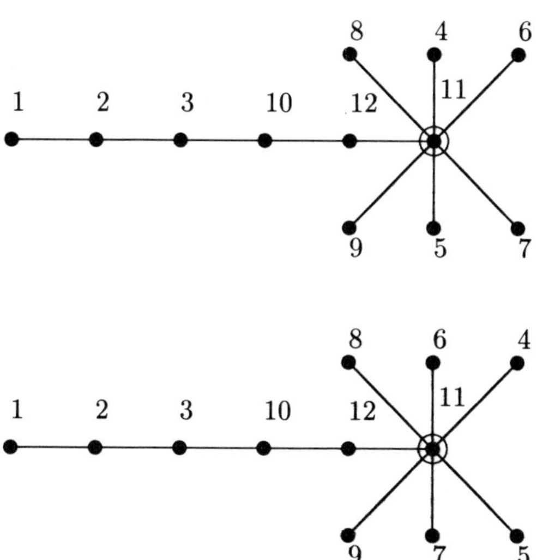

To exclude the second tree, we use the following tensor products:

$$\chi_{185} \otimes \chi_{186} \approx_B \chi_2 + \chi_7 + \chi_{27}$$

in terms of nodes

$$1_C \otimes 1_D \approx_B 1_B + 2_B + 3_B$$

and

$$\chi_{185} \otimes \chi_{199} \approx_B \chi_{117}^2 + \chi_{136}^2 + \chi_{147}^3$$

in terms of nodes

$$1_B \otimes 4_D \approx_B 10_B^2 + 11_B^2 + 12_B^3.$$

The first tensor product tells us that $1C0 \otimes 1D0 \approx 1B0$ or $1C0 \otimes 1D0 \approx 1B1$. Because χ_{185} and χ_{186} are self dual leaves on the trees they belong to, their Green correspondents $1C0$ and $1D0$ are self dual. It follows that the tensor product $1C0 \otimes 1D0$ is again self dual. This implies that $1B1$ cannot appear in this tensor product, because otherwise its dual $1B10$ had to appear in the tensor product, too. It follows that $1C0 \otimes 1D0 \approx 1B0$. In the second possible tree above, the node 4_B (χ_{199}) has the Green correspondent $1D5$. This implies that in the tensor product $\chi_{185} \otimes \chi_{199}$ the character χ_{88}, which has the Green correspondent $1B5 = 1C0 \otimes 1D5$, should appear. This contradiction shows that the first tree is correct.

Group: 2BM Prime: 31 Block: 1

Nr.	CAS-Nr.	Degree	CC	N&C
1	1	1	r	×
2	5	9458750	r	∘
3	24	156871952640	r	×
4	45	8379477898506	46	∘
4	46	8379477898506	45	∘
5	64	38658236225625	r	∘
6	115	2123804897280000	r	×
7	119	2642676197359616	r	×
8	121	2788834713600000	r	×
9	129	3669877711500000	r	∘
10	137	4331775591000000	r	∘
11	138	4331775591000000	r	∘
12	145	5191701169700864	r	×
13	173	10134024677130240	r	∘
14	176	11854109736960000	r	×
15	177	11854109736960000	r	×
16	181	13940902029600000	r	∘

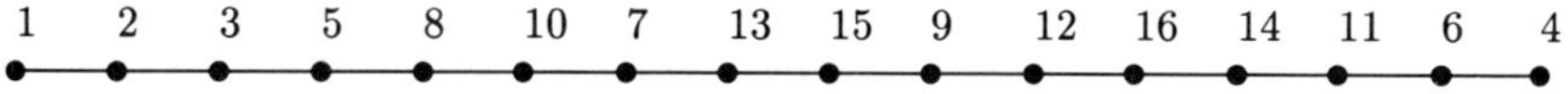

1 — 2 — 3 — 5 — 8 — 10 — 7 — 13 — 15 — 9 — 12 — 16 — 14 — 11 — 6 — 4

PROJECTIVES:

Nr.	1	2	3	4	5	6	7	8	9	10	11	12	13	14	15	16
CC	r	r	r	r	r	r	r	r	r	r	r	r	r	r	r	r
N&C	×	∘	×	∘	∘	×	×	×	∘	∘	∘	×	∘	×	×	∘
$\chi_2 \otimes \chi_2$	1	1														
$\chi_7 \otimes \chi_2$		1	1													
$\chi_{69} \otimes \chi_2$			1		4	2		6		3	3			1		
$\chi_{16} \otimes \chi_2$			1		1											
$\chi_{84} \otimes \chi_2$						4	1	6	1	7	7		2	8	3	5
$\chi_{90} \otimes \chi_2$							2	1	2	2	2	1	10	7	11	6
$\chi_{97} \otimes \chi_2$							1	1	4	2	2	2	10	7	13	6
$\chi_{80} \otimes \chi_2$							1						6	1	5	1
$\chi_{28} \otimes \chi_4$							1						2		1	
$\chi_{78} \otimes \chi_2$									1			1	3	2	4	3
$\chi_{75} \otimes \chi_2$									1				2		3	

Observing that the characters χ_{137} and χ_{138} are algebraic conjugate characters, we can assume that the node 10 is the first on the real stem.

Group: 2BM Prime: 31 Block: 2

Nr.	CAS-Nr.	Degree	CC	N&C
1	185	96256	r	×
2	186	10506240	r	∘
3	190	53936390144	191	×
4	191	53936390144	190	×
5	198	21400636907520	r	×
6	214	261841888000000	215	×
7	215	261841888000000	214	×
8	216	312199319900160	r	∘
9	217	447224633856000	218	×
10	218	447224633856000	217	×
11	223	2642676197359616	224	×
12	224	2642676197359616	223	×
13	237	5191701169700864	r	×
14	238	8580585368070144	239	∘
14	239	8580585368070144	238	∘
15	243	21065543276560384	r	×
16	244	24089453696000000	r	∘

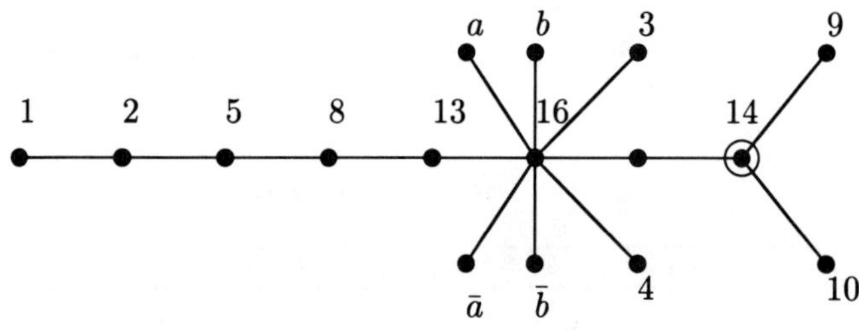

$\{a, b\} = \{6, 11\}$

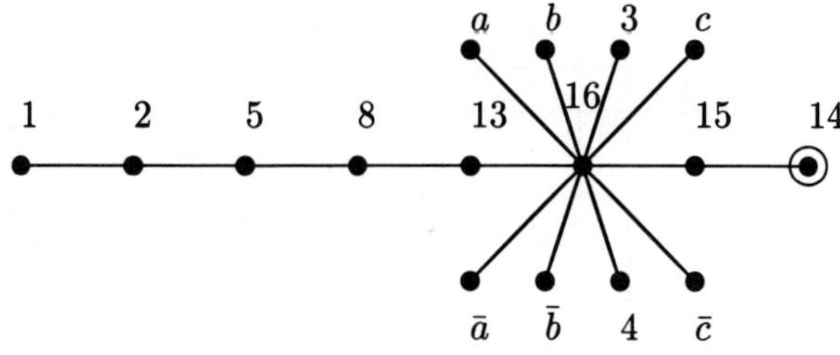

$\{a, b, c\} = \{6, 9, 11\}$

PROJECTIVES:

Nr.	1	2	3	4	5	6	7	8	9	10	11	12	13	14	15	16
CC	r	r	4	3	r	7	6	r	10	9	12	11	r	r	r	r
N&C	×	∘	×	×	×	×	×	∘	×	×	×	×	×	∘	×	∘
$\chi_{185} \otimes \chi_2$	1	1														
$\chi_{188} \otimes \chi_2$		1			1											
$\chi_{212} \otimes \chi_2$			1										1	2	4	4
$\chi_{195} \otimes \chi_2$					2			3					1			
$\chi_{214} \otimes \chi_2$						1	1				1	1	1	2	6	9
$\chi_{185} \otimes \chi_{25}$							1									1
$\chi_{217} \otimes \chi_2$									1	1	1	1	1	4	10	11
$\chi_{205} \otimes \chi_2$											1	1		1	2	3
$\chi_{201} \otimes \chi_2$													1			1

The above projectives determine the real stem which is as follows.

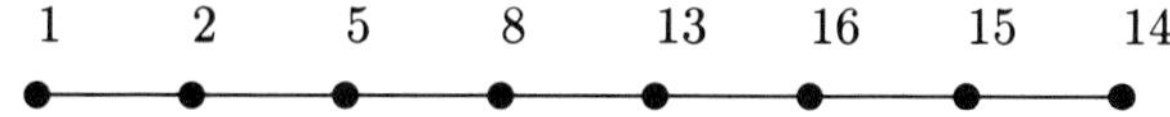

We now show that the nodes 3_B and 4_B belonging to the complex conjugate pair χ_{190}, χ_{191} cannot lie at node 14_B.

We already know by the above projectives that the nodes 6_B, 7_B, 11_B and 12_B lie at the node 16_B and that the nodes 3_B, 4_B, 9_B and 10_B can lie at either of the nodes 14_B or 16_B. Let us define the Green correspondent of the node 1_B (χ_{185}) to be $1B0$. Because

$$\chi_{185} \otimes \chi_{185} \approx_A \chi_1$$

we conclude for the Green correspondents,

$$1B0 \otimes 1B0 \approx_A 1A0.$$

If 3_B, 4_B lay at node 14_B, the Green correspondents would be either $1B7$, $1B8$ or $1B6$, $1B9$. Green correspondence tells us that $\chi_{185} \otimes \chi_{190}$ should contain the Green correspondent of $1A7$, $1A8$ or $1A6$, $1A9$. But in reality this tensor product is

$$\chi_{185} \otimes \chi_{190} \approx_A \chi_{145}.$$

The node 12_A belonging to χ_{145} has Green correspondents $1A5$ and $1A10$. This is a contradiction. So we are left with the possible embedded trees given above, if we assume without loss of generality that the nodes 3_B, 6_B, 9_B and 11_B lie on the upper half of the embedded tree.

Group: 2BM Prime: 47 Block: 1

Nr.	CAS-Nr.	Degree	CC	N&C
1	1	1	r	×
2	3	96255	r	○
3	14	4622913750	r	×
4	21	90807234375	22	○
4	22	90807234375	21	○
5	25	252984703125	26	○
6	26	252984703125	25	○
7	31	641238778125	r	○
8	35	4097337118875	36	○
9	36	4097337118875	35	○
10	91	662947625486250	r	×
11	100	835517991997440	r	×
12	101	844994666880000	r	×
13	119	2642676197359616	r	○
14	137	4331775591000000	r	○
15	138	4331775591000000	r	○
16	158	7744013981383680	r	×
17	159	7853379563452500	r	×
18	169	9573585768140625	r	○
19	170	9573585768140625	r	○
20	175	11381560819200000	r	×
21	176	11854109736960000	r	○
22	177	11854109736960000	r	○
23	178	12424315904000000	r	×
24	179	12424315904000000	r	×

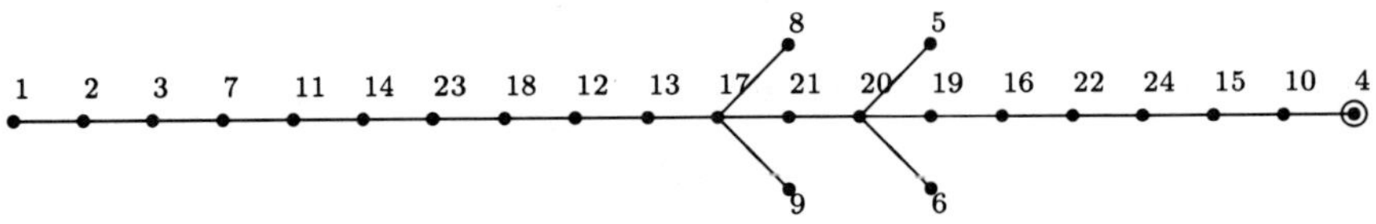

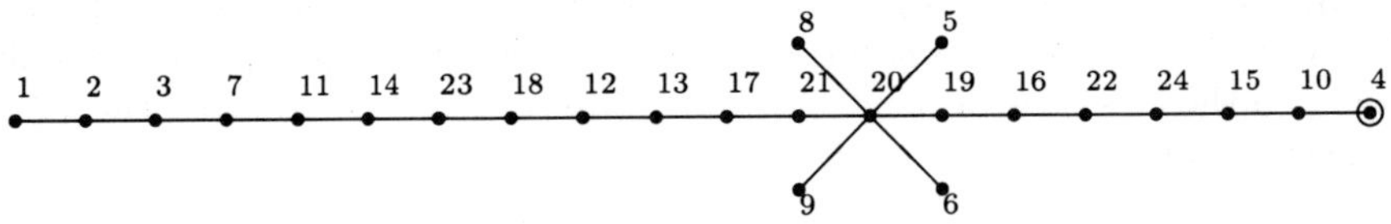

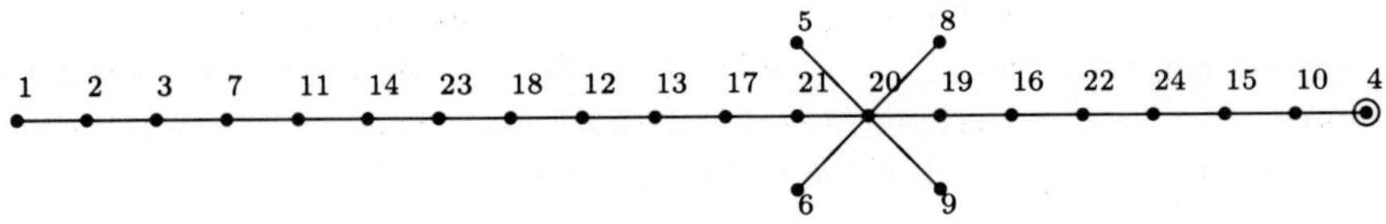

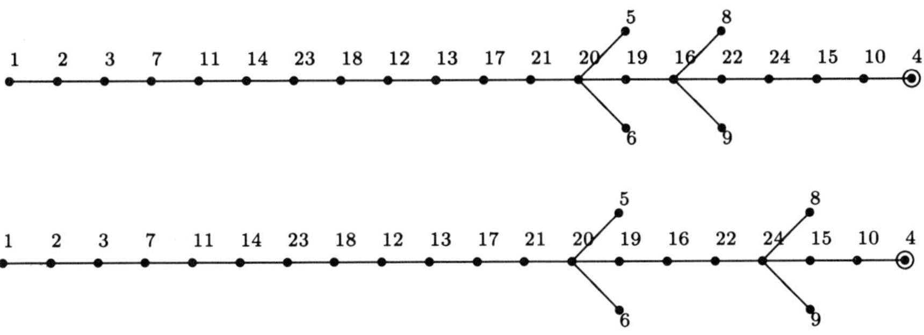

PROJECTIVES:

Nr.	1	2	3	4	5	6	7	8	9	10	11	12	13	14	15	16	17	18	19	20	21	22	23	24
CC	r	r	r	r	6	5	r	9	8	r	r	r	r	r	r	r	r	r	r	r	r	r	r	r
N&C	×	∘	×	∘	∘	∘	∘	∘	∘	×	×	×	∘	∘	∘	×	×	∘	∘	×	∘	∘	×	×
$\chi_2 \otimes \chi_2$	1	1																						
$\chi_9 \otimes \chi_2$		1	2				1																	
$\chi_{32} \otimes \chi_2$			1				2				1													
$\chi_{84} \otimes \chi_2$				1						5	3		1	7	7		4	2	2	7	8	3	6	6
$\chi_{52} \otimes \chi_2$					1															1				
$\chi_{69} \otimes \chi_2$							2			2	5			3	3			1	1	2	1		1	1
$\chi_{73} \otimes \chi_2$							1				3			2	2	1	1	2	2	2	2		2	2
$\chi_{161} \otimes \chi_2$								1	1	4	4	4	21	33	33	66	71	79	79	99	107	104	105	105
$\chi_3 \otimes \chi_{37}$										3	3			5	5		1			1	2		2	2
$\chi_{63} \otimes \chi_2$											1			1	1			1	1	1			1	1
$\chi_{91} \otimes \chi_2$												2	2	1	1	10	3	8	8	6	5	12	8	8
$\chi_{67} \otimes \chi_2$												1				2		1	1			1		
$\chi_{65} \otimes \chi_2$																1						1		

There are three pairs of algebraic conjugate characters in this block, namely the characters $\chi_{137}, \chi_{138}, \chi_{169}, \chi_{170}$ and χ_{178}, χ_{179} corresponding to the nodes 14, 15, 18, 19, 23 and 24. Because the irrationalities of these pairs lie in disjoint fields, we can assume without loss of generality that the nodes 14, 18 and 23 come first on the real stem. We can also assume that the nodes 5 and 8 lie on the upper half of the tree. Under these assumptions we get the following embedded trees consistent with the projectives given above:

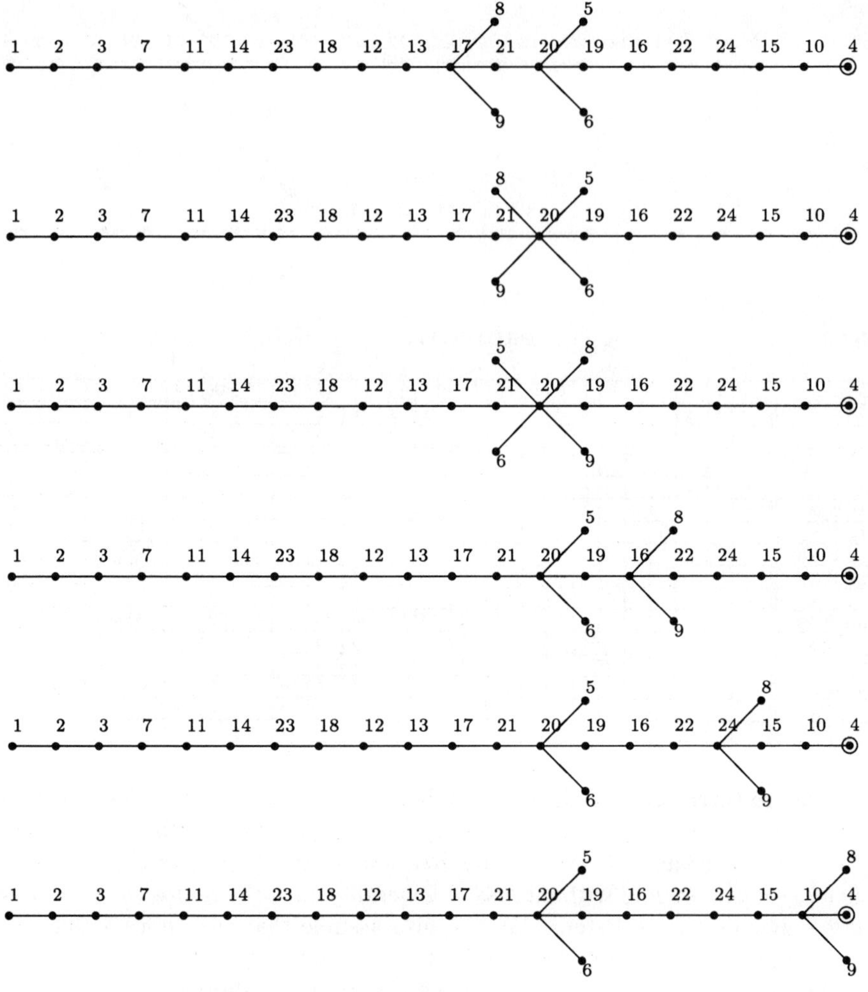

We rule out the last tree by the following argument: The tensor product

$$\chi_3 \otimes \chi_{35} \approx \chi_{158} + \chi_{159} + \chi_{169} + \chi_{170} + \chi_{175} + \chi_{176} + \chi_{177} + \chi_{178} + \chi_{179}$$

or in term of nodes

$$2 \otimes 8 \approx 16 + 17 + 18 + 19 + 20 + 21 + 22 + 23 + 24$$

should contain the node which has Green correspondent $46A1 \otimes 46A11 = 1A11$. This is the node 10, which obviously does not occur in that tensor product. So we are finally left with the five trees given above.

Group: 2BM Prime: 47 Block: 2

Nr.	CAS-Nr.	Degree	CC	N&C
1	186	10506240	r	×
2	187	410132480	r	∘
3	190	53936390144	191	×
4	191	53936390144	190	×
5	193	772566552576	r	×
6	199	23459577856000	200	∘
7	200	23459577856000	199	∘
8	204	60863961600000	r	∘
9	205	77683916800000	206	×
10	206	77683916800000	205	×
11	210	168430710030336	211	×
12	211	168430710030336	210	×
13	217	447224633856000	218	×
14	218	447224633856000	217	×
15	223	2642676197359616	224	∘
16	224	2642676197359616	223	∘
17	229	4215392896000000	230	∘
17	230	4215392896000000	229	∘
18	231	4250243698421760	r	×
19	232	4490309824000000	233	×
20	233	4490309824000000	232	×
21	235	5191647233310720	236	∘
22	236	5191647233310720	235	∘
23	243	21065543276560384	r	∘
24	246	26438944243712000	r	×

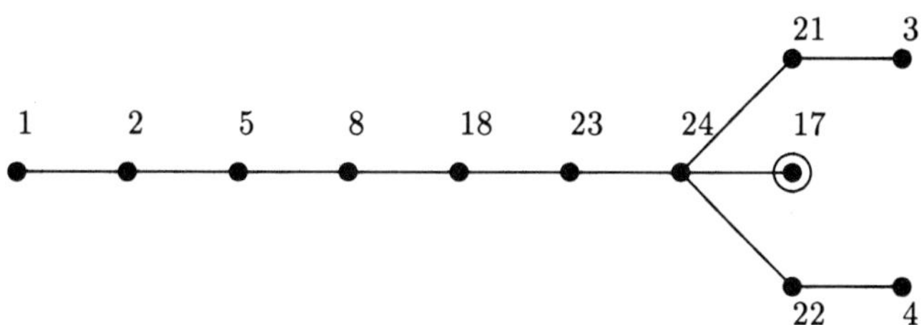

Warning! This is not the complete Brauer tree, but the maximal subtree we know.

PROJECTIVES:

Nr.	1	2	3	4	5	6	7	8	9	10	11	12	13	14	15	16	17	18	19	20	21	22	23	24
CC	r	r	4	3	r	7	6	r	10	9	12	11	14	13	16	15	r	r	20	19	22	21	r	r
N&C	×	∘	×	×	×	∘	∘	∘	×	×	×	×	×	×	∘	∘	∘	×	×	×	∘	∘	∘	×
$\chi_{187} \otimes \chi_2$	1	2			1																			
$\chi_{212} \otimes \chi_2$			1														2		2	2	5	4	4	10
$\chi_{196} \otimes \chi_2$					2			3										1						
$\chi_{222} \otimes \chi_2$						1	1						2	2	5	5	10	7	8	8	9	9	40	53
$\chi_{205} \otimes \chi_2$									1	1					1	1		1					2	1
$\chi_{210} \otimes \chi_2$											1				1	1	1		1	1	1	1	4	6
$\chi_{217} \otimes \chi_2$													1	1	1	1	3	1	2	2	4	4	10	16
$\chi_{214} \otimes \chi_2$															1	1	1	1	1	2	1	1	6	7

There are 7200 trees which are consistent with the projectives given above. The following part of the tree is nevertheless determined by them:

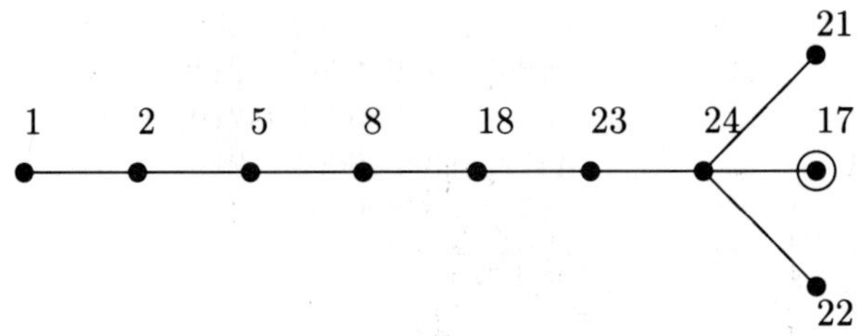

If the node 3_B (χ_{190}) has the Green correspondent $1Bx$, we derive from the tensor product

$$\chi_3 \otimes \chi_{190} \approx_B \chi_{235}$$

in terms of nodes

$$2_A \otimes 3_B \approx_B 21_B,$$

that the node 21_B has the Green correspondent $46Bx = 46A0 \otimes 1Bx$. This shows that the node 3_B must be joint to the node 21_B.

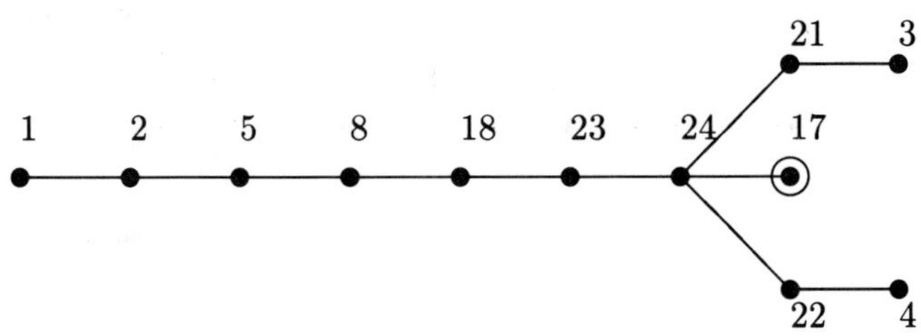

6.26 The Monster

Group: M Prime: 3 Block: 2

Nr.	CAS-Nr.	Degree	CC	N&C
1	26	3503434660075044981	27	×
2	105	6897632227448950059492 42	106	×
3	107	6897667261795550809942 23	108	∘

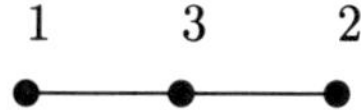

Group: M Prime: 3 Block: 3

Complex conjugate to Block 2.

Group: M Prime: 3 Block: 5

Nr.	CAS-Nr.	Degree	CC	N&C
1	101	600020772685064502392907	r	×
2	152	4200145408795451516750349 0	r	×
3	153	4260147486063957966989639 7	r	∘

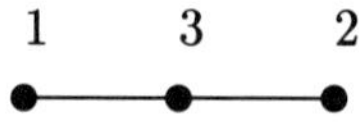

Group: M Prime: 3 Block: 6

Nr.	CAS-Nr.	Degree	CC	N&C
1	122	5334046162969208352215625	r	×
2	160	6375081284503582807900841	r	×
3	163	6908485900800503643122406	r	○

1 3

Group: M Prime: 3 Block: 7

Nr.	CAS-Nr.	Degree	CC	N&C
1	164	7461221352972038365477935	r	×
2	169	10335410424391272776309145	r	×
3	187	17796631777363311141787081	r	○

1 3

Group: M Prime: 5 Block: 5

Nr.	CAS-Nr.	Degree	CC	N&C
1	49	77609719227713750000	r	×
2	89	26079952410708376796875	90	×
2	90	26079952410708376796875	89	×
3	91	26157562129936090546875	r	○

1 3 2

Group: M Prime: 5 Block: 3

Nr.	CAS-Nr.	Degree	CC	N&C
1	36	77316619273928125000	r	×
2	97	392611651975065600000000	r	×
3	119	4004308274823270400000000	r	∘
4	120	4239315652979009728125000	r	∘
5	130	7850934959207940600000000	r	×

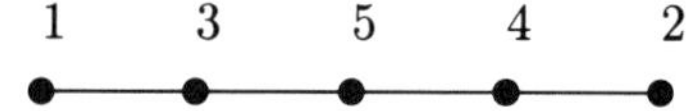

PROJECTIVES:

Nr.	1	2	3	4	5
CC	r	r	r	r	r
N&C	×	×	∘	∘	×
$\chi_2 \otimes \chi_{111}$		31		50	19
$\chi_2 \otimes \chi_{128}$			7	1	8

Group: M Prime: 5 Block: 4

Nr.	CAS-Nr.	Degree	CC	N&C
1	37	1304153504203429687 50	r	×
2	93	30333790382470158117 18750	r	×
3	134	94794957458053056531 25000	r	∘
4	190	20331426126115785227 4218750	r	∘
5	192	21249024755336572177 2656250	r	×

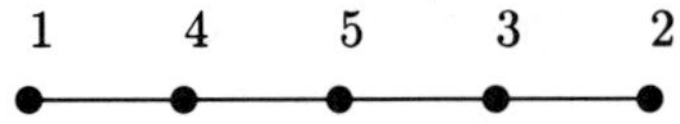

PROJECTIVES:

Nr.	1	2	3	4	5
CC	r	r	r	r	r
N&C	×	×	∘	∘	×
$\chi_2 \otimes \chi_{70}$			1	1	2

We have two trees which are consistent with the projective above. Unfortunately, we cannot apply Green correspondence, since 5 divides the order of the Monster more than once. So we have to use more elementary tricks.

We shall assume that the first tree is correct and arrive at a contradiction. Let 2_A denote the character χ_2 (although we do not know whether it lies in the principal block). The following tensor products will be used:

$$\begin{aligned} \chi_2 \otimes \chi_{49} &\approx_D \chi_{37} + 2\chi_{93} + \chi_{134} \\ \chi_2 \otimes \chi_{37} &\approx_E \chi_{49} \\ \chi_2 \otimes \chi_{134} &\approx_E \chi_{49} + (\chi_{89} + \chi_{90}) + 3\chi_{91} \end{aligned}$$

In terms of nodes these tensor products are as follows:

$$\begin{aligned} 2_A \otimes 1_E &\approx_D 1_D + 2_D^2 + 3_D \\ 2_A \otimes 1_D &\approx_E 1_E \\ 2_A \otimes 3_D &\approx_E 1_E + 2_E + 3_E^3 \end{aligned}$$

Here, as usual, the letters D and E are used to denote the Blocks 4 and 5 respectively.

Let M denote an indecomposable module with Brauer character $\hat{\chi}_2$, and let M^* be its dual. The irreducible modules in the present block are denoted by T, U, V and W according to the following picture:

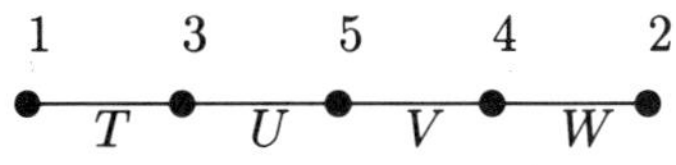

$$\overset{1}{\bullet} \underset{T}{\text{———}} \overset{3}{\bullet} \underset{U}{\text{———}} \overset{5}{\bullet} \underset{V}{\text{———}} \overset{4}{\bullet} \underset{W}{\text{———}} \overset{2}{\bullet}$$

Finally, the two irreducibles modules in Block 5 are called X and Y, where X corresponds to node 1_E.

We are going to find the structure of the component of $M^* \otimes X$ in the present block. From the second tensor product we know that $M \otimes T \approx_E X$. Thus

$$[M \otimes T, X] = [X, M \otimes T] = 1,$$

where $[L, N]$ denotes the dimension of the space of homomorphisms from L to N. It follows that

$$[T, M^* \otimes X] = [M^* \otimes X, T] = 1.$$

The knowledge of the indecomposable modules and the first tensor product now allows us to conclude

$$M^* \otimes X \approx_D \begin{matrix} T \\ U \\ T \end{matrix} \oplus W \oplus W,$$

where the first module on the right-hand side is the projective cover of T. In particular, U does not occur in the socle or in the head of $M^* \otimes X$. As above, it follows that X does not occur in the head or in the socle of $U \otimes M$.

The indecomposable modules in a block with a cyclic defect group have heads and socles which are multiplicity free. Hence every indecomposable summand of $U \otimes M$ in Block 5 is a factor module of the projective cover of Y. If in addition it contains X as a composition factor, it has to be the projective cover itself. Now the last two tensor products show that X is contained 3 times and Y is contained 5 times as a composition factor in $U \otimes M$ (note that node 2_E yields Y twice). This contradicts the structure of the projective cover of Y, which contains X just once but Y 3 times as a composition factor. Hence the second of the above trees must be correct.

Group: M Prime: 7 Block: 3

Nr.	CAS-Nr.	Degree	CC	N&C
1	19	39660520552077425	r	×
2	52	1353006807137391674268	r	×
3	76	8655148946923327384900	r	∘
4	151	41762322738385820195625000	r	×
5	168	889438206202883432616723393	r	∘
6	173	1249821560727476472572928000	r	∘
7	184	1722488523976517456534375000	r	×

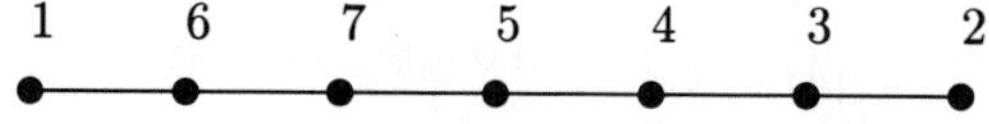

PROJECTIVES:

Nr.	1	2	3	4	5	6	7
CC	r	r	r	r	r	r	r
N&C	×	×	∘	×	∘	∘	×
$\chi_2 \otimes \chi_{65}$		8	11	3			
$\chi_2 \otimes \chi_{73}$			1	7	6		
$\chi_2 \otimes \chi_{83}$					1	8	9

Group: M Prime: 7 Block: 4

Nr.	CAS-Nr.	Degree	CC	N&C
1	20	60359800576579350	r	×
2	46	643356925889917747200	r	×
3	80	146575737439884098045700	r	∘
4	96	391009081837477378329600	r	∘
5	119	4004308274823270400000000	r	×
6	121	4926670174323484069683200	r	×
7	131	8394037047155083487634450	r	∘

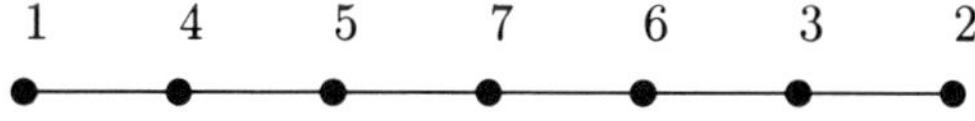

PROJECTIVES:

Nr.	1	2	3	4	5	6	7
CC	r	r	r	r	r	r	r
N&C	×	×	∘	∘	×	×	∘
$\chi_2 \otimes \chi_{65}$		5	10			9	4
$\chi_2 \otimes \chi_{109}$			4	1	2	15	12
$\chi_2 \otimes \chi_{73}$			1			3	2
$\chi_2 \otimes \chi_{87}$					1		1

Group: M | Prime: 7 | Block: 5

Nr.	CAS-Nr.	Degree	CC	N&C
1	22	290568421805921077	r	×
2	31	28585990950721640625	r	×
3	95	351532203382732066094400	r	∘
4	141	16109407269221032565630370	r	×
5	154	43527130990147981755651072	r	∘
6	158	58437394633227526183321600	r	∘
7	167	86206621680977834911875000	r	×

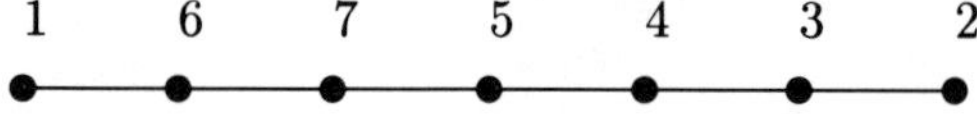

PROJECTIVES:

Nr.	1	2	3	4	5	6	7
CC	r	r	r	r	r	r	r
N&C	×	×	∘	×	∘	∘	×
$\chi_2 \otimes \chi_{65}$		2	3	1		2	2
$\chi_2 \otimes \chi_{73}$			1	4	4	3	4
$\chi_2 \otimes \chi_{51}$					1		1
$\chi_2 \otimes \chi_{83}$						1	1

We have the following two trees consistent with the above set of projectives:

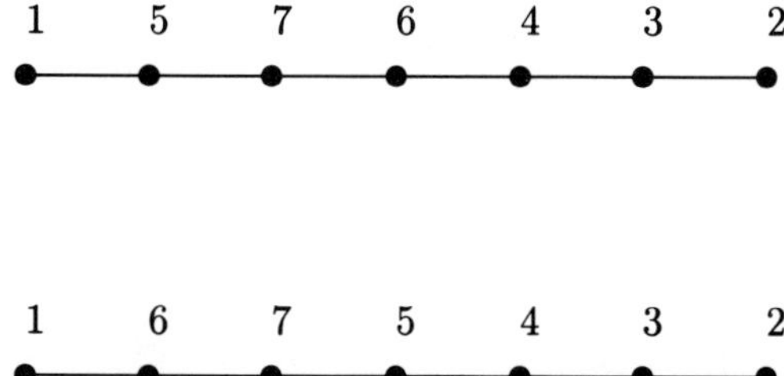

To conclude that the second tree is correct we use the information of the next block. In Block 6, node 1 is joined to node 3. In other words, $\chi_{32} + \chi_{62}$ is a projective character. We have

$$\chi_2 \otimes (\chi_{32} + \chi_{62}) \approx 2\chi_{22} + 6\chi_{31} + 7\chi_{95} + \chi_{141} + 2\chi_{158}.$$

In terms of nodes

$$1^2 + 2^6 + 3^7 + 4 + 6^2$$

is projective. This excludes the first tree.

Group: M Prime: 7 Block: 6

Nr.	CAS-Nr.	Degree	CC	N&C
1	32	30815545786259524745	r	×
2	47	691170144025469730622	48	∘
2	48	691170144025469730622	47	∘
3	62	6566555764392010419123	r	∘
4	63	7226910362631220625000	r	×

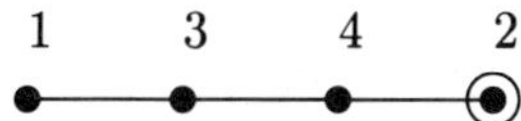

PROJECTIVES:

Nr.	1	2	3	4
CC	r	r	r	r
N&C	×	∘	∘	×
$\chi_3 \otimes \chi_{65}$	3		44	41

Group: M Prime: 11 Block: 2

Nr.	CAS-Nr.	Degree	CC	N&C
1	5	18538750076	r	×
2	23	336041615485626050	r	×
3	30	875419382211257812 5	r	∘
4	92	277540481294528814140625	r	×
5	99	597787522207315571077947	100	∘
6	100	597787522207315571077947	99	∘
7	116	328251054028363144217510 4	r	∘
8	117	353729279653874141507490 0	r	∘
9	130	785093495920794060000000 0	r	×
10	134	947949574580530565312500 0	r	×
11	137	959258438691858297965772 8	r	∘

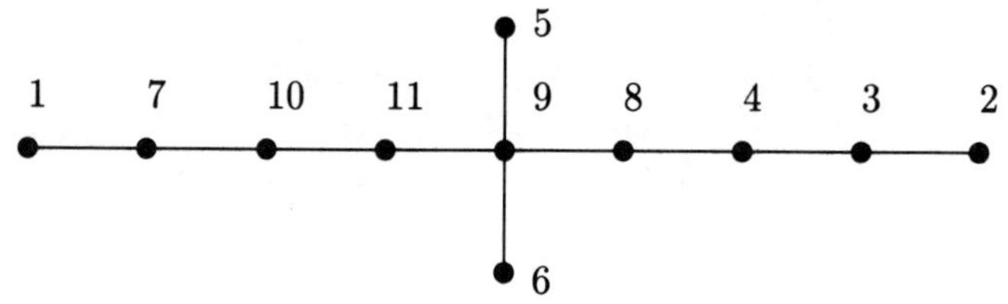

PROJECTIVES:

Nr.	1	2	3	4	5	6	7	8	9	10	11
CC	r	r	r	r	6	5	r	r	r	r	r
N&C	×	×	∘	×	∘	∘	∘	∘	×	×	∘
$\chi_3 \otimes \chi_{25}$		6	8	2			1	1	1	1	
$\chi_2 \otimes \chi_{52}$		3	5	4			2	2		2	
$\chi_2 \otimes \chi_{61}$		1	1	3			2	3		3	1
$\chi_2 \otimes \chi_{122}$					1	1	3	4	10	10	11
$\chi_2 \otimes \chi_{124}$					1	1	3	1	4	4	2
$\chi_2 \otimes \chi_{50}$							1	1	1	1	
$\chi_2 \otimes \chi_{69}$								1	1		
$\chi_2 \otimes \chi_{102}$									1	1	2

Group: M Prime: 11 Block: 3

Nr.	CAS-Nr.	Degree	CC	N&C
1	7	293553734298	r	×
2	19	39660520552077425	r	×
3	33	31569817307122699605	r	∘
4	66	19795913912408993711352	r	×
5	95	351532203382732066094400	r	∘
6	120	4239315652979009728125000	r	×
7	141	16109407269221032565630370	r	∘
8	151	41762322738385820195625000	r	∘
9	171	121170799240938738783416925	r	×
10	178	136107644194473772613203125	r	×
11	190	203314261261157852274218750	r	∘

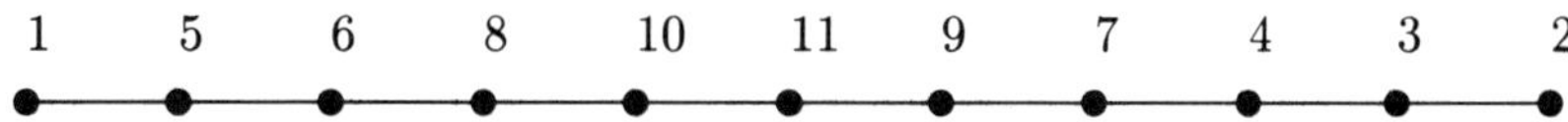

PROJECTIVES:

Nr.	1	2	3	4	5	6	7	8	9	10	11
CC	r	r	r	r	r	r	r	r	r	r	r
N&C	×	×	∘	×	∘	×	∘	∘	×	×	∘
$\chi_3 \otimes \chi_{25}$	1	5	9	4	2	1					
$\chi_2 \otimes \chi_{25}$		2	3	1							
$\chi_2 \otimes \chi_{80}$			2	7	12	22	10	14	6	4	1
$\chi_2 \otimes \chi_{79}$			1	5	5	15	9	14	5	4	
$\chi_2 \otimes \chi_{88}$							1	3	8	9	13
$\chi_2 \otimes \chi_{81}$								1	4	5	8
$\chi_2 \otimes \chi_{74}$									2	2	4

Group: M Prime: 11 Block: 4

Nr.	CAS-Nr.	Degree	CC	N&C
1	8	3879214937598	r	×
2	15	2374124840062976	r	×
3	37	130415350420342968750	r	∘
4	64	10145274012943412428800	r	∘
5	65	12810005542623250817856	r	×
6	97	392611651975065600000000	r	×
7	133	9416031858681585751556096	r	∘
8	140	14930164283563048960000000	r	∘
9	145	29734941419909382162874368	r	×
10	155	50572542024949598403750000	r	×
11	157	56356433273146675005489152	r	∘

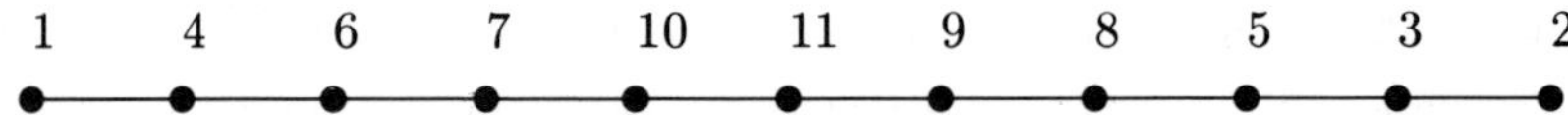

PROJECTIVES:

Nr.	1	2	3	4	5	6	7	8	9	10	11
CC	r	r	r	r	r	r	r	r	r	r	r
N&C	×	×	∘	∘	×	×	∘	∘	×	×	∘
$\chi_3 \otimes \chi_{25}$	1	3	8	4	5	3					
$\chi_2 \otimes \chi_{25}$		1	2		1						
$\chi_2 \otimes \chi_{52}$			7	5	8	7	2	2	1		
$\chi_2 \otimes \chi_{101}$			1	6	4	16	25	30	29	19	6
$\chi_2 \otimes \chi_{79}$			1	2	6	12	14	16	11	4	
$\chi_2 \otimes \chi_{94}$				1	7	13	22	25	18	11	1
$\chi_2 \otimes \chi_{81}$									1	1	2

Group: M | Prime: 11 | Block: 5

Nr.	CAS-Nr.	Degree	CC	N&C
1	11	190292345709543	r	×
2	13	1044868466775133	r	×
3	28	36057187535969531 25	r	∘
4	29	84568363435803104 00	r	×
5	58	2382987417506242421875	r	∘
6	67	21803647757861753437500	r	×
7	70	41209556844092914062500	r	∘
8	87	218028402153522030021875	r	×
9	111	1599110387863558882812500	r	∘
10	114	2216343020913351966796875	r	∘
11	118	3619209050774375426792424	r	×

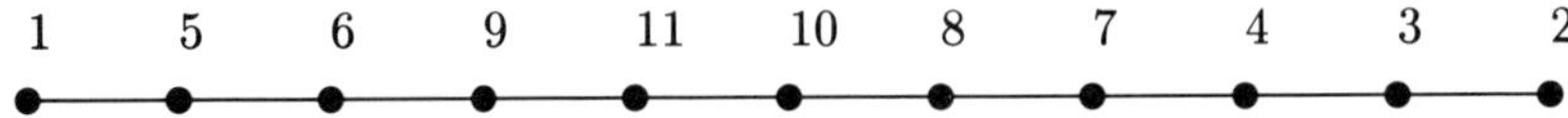

PROJECTIVES:

Nr.	1	2	3	4	5	6	7	8	9	10	11
CC	r	r	r	r	r	r	r	r	r	r	r
N&C	×	×	∘	×	∘	×	∘	×	∘	∘	×
$\chi_3 \otimes \chi_{25}$	2	1	6	5	10	9			2		1
$\chi_2 \otimes \chi_{25}$	1		1	1	3	2					
$\chi_3 \otimes \chi_{52}$		1	18	17	48	73			64	1	40
$\chi_2 \otimes \chi_{52}$			4	4	9	13			7		3
$\chi_2 \otimes \chi_{61}$			2	2	3	7			9		5
$\chi_2 \otimes \chi_{94}$				1	2	6	1		24	6	26
$\chi_4 \otimes \chi_{35}$				1	1	1	1		8	6	14
$\chi_2 \otimes \chi_{50}$					2	2			1		1
$\chi_2 \otimes \chi_{109}$							1	2	11	8	18
$\chi_2 \otimes \chi_{35}$										1	1

Group: M Prime: 11 Block: 6

Nr.	CAS-Nr.	Degree	CC	N&C
1	12	222879856734249	r	×
2	43	37991382469431237017 6	r	○
3	57	2351753641814605348320	r	×
4	98	4335286945605989785251 84	r	○
5	135	9592298143650890255171584	136	○
5	136	9592298143650890255171584	135	○
6	138	10023854998171489083984375	r	×

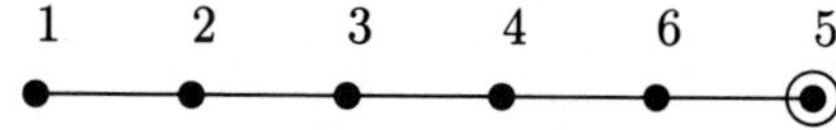

PROJECTIVES:

Nr.	1	2	3	4	5	6
CC	r	r	r	r	r	r
N&C	×	○	×	○	○	×
$\chi_2 \otimes \chi_{25}$	1	2	1			
$\chi_2 \otimes \chi_{61}$		4	5	1		

Group: M Prime: 11 Block: 7

Nr.	CAS-Nr.	Degree	CC	N&C
1	20	60359800576579350	r	×
2	24	2500435234254428856	r	∘
3	38	15594307673918258285 0	r	×
4	41	28624326769272448614 4	42	∘
4	42	28624326769272448614 4	41	∘
5	46	64335692588991774720 0	r	∘
6	49	77609719227713750000 0	r	×

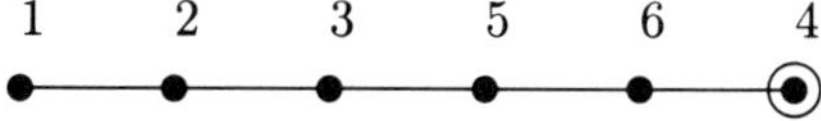

PROJECTIVES:

Nr.	1	2	3	4	5	6
CC	r	r	r	r	r	r
N&C	×	∘	×	∘	∘	×
$\chi_2 \otimes \chi_{52}$	1	4	10		9	2
$\chi_2 \otimes \chi_{25}$	1	3	3		2	1
$\chi_2 \otimes \chi_{80}$		1	5		6	2

Group: M | Prime: 13 | Block: 2

Nr.	CAS-Nr.	Degree	CC	N&C
1	4	842609326	r	×
2	15	2374124840062976	r	×
3	22	290568421805921077	r	∘
4	64	10145274012943412428800	r	×
5	66	19795913912408993711352	r	∘
6	79	115192831837135016250000	r	×
7	95	351532203382732066094400	r	∘
8	123	5514132424881463208443904	r	×
9	128	7567151576542452425781250	129	×
10	129	7567151576542452425781250	128	×
11	142	22626621365160537099927552	r	∘
12	161	64326163427522624205703125	r	∘
13	162	66550339514356152000000000	r	×

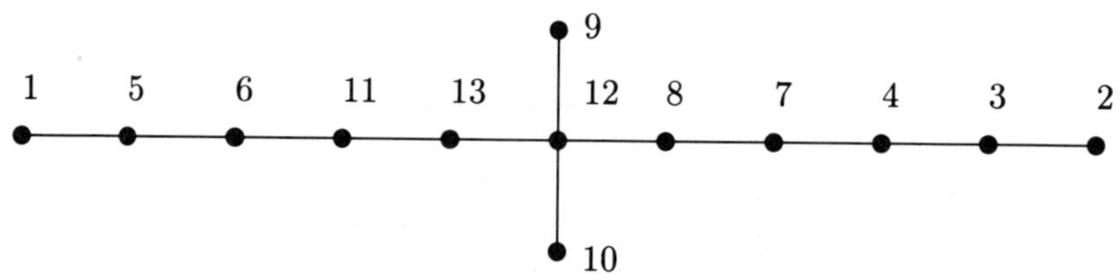

PROJECTIVES:

Nr.	1	2	3	4	5	6	7	8	9	10	11	12	13
CC	r	r	r	r	r	r	r	r	10	9	r	r	r
N&C	×	×	∘	×	∘	×	∘	×	×	×	∘	∘	×
$\chi_3 \otimes \chi_{18}$	1	8	9	1	1								
$\chi_2 \otimes \chi_{18}$		3	3										
$\chi_2 \otimes \chi_{32}$			1	2			1						
$\chi_2 \otimes \chi_{68}$				3	4	7	5	2			7	1	5
$\chi_2 \otimes \chi_{57}$				2	4	5	2				2		1
$\chi_3 \otimes \chi_{34}$				1	2	4	1				3		1
$\chi_2 \otimes \chi_{96}$				1			2	1	2	2	1	8	5
$\chi_2 \otimes \chi_{91}$							1	2			5	4	8

Group: M Prime: 13 Block: 3

Nr.	CAS-Nr.	Degree	CC	N&C
1	6	19360062527	r	×
2	9	36173193327999	r	×
3	33	31569817307122699605	r	∘
4	58	238298741750624242187 5	r	∘
5	62	656655576439201041912 3	r	×
6	63	722691036263122062500 0	r	×
7	109	10376058869846974817553 04	r	∘
8	118	36192090507743754267924 24	r	∘
9	151	41762322738385820195625 000	r	×
10	168	88943820620288343261672 393	r	×
11	172	12405838559302147118832 0256	r	∘
12	185	17386530525197214044726 5625	r	×
13	186	17586762698879416222700 8203	r	∘

1 — 4 — 5 — 7 — 10 — 11 — 12 — 13 — 9 — 8 — 6 — 3 — 2

PROJECTIVES:

Nr.	1	2	3	4	5	6	7	8	9	10	11	12	13
CC	r	r	r	r	r	r	r	r	r	r	r	r	r
N&C	×	×	∘	∘	×	×	∘	∘	×	×	∘	×	∘
$\chi_3 \otimes \chi_{18}$	1	4	6	3	2	2							
$\chi_2 \otimes \chi_{23}$		1	2	1	1	1							
$\chi_2 \otimes \chi_{68}$			1	5	7	7	3	10	4	2	1		
$\chi_2 \otimes \chi_{32}$			1	2	2	1							
$\chi_2 \otimes \chi_{73}$							3	6	7	6	3		1
$\chi_2 \otimes \chi_{88}$							1		3	8	8	5	7
$\chi_2 \otimes \chi_{74}$										1	2	4	3

Group: M Prime: 13 Block: 4

Nr.	CAS-Nr.	Degree	CC	N&C
1	8	3879214937598	r	×
2	25	2986480825407204125	r	∘
3	61	5578077210155766091776	r	×
4	94	331150814995116217581480	r	∘
5	124	5514132424881463208443904	125	×
5	125	5514132424881463208443904	124	×
6	143	24546384719289825598186695	r	×
7	145	29734941419909382162874368	r	∘

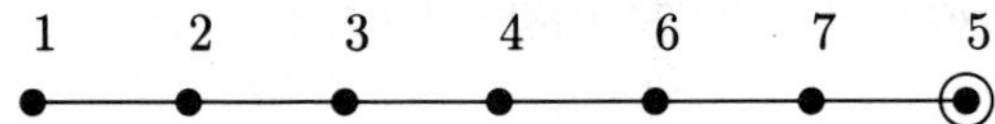

PROJECTIVES:

Nr.	1	2	3	4	5	6	7
CC	r	r	r	r	r	r	r
N&C	×	∘	×	∘	×	×	∘
$\chi_2 \otimes \chi_{23}$	1	2	1				
$\chi_2 \otimes \chi_{65}$		1	6	7		5	3
$\chi_2 \otimes \chi_{57}$		1	5	5		2	1
$\chi_2 \otimes \chi_{73}$			1	5		6	2

Group: M　　　　Prime: 13　　　　Block: 5

Nr.	CAS-Nr.	Degree	CC	N&C
1	12	222879856734249	r	×
2	13	1044868466775133	r	×
3	19	39660520552077425	r	∘
4	21	251098487132187500	r	×
5	30	8754193822112578125	r	∘
6	35	49609712911192813665	r	∘
7	52	1353006807137391674268	r	×
8	70	41209556844092914062500	r	×
9	77	91068387388302451493925	r	∘
10	117	3537292796538741415074900	r	×
11	138	10023854998171489083984375	r	∘
12	175	129572518017902934396764160	r	∘
13	178	136107644194473772613203125	r	×

1 — 3 — 4 — 6 — 8 — 11 — 13 — 12 — 10 — 9 — 7 — 5 — 2

PROJECTIVES:

Nr.	1	2	3	4	5	6	7	8	9	10	11	12	13
CC	r	r	r	r	r	r	r	r	r	r	r	r	r
N&C	×	×	∘	×	∘	∘	×	×	∘	×	∘	∘	×
$\chi_2 \otimes \chi_{18}$	2	1	3	1	2		1						
$\chi_2 \otimes \chi_{23}$	1	1	3	2	4		3						
$\chi_2 \otimes \chi_{32}$		1			1		1		1				
$\chi_2 \otimes \chi_{34}$				1		1	1		1				
$\chi_2 \otimes \chi_{65}$					2		8		9	4		1	
$\chi_2 \otimes \chi_{73}$						1		1	1	1		3	3
$\chi_2 \otimes \chi_{115}$						1		1		7	2	98	93
$\chi_2 \otimes \chi_{147}$								2		38	77	1120	1157
$\chi_2 \otimes \chi_{98}$								1	2	5	2	18	16
$\chi_2 \otimes \chi_{96}$									1	2	2	10	11
$\chi_2 \otimes \chi_{69}$										1	1	1	1
$\chi_2 \otimes \chi_{102}$											2	19	21

Group: M Prime: 17 Block: 1

Nr.	CAS-Nr.	Degree	CC	N&C
1	1	1	r	×
2	14	1109944460516150	r	×
3	21	251098487132187500	r	○
4	53	1480279477146615234375	54	×
5	54	1480279477146615234375	53	×
6	63	7226910362631220625000	r	×
7	71	42940402913709544921875	72	×
8	72	42940402913709544921875	71	×
9	80	146575737439884098045700	r	○
10	96	391009081837477378329600	r	○
11	133	9416031858681585751556096	r	×
12	152	42001454087954515167503490	r	○
13	155	50572542024949598403750000	r	×
14	156	51324350389558097414062500	r	×
15	170	115165062362004433625000000	r	○
16	175	129572518017902934396764160	r	○
17	186	175867626988794162227008203	r	×

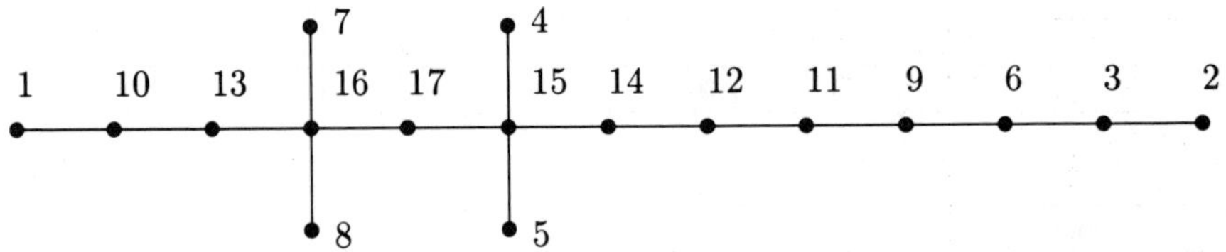

PROJECTIVES:

Nr.	1	2	3	4	5	6	7	8	9	10	11	12	13	14	15	16	17
CC	r	r	r	5	4	r	8	7	r	r	r	r	r	r	r	r	r
N&C	×	×	○	×	×	×	×	×	○	○	×	○	×	×	○	○	×
$\chi_2 \otimes \chi_{24}$		1	2			2			1								
$\chi_2 \otimes \chi_9$		1	1														
$\chi_2 \otimes \chi_{166}$				1	1		2	2		5	193	867	1052	1036	2283	2687	3555
$\chi_2 \otimes \chi_{64}$						5			8	1	4	1	2			1	
$\chi_2 \otimes \chi_{147}$							2	2	1		94	363	445	397	839	1120	1383
$\chi_2 \otimes \chi_{115}$							1	1	2		17	31	45	21	39	98	85
$\chi_2 \otimes \chi_{126}$										4	16	70	84	92	231	212	325
$\chi_2 \otimes \chi_{70}$											1	1	1			2	1
$\chi_2 \otimes \chi_{69}$												1		1	1	1	2
$\chi_2 \otimes \chi_{74}$													1		2	2	3
$\chi_2 \otimes \chi_{55}$														1	1		
$\chi_2 \otimes \chi_{59}$															1		1

The above projectives determine the real stem of the tree. It is as follows:

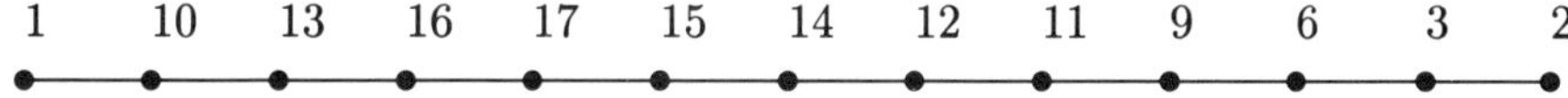

The tensor product of an ordinary irreducible character of the second block with an ordinary irreducible character of the principal block is projective when restricted to the principal block. This follows from the fact that the tensor product of the Green correspondents has no non-projective component in the principal block. Thus

$$\chi_2 \otimes \chi_{71} \approx \chi_{72} + 2\chi_{175} + \chi_{186}$$

is a projective character. In terms of nodes

$$8 + 16^2 + 17$$

is projective, showing that node 8 is joined to node 16. We have four possibilities for the location of nodes 4 and 5, leading to the following trees:

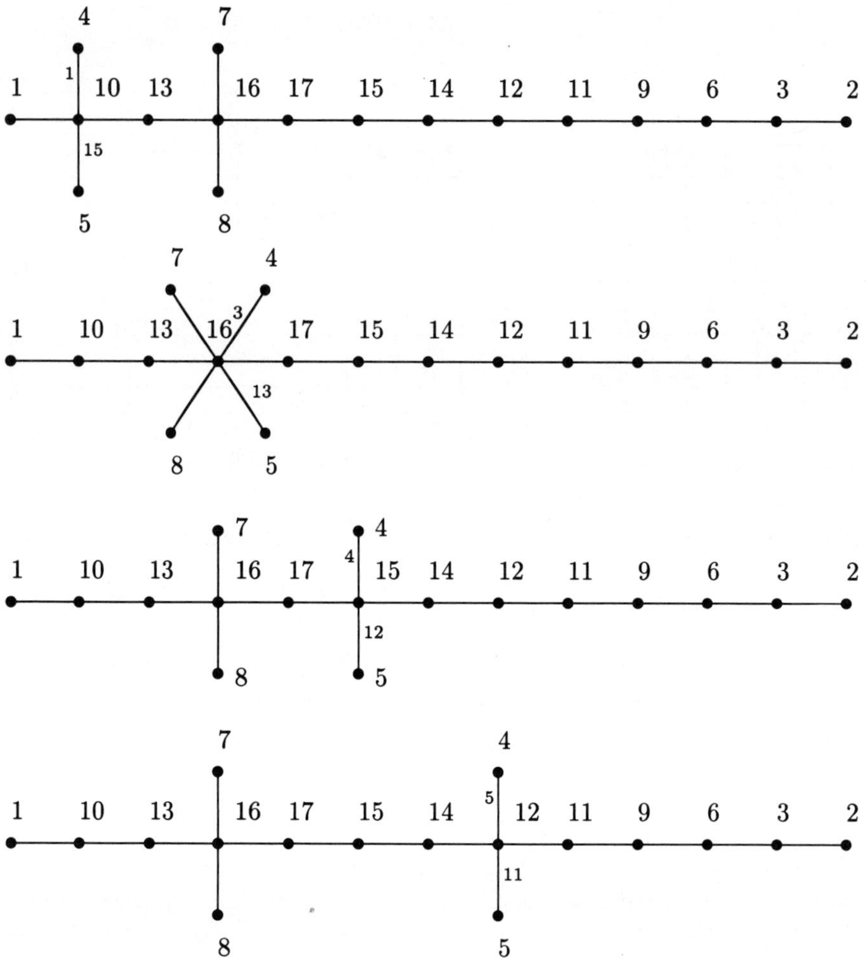

To show that the third tree is correct, we use the tensor product

$$\chi_2 \otimes \chi_{53} \approx_B \chi_{191}$$

and the tree for Block 2. The latter is obtained without using any information about the principal block. Now χ_{191} is node 17 in the second block. Let $1B0$ denote the Green correspondent of the irreducible module with Brauer character χ_2. Then the two edges incident to node 17 in Block 2 have labels 4 and 12. Let x label the edge incident to node 4 in the principal block. Then $1B0 \otimes 1Ax = 1Bx$. It follows that $x \in \{4, 12\}$. Hence the second of the above trees is correct. We remark that the irrationalities of nodes 4 and 7 are independent, so we can choose on ordering of 7 and 8.

Group: M Prime: 17 Block: 2

Nr.	CAS-Nr.	Degree	CC	N&C
1	2	196883	r	×
2	5	18538750076	r	×
3	20	60359800576579350	r	∘
4	25	2986480825407204125	r	×
5	32	30815545786259524745	r	∘
6	49	776097192277137500000	r	×
7	50	918438233727730974720	r	∘
8	91	261575621299360905468750	r	×
9	119	4004308274823270400000000	r	∘
10	122	5334046162969208352215625	r	×
11	124	5514132424881463208443904	125	×
12	125	5514132424881463208443904	124	×
13	130	7850934959207940600000000	r	∘
14	163	69084859008005036431224066	r	×
15	169	103354104243912727763091456	r	∘
16	187	177966317773633111417870812	r	∘
17	191	207467089840006711558593750	r	×

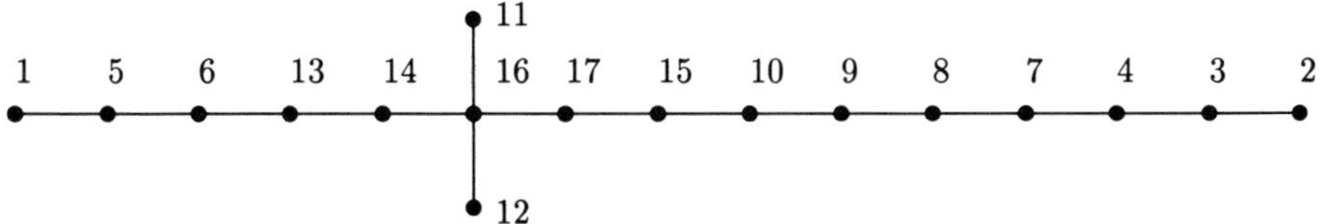

PROJECTIVES:

Nr.	1	2	3	4	5	6	7	8	9	10	11	12	13	14	15	16	17
CC	r	r	r	r	r	r	r	r	r	r	12	11	r	r	r	r	r
N&C	×	×	○	×	○	×	○	×	○	×	×	×	○	×	○	○	×
$\chi_4 \otimes \chi_9$	1	3	6	3	1												
$\chi_2 \otimes \chi_9$		1	1														
$\chi_3 \otimes \chi_{36}$			3	5	3	4	3	1					1				
$\chi_2 \otimes \chi_{24}$			2	3	1	1	1										
$\chi_2 \otimes \chi_{30}$			2	2	1	1											
$\chi_2 \otimes \chi_{76}$				1	1	3	1						4	2	1	1	2
$\chi_2 \otimes \chi_{66}$				1		1	1						2	1			
$\chi_2 \otimes \chi_{64}$					2	2											
$\chi_2 \otimes \chi_{68}$						1	1	1					2	1			
$\chi_2 \otimes \chi_{94}$							1	1					8	12	7	7	10
$\chi_2 \otimes \chi_{83}$								1	1					3	5	9	11
$\chi_2 \otimes \chi_{87}$									1	1			2	6	5	9	10
$\chi_3 \otimes \chi_{59}$										1	2	2		7	13	23	24
$\chi_2 \otimes \chi_{98}$										1			6	14	11	15	17
$\chi_2 \otimes \chi_{70}$													1	2	1	1	1
$\chi_2 \otimes \chi_{69}$													1	1	1	1	2
$\chi_2 \otimes \chi_{74}$														1	2	3	4

The following four trees are consistent with the above set of projectives.

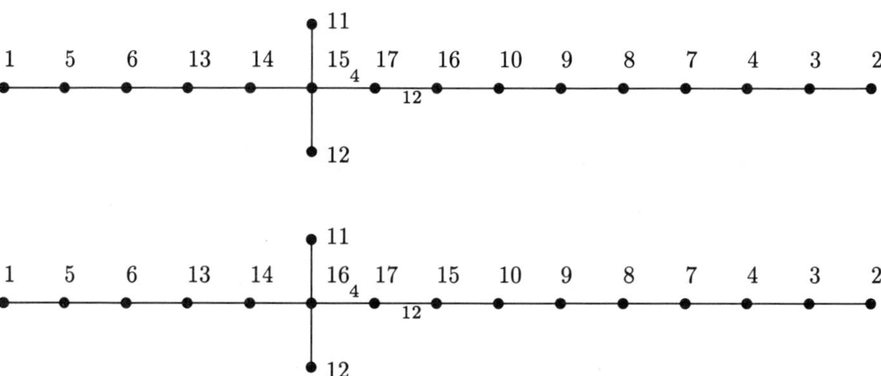

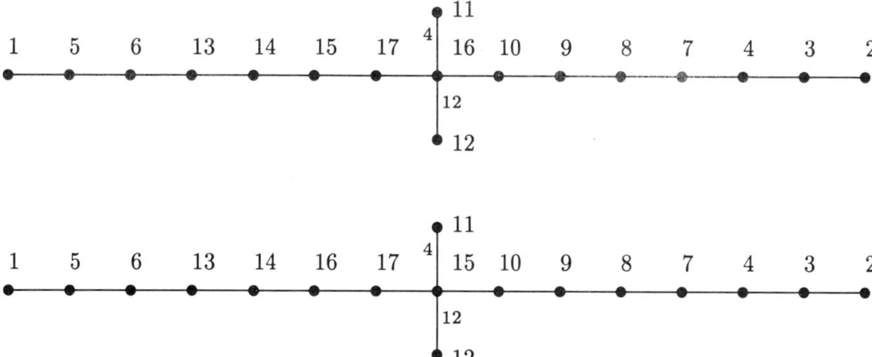

To show that the second possibility is correct we use the following tensor products:

$$\begin{array}{lll} \chi_3 \otimes \chi_8 & \approx_B & \chi_2 + 2\chi_5 + 2\chi_{20} + \chi_{25} \\ \chi_3 \otimes (\chi_{47} + \chi_{48}) & \approx_B & 2\chi_{130} + 6\chi_{163} + 2\chi_{169} + 6\chi_{187} + 6\chi_{191} \end{array}$$

In terms of nodes these tensor products are:

$$\begin{array}{lll} 1_C \otimes 1_E & \approx_B & 1_B + 2_B^2 + 3_B^2 + 4_B \\ 1_C \otimes 4_E & \approx_B & 13_B^2 + 14_B^6 + 15_B^2 + 16_B^6 + 17_B^6 \end{array}$$

Let $1B0$, $1C0$ resp. $1E0$ denote the Green correspondents of the first nodes of Blocks 2, 3 and 5. We want to determine the non-projective component of $1C0 \otimes 1E0$ in the (Brauer correspondent) of the second block. Since every direct summand of $1_C \otimes 1_E$ is a trivial source module, it follows from the first tensor product above that $1C0 \otimes 1E0 \approx_B 1B0 \oplus 1B8$. Now the indecomposable trivial source module in Block 5 with Brauer character $\chi_{47} + \chi_{48}$ has Green correspondent $1E4$. We thus have $1C0 \otimes 1E4 = (1C0 \otimes 1E0) \otimes 1A4 \approx_B 1B4 \oplus 1B12$. Since the nodes 11_B and 12_B do not occur at all in the above tensor product this immediately excludes the last two possibilities. Furthermore it shows that

$$13_B^2 + 14_B^6 + 15_B^2 + 16_B^6 + 17_B^4$$

is projective, so the second tree is correct.

Group: M Prime: 17 Block: 3

Nr.	CAS-Nr.	Degree	CC	N&C
1	3	21296876	r	×
2	7	293553734298	r	×
3	22	290568421805921077	r	∘
4	31	2858599095072164062 5	r	×
5	37	13041535042034296875 0	r	∘
6	52	135300680713739167426 8	r	×
7	78	114212876389603002704448	r	∘
8	79	115192831837135016250000	r	∘
9	140	1493016428356304896000000 0	r	×
10	141	16109407269221032565630370	r	×
11	144	27501917609709102247187500	r	∘
12	151	41762322738385820195625000	r	∘
13	153	42601474860639579669896397	r	×
14	154	43527130990147981755651072	r	×
15	172	124058385593021471188320256	r	∘
16	178	136107644194473772613203125	r	∘
17	192	212490247553365721772656250	r	×

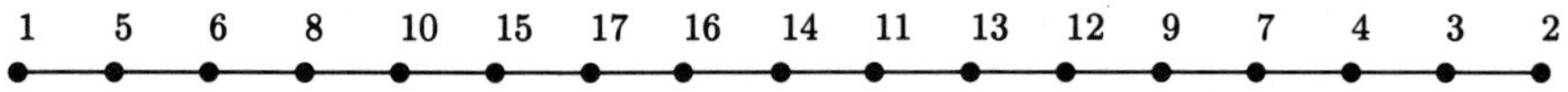

1 — 5 — 6 — 8 — 10 — 15 — 17 — 16 — 14 — 11 — 13 — 12 — 9 — 7 — 4 — 3 — 2

PROJECTIVES:

Nr.	1	2	3	4	5	6	7	8	9	10	11	12	13	14	15	16	17
CC	r	r	r	r	r	r	r	r	r	r	r	r	r	r	r	r	r
N&C	×	×	∘	×	∘	×	∘	∘	×	×	∘	∘	×	×	∘	∘	×
$\chi_3 \otimes \chi_9$	1	2	3	1	1												
$\chi_2 \otimes \chi_{18}$		1	3	2	1	1											
$\chi_3 \otimes \chi_{43}$			8	17	21	34	19	17	16	5	1	6		1	1		
$\chi_3 \otimes \chi_{36}$			6	11	14	19	10	6	5	1							
$\chi_2 \otimes \chi_{30}$			4	5	5	5	1										
$\chi_2 \otimes \chi_{43}$			2	4	5	7	3	2	1								
$\chi_2 \otimes \chi_{57}$				2	3	7	3	5	2	1		1					
$\chi_2 \otimes \chi_{68}$				1	2	6	5	7	8	4	2	4	1	1	1		
$\chi_2 \otimes \chi_{92}$				1	2	4	8	5	20	8	4	14	3	5	10	9	11
$\chi_2 \otimes \chi_{73}$								2	4	4	3	7	4	4	3	3	2
$\chi_2 \otimes \chi_{87}$									1	2	4	5	6	6	8	9	11
$\chi_2 \otimes \chi_{70}$									1	1	2	3	2	3	2	2	2
$\chi_2 \otimes \chi_{105}$										1	3	3	5	4	19	21	36
$\chi_2 \otimes \chi_{51}$											1			1			
$\chi_2 \otimes \chi_{69}$															1	1	2

We have the following two consistent trees:

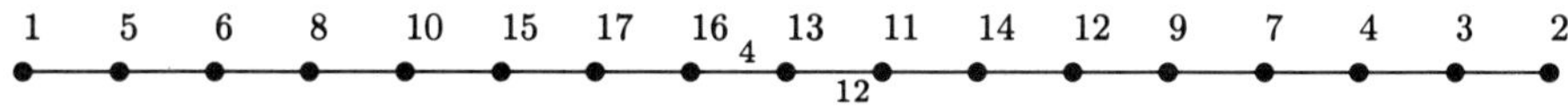

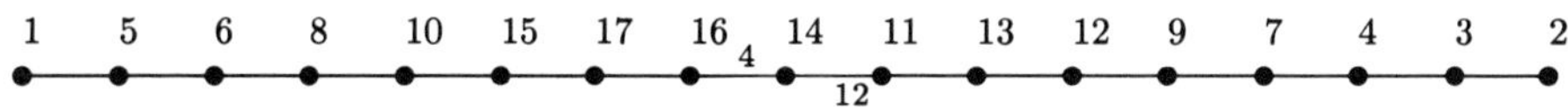

We use the same notation as in the proof for Block 2, and a similar argument involving the following tensor products:

$$\begin{array}{rcl} \chi_2 \otimes \chi_8 & \approx_C & \chi_3 + 2\chi_7 + \chi_{22} \\ \chi_2 \otimes (\chi_{47} + \chi_{48}) & \approx_C & 2\chi_{154} \end{array}$$

Again we list these tensor products in terms of nodes:

$$\begin{array}{rcl} 1_B \otimes 1_E & \approx_C & 1_C + 2_C^2 + 3_C \\ 1_B \otimes 4_E & \approx_C & 14_C^2 \end{array}$$

The first tensor product now shows that $1B0 \otimes 1E0 \approx_C 1C0 \oplus 1C8$. From this and the second tensor product it follows that the two edges incident to node 14_C must have the labels 4 and 12. Hence the second tree is correct.

Group: M Prime: 17 Block: 4

Nr.	CAS-Nr.	Degree	CC	N&C
1	4	842609326	r	×
2	6	19360062527	r	×
3	13	1044868466775133	r	∘
4	28	3605718753596953125	r	∘
5	38	155943076739182582850	r	×
6	46	643356925889917747200	r	×
7	62	6566555764392010419123	r	∘
8	77	91068387388302451493925	r	∘
9	101	600020772685064502392907	r	×
10	102	626877403613887304040448	103	×
11	103	626877403613887304040448	102	×
12	104	655159231073705404921875	r	×
13	131	8394037047155083487634450	r	∘
14	139	12650882100466187033706250	r	×
15	146	33684388830359981044531200	r	∘
16	148	37310715211546624000000000	r	∘
17	161	64326163427522624205703125	r	×

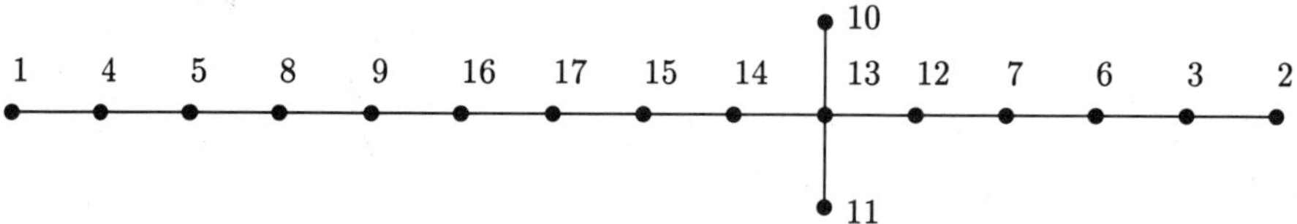

PROJECTIVES:

Nr.	1	2	3	4	5	6	7	8	9	10	11	12	13	14	15	16	17
CC	r	r	r	r	r	r	r	r	r	11	10	r	r	r	r	r	r
N&C	×	×	∘	∘	×	×	∘	∘	×	×	×	×	∘	×	∘	∘	×
$\chi_2 \otimes \chi_9$	1	1	1	1													
$\chi_2 \otimes \chi_{12}$		1	1														
$\chi_2 \otimes \chi_{18}$			1	2	2	1											
$\chi_2 \otimes \chi_{33}$				3	5	4	4	3	1								
$\chi_2 \otimes \chi_{57}$				2	6	5	6	7	3			1	1	1			
$\chi_2 \otimes \chi_{65}$				1	6	5	6	9	4			2	4	4	1		
$\chi_2 \otimes \chi_{64}$				1	2	4	7	7	6			5	2		1		1
$\chi_2 \otimes \chi_{92}$					1	3	6	10	11			9	12	8	9	2	7
$\chi_2 \otimes \chi_{109}$							1	3	5			3	12	17	16	12	19
$\chi_2 \otimes \chi_{114}$								1	3			2	14	21	25	24	38
$\chi_2 \otimes \chi_{73}$								1	1				2	3	1	1	1
$\chi_2 \otimes \chi_{126}$									1	1	1	1	10	17	63	64	116
$\chi_2 \otimes \chi_{70}$													1	1		1	1
$\chi_2 \otimes \chi_{69}$															1		1

There are the following three trees consistent with the set of pojectives given above:

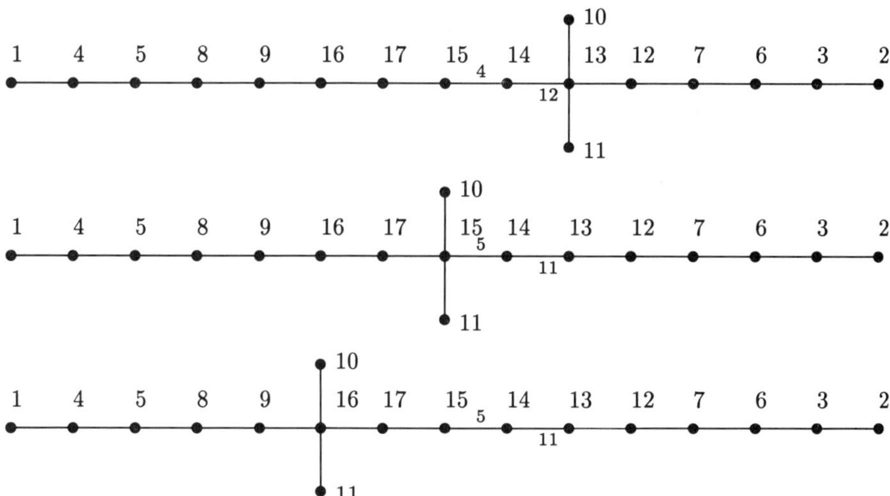

The argument used to rule out the last two trees is exactly the same as the one used in Block 3, so we only give the tensor products:

$$\begin{array}{lcl} \chi_2 \otimes \chi_8 & \approx_D & \chi_4 + 2\chi_6 + \chi_{13} \\ \chi_2 \otimes (\chi_{47} + \chi_{48}) & \approx_D & 2\chi_{139} \end{array}$$

These are, in terms of nodes,

$$\begin{array}{lcl} 1_B \otimes 1_E & \approx_D & 1_D + 2_D^2 + 3_D \\ 1_B \otimes 4_E & \approx_D & 14_D^2 \end{array}$$

Group: M Prime: 17 Block: 5

Nr.	CAS-Nr.	Degree	CC	N&C
1	8	3879214937598	r	×
2	15	2374124840062976	r	∘
3	29	8456836343580310400	r	×
4	47	691170144025469730622	48	×
4	48	691170144025469730622	47	×
5	61	5578077210155766091776	r	∘
6	111	1599110387863558882812500	r	×
7	113	2181694185821505680397072	r	∘
8	121	4926670174323484069683200	r	∘
9	123	5514132424881463208443904	r	×

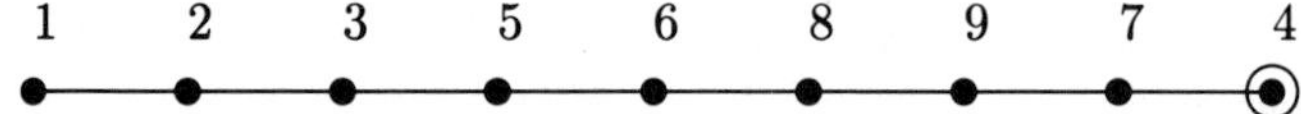

PROJECTIVES:

Nr.	1	2	3	4	5	6	7	8	9
CC	r	r	r	r	r	r	r	r	r
N&C	×	∘	×	×	∘	×	∘	∘	×
$\chi_2 \otimes \chi_{10}$	2	3	1						
$\chi_2 \otimes \chi_{57}$			3		5	6		4	
$\chi_2 \otimes \chi_{43}$			2		4	3		1	
$\chi_2 \otimes \chi_{66}$			1		3	8		7	1
$\chi_2 \otimes \chi_{34}$			1		2	1			
$\chi_2 \otimes \chi_{115}$				1		10	3	14	6
$\chi_2 \otimes \chi_{98}$					1	5	1	7	4
$\chi_2 \otimes \chi_{105}$							1		1
$\chi_2 \otimes \chi_{89}$								1	1

Group: M Prime: 19 Block: 1

Nr.	CAS-Nr.	Degree	CC	N&C
1	1	1	r	×
2	10	125510727015275	r	×
3	15	2374124840062976	r	○
4	26	35034346600750449 81	27	○
5	27	35034346600750449 81	26	○
6	31	285859909507216406 25	r	×
7	39	1723994342015933547 56	40	×
8	40	1723994342015933547 56	39	×
9	46	6433569258899177472 00	r	○
10	49	7760971922771375000 00	r	○
11	64	10145274012943412428 800	r	×
12	78	11421287638960300270 4448	r	○
13	79	11519283183713501625 0000	r	×
14	121	49266701743234840696 83200	r	×
15	133	94160318586815857515 56096	r	○
16	140	14930164283563048960 000000	r	○
17	158	58437394633227526183 321600	r	×
18	161	64326163427522624205 703125	r	×
19	169	10335410424391272776 3091456	r	○

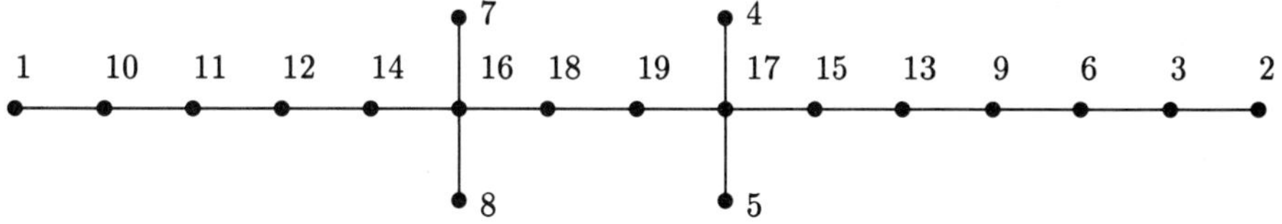

PROJECTIVES:

Nr.	1	2	3	4	5	6	7	8	9	10	11	12	13
CC	r	r	r	5	4	r	8	7	r	r	r	r	r
N&C	×	×	∘	∘	∘	×	×	×	∘	∘	×	∘	×
$\chi_7 \otimes \chi_7$	1	9	12			6			3	1			
$\chi_2 \otimes \chi_{22}$		1	3			4			2	1	1		
$\chi_2 \otimes \chi_7$		1	1										
$\chi_4 \otimes \chi_{171}$				1	1	3	25	25	104	117	1410	15065	15408
$\chi_4 \otimes \chi_{180}$				1	1	1	24	24	74	102	1404	16156	16289
$\chi_2 \otimes \chi_{43}$						4			5	1	4	3	2
$\chi_2 \otimes \chi_{65}$						2			5	1	3	5	6
$\chi_2 \otimes \chi_{92}$						1			3	2	6	8	5
$\chi_2 \otimes \chi_{147}$							1	1					1
$\chi_2 \otimes \chi_{50}$												1	
$\chi_2 \otimes \chi_{73}$													2
$\chi_2 \otimes \chi_{71}$													
$\chi_2 \otimes \chi_{69}$													

PROJECTIVES (continued):

Nr.	14	15	16	17	18	19
CC	r	r	r	r	r	r
N&C	×	∘	∘	×	×	∘
$\chi_7 \otimes \chi_7$						
$\chi_2 \otimes \chi_{22}$						
$\chi_2 \otimes \chi_7$						
$\chi_4 \otimes \chi_{171}$	630057	1198478	1897750	7395004	8129309	13059725
$\chi_4 \otimes \chi_{180}$	701141	1340544	2125746	8322276	9161342	14719877
$\chi_2 \otimes \chi_{43}$	1	1	1			
$\chi_2 \otimes \chi_{65}$	9	5	6	2		
$\chi_2 \otimes \chi_{92}$	19	16	20	17	7	6
$\chi_2 \otimes \chi_{147}$	53	94	148	511	519	844
$\chi_2 \otimes \chi_{50}$	1					
$\chi_2 \otimes \chi_{73}$	3	4	4	3	1	1
$\chi_2 \otimes \chi_{71}$				1		1
$\chi_2 \otimes \chi_{69}$					1	1

The projectives above determine the following real stem:

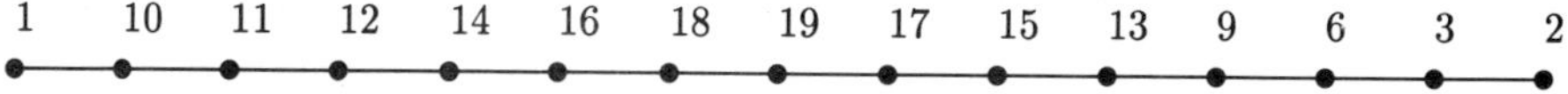

In addition they show that node 4 is joined to one of $\{7, 8, 11, 13, 14, 17, 18\}$ and node 7 to one of $\{15, 16, 19\}$. The proof that node 7 is indeed joined to node 16 and node 4 to 17 is given simultaneously with the proof for Block 2. Note that the irrationalities of the two pairs of complex conjugate characters are independent, so we do not have to worry about the planar embedding. Notice also that the edge incident to node 7 carries label 3, and the edge incident to node 4 has label 6.

Group: M Prime: 19 Block: 2

Nr.	CAS-Nr.	Degree	CC	N&C
1	2	196883	r	×
2	4	842609326	r	×
3	11	190292345709543	r	∘
4	25	2986480825407204125	r	∘
5	32	30815545786259524745	r	×
6	37	130415350420342968750	r	×
7	63	722691036263122062500	r	∘
8	66	19795913912408993711352	r	×
9	70	41209556844092914062500	r	∘
10	88	220326476909636307378168	r	×
11	104	655159231073705404921875	r	∘
12	107	689766726179555080994223	108	∘
13	108	689766726179555080994223	107	∘
14	145	29734941419909382162874368	r	×
15	172	124058385593021471188320256	r	∘
16	175	129572518017902934396764160	r	∘
17	187	177966317773633111417870812	r	∘
18	189	200390867219082687273984375	r	×
19	190	203314261261157852274218750	r	×

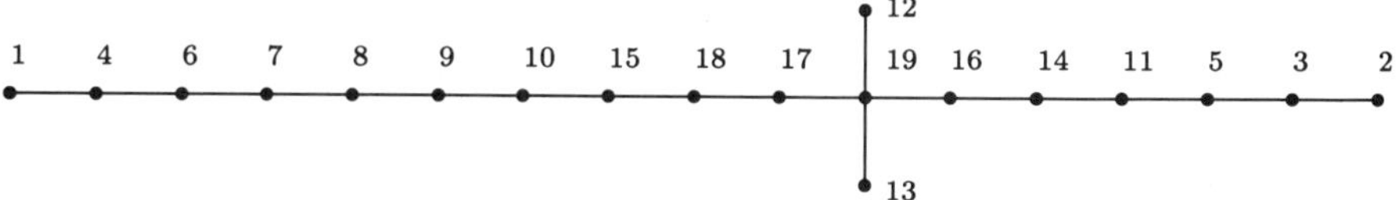

PROJECTIVES:

Nr.	1	2	3	4	5	6	7	8	9	10	11	12	13	14	15	16	17	18	19
CC	r	r	r	r	r	r	r	r	r	r	r	13	12	r	r	r	r	r	r
N&C	×	×	∘	∘	×	×	∘	×	∘	×	∘	∘	∘	×	∘	∘	∘	×	×
$\chi_3 \otimes \chi_7$	1	1	1	1															
$\chi_2 \otimes \chi_7$		1	1																
$\chi_2 \otimes \chi_{24}$			1	3	1	4	2	1											
$\chi_2 \otimes \chi_{20}$			1	1	1	1													
$\chi_2 \otimes \chi_{52}$				2	1	7	9	4			2			1					
$\chi_2 \otimes \chi_{62}$				1	2	6	9	4			5			3					
$\chi_2 \otimes \chi_{65}$				1		3	6	4			2			3		1			
$\chi_2 \otimes \chi_{96}$					1						2			4	10	10	18	19	16
$\chi_2 \otimes \chi_{73}$								1	1					2	3	3	1	4	1
$\chi_2 \otimes \chi_{147}$									2	7	6	5	5	275	1066	1120	1432	1676	1678
$\chi_2 \otimes \chi_{115}$									1	3	3			33	90	98	95	129	122
$\chi_3 \otimes \chi_{51}$										1				3	6	7	4	7	6
$\chi_2 \otimes \chi_{113}$											1	1	1	14	64	66	97	103	113
$\chi_2 \otimes \chi_{83}$															3	3	8	7	7
$\chi_2 \otimes \chi_{71}$															2	2	1	2	3
$\chi_2 \otimes \chi_{69}$															1	1	1	2	1

The following three trees are consistent with the above set of projectives:

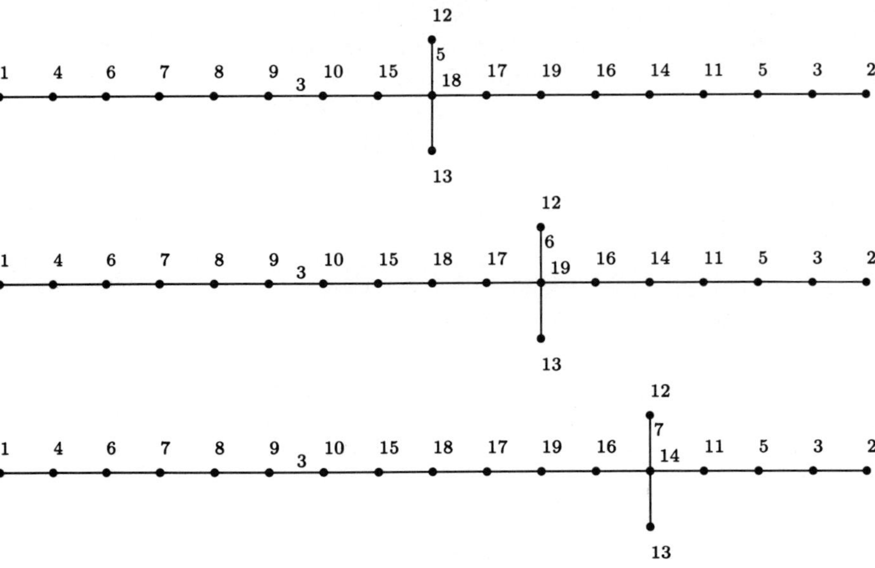

To rule out the first and the last of these and to prove the trees for the principal block, we use the following tensor products:

$$\begin{array}{lll} \chi_2 \otimes \chi_{39} & \approx_B & \chi_{88} \\ \chi_2 \otimes \chi_{26} & \approx_B & \chi_{107} \\ \chi_3 \otimes \chi_{26} & \approx_C & \chi_{164} \end{array}$$

In terms of nodes these tensor products are as follows:

$$\begin{array}{lll} 1_B \otimes 7_A & \approx_B & 10_B \\ 1_B \otimes 4_A & \approx_B & 12_B \\ 1_C \otimes 4_A & \approx_C & 16_C \end{array}$$

Here, the capital letters A, B, C denote the blocks 1, 2 and 3 resp., and $\approx_X$ as usual indicates the restriction to a particular block with letter X. Let $1B0$ denote the Green correspondent of the irreducible module belonging to node 1_B. Similarly, $1C0$ is the Green correspondent of the leaf 1_C.

Let $1Ay$ be the Green correspondent of the edge joining node 7_A with the real stem in the principal block. We assume without loss of generality that y is less than 9, i.e. that 7_A sits on the upper half of the tree. We have $1B0 \otimes 1Ay = 1By$. The first tensor product now implies that y equals 3, since this is the only number less than 9 labelling an edge incident to node 10_B.

Let $18Ax$ be the Green correspondent of the edge joining node 4_A with the real stem. As above we assume that x is smaller than 9. We have $1B0 \otimes 18Ax = 18Bx$. The second tensor product now implies that $x \in \{5, 6, 7\}$ according to the three possibilities for the Brauer tree of Block 2. In particular we see that node 4_A can be joined neither to node 11_A nor to node 14_A.

It now follows that node 7_A is linked to node 16_A and node 4_A to one of 18_A, 17_A or 13_A. To exclude the last possibility we use the third tensor product to show that x cannot equal 7. We have $1C0 \otimes 18Ax = 18Cx$. If x were 7, then in the tree for Block 3 we had an edge incident to node 16_C with label 7 by the above tensor product. The projectives given below determine the real stem of that tree and restrict the possibilities for the location of the complex conjugate characters. In any one of those possibilities, the edge with label 7 joins nodes 8_C and 10_C on the real stem. Hence there can be no edge with this label incident to node 16_C.

We finally rule out the possibility that x equals 5. Suppose $x = 5$. Then the edge joining the complex conjugate node 5_A with the real stem has label 14. We have $18A5 \otimes 1A9 = 18A14$, so 5_A has to occur in the tensor product $4_A \otimes 2_A$. However, $(\chi_{10} \otimes \chi_{26}, \chi_{27}) = 0$, where $(\ ,\ )$ denotes the usual scalar product. This contradiction shows that x equals 6, concluding the proof for Blocks 1 and 2.

Group: M Prime: 19 Block: 3

Nr.	CAS-Nr.	Degree	CC	N&C
1	3	21296876	r	×
2	5	18538750076	r	×
3	12	222879856734249	r	∘
4	13	1044868466775133	r	×
5	19	39660520552077425	r	∘
6	30	8754193822112578125	r	×
7	67	21803647757861753437500	r	∘
8	77	91068387388302451493925	r	×
9	91	261575621299360905468750	r	∘
10	98	433528694560598978525184	r	∘
11	102	626877403613887304040448	103	×
12	103	626877403613887304040448	102	×
13	105	689763222744895005949242	106	∘
14	106	689763222744895005949242	105	∘
15	132	8874260875527017936065100	r	×
16	164	74612213529720383654779356	r	∘
17	178	136107644194473772613203125	r	×
18	184	172248852397651745653437500	r	×
19	193	241866941438795926688759808	r	∘

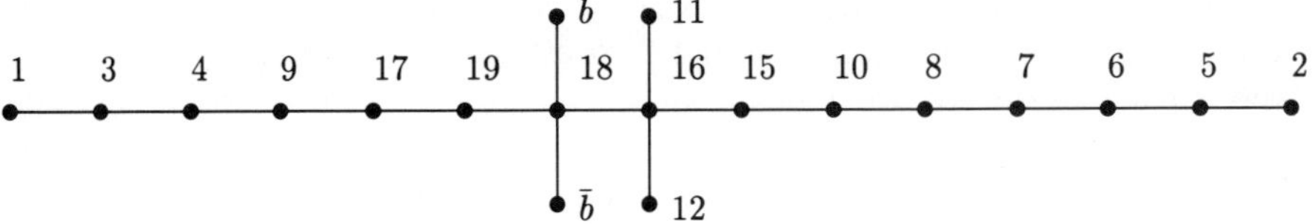

$$\{b, \bar{b}\} = \{13, 14\}$$

We do not know the embedding of nodes 13 and 14 consistent with Blocks 1 and 2.

PROJECTIVES:

Nr.	1	2	3	4	5	6	7	8	9	10	11	12	13	14	15	16	17	18	19
CC	r	r	r	r	r	r	r	r	r	r	12	11	14	13	r	r	r	r	r
N&C	×	×	∘	×	∘	×	∘	×	∘	∘	×	×	∘	∘	×	∘	×	×	∘
$\chi_2 \otimes \chi_8$	1	1	2	1	1														
$\chi_2 \otimes \chi_7$	1	1	1		1														
$\chi_2 \otimes \chi_9$		1	1	1	1														
$\chi_2 \otimes \chi_{14}$		1			1														
$\chi_4 \otimes \chi_{33}$			5	5	20	46	103	81	1	4							3		2
$\chi_3 \otimes \chi_{36}$			1	2	3	11	22	14	1										
$\chi_2 \otimes \chi_{20}$			1	1	2	2													
$\chi_2 \otimes \chi_{127}$							1	5	2	6	1	1			16	108	265	261	432
$\chi_2 \otimes \chi_{73}$								1		2					1		3		3
$\chi_2 \otimes \chi_{34}$								1		1									
$\chi_2 \otimes \chi_{113}$									1	1			1	1	6	43	70	99	128
$\chi_2 \otimes \chi_{122}$										1	1	1	1	1	14	102	176	227	314
$\chi_2 \otimes \chi_{69}$										1					1		1	1	2

The following six trees are consistent with the above set of projectives. Note that in any of these possibilities the edge joining nodes 7 and 10 on the real stem has label 7 as claimed in the proof for Block 2.

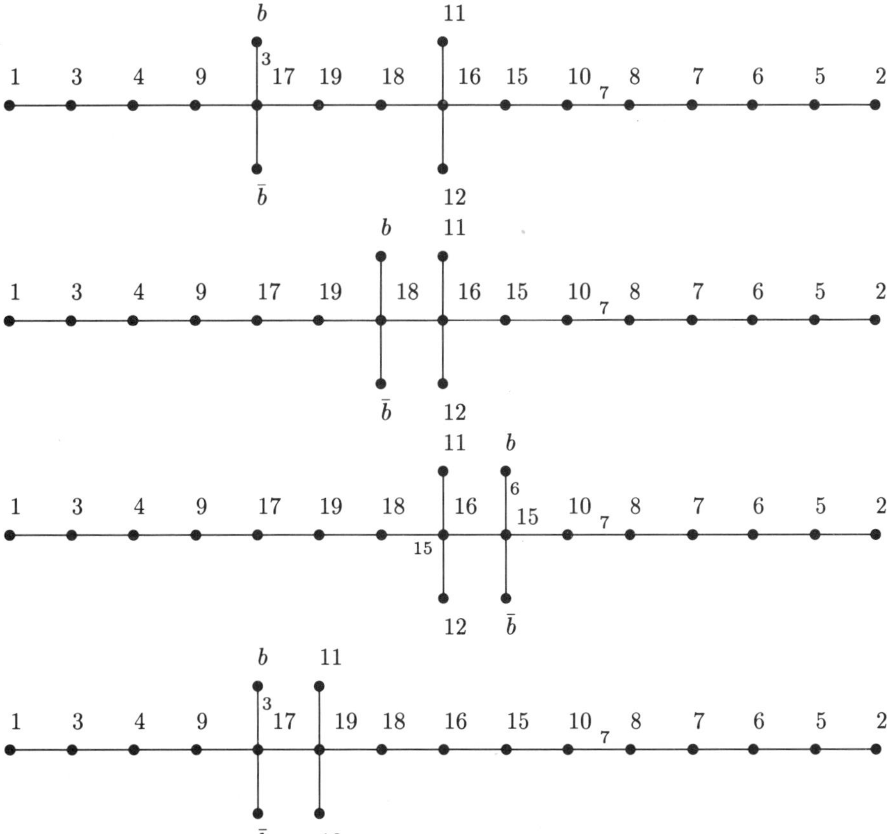

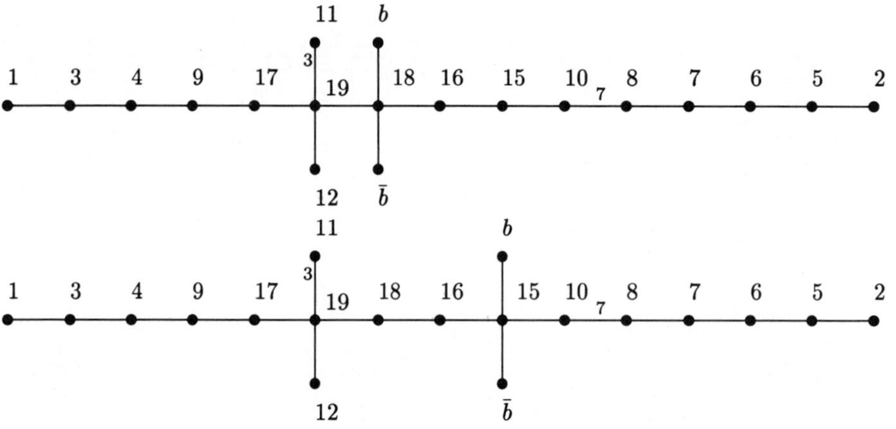

$\{b, \bar{b}\} = \{13, 14\}$
We use the following tensor products:

$$\begin{array}{lll} \chi_2 \otimes \chi_2 & \approx_C & \chi_3 + \chi_5 \\ \chi_2 \otimes \chi_{70} & \approx_C & \chi_{98} + 2\chi_{178} + 3\chi_{193} \\ \chi_3 \otimes \chi_{39} & \approx_C & \chi_{178} + \chi_{184} + \chi_{193} \\ \chi_2 \otimes \chi_{107} & \approx_C & 2\chi_{105} + \chi_{106} + \chi_{132} + 15\chi_{164} + 21\chi_{178} + 34\chi_{184} + 40\chi_{193} \end{array}$$

Again we give these tensor products in terms of nodes, where we use the same convention as in the proof for Block 2:

$$\begin{array}{lll} 1_B \otimes 1_B & \approx_C & 1_C + 2_C \\ 1_B \otimes 9_B & \approx_C & 10_C + 17_C^2 + 19_C^3 \\ 1_C \otimes 7_A & \approx_C & 17_C + 18_C + 19_C \\ 1_B \otimes 12_B & \approx_C & 13_C^2 + 14_C + 15_C + 16_C^{15} + 17_C^{21} + 18_C^{34} + 19_C^{40} \end{array}$$

It is clear from the first tensor product that $1B0 \otimes 1B0 \approx_C 1C0 \oplus 1C9$.

In Block 2, node 9_B is of type nought and is incident to an edge with label 3. From the second tensor product above and $1B0 \otimes 18B3 \approx 18C3 \oplus 18C12$, we see that the edge joining node 13_C with the real stem does not carry label 3 since 13_C does not occur in that product. Of course the same is true for node 14_C. This excludes the first and the fourth possibility.

Now consider the third tensor product. We have $1C0 \otimes 1A3 = 1C3$. Hence one of nodes 17_C or 18_C is incident with an edge carrying label 3. This excludes the last two possibilities.

Finally consider the last tensor product. Remember that the edge joining node 12_B to the real stem in Block 2 has label 6. We have $1B0 \otimes 18B6 \approx 18C6 \oplus 18C15$. In the third of the above trees, at most one of 13_C, 14_C can be among the Green correspondents of $18C6 \oplus 18C15$. Hence the multiplicity of 13_C and 14_C in the projective component of this tensor product is at least 2, whereas the multiplicity of 15_C in this projective is at most 1. So 13_C is not joined to 15_C proving that the second tree is correct.

Group: M   Prime: 19   Block: 4

Nr.	CAS-Nr.	Degree	CC	N&C
1	6	19360062527	r	×
2	18	15178147608537368	r	∘
3	28	3605718753596953125	r	×
4	68	24670833602960142274950	r	∘
5	89	260799524107083767968750	90	∘
5	90	260799524107083767968750	89	∘
6	118	3619209050774375426792424	r	×
7	126	7118465328761788475375616	r	×
8	176	13028713526683728923731674 3	r	∘
9	181	13898854987658452014832025 6	r	∘
10	194	25882347753105506404523437 5	r	×

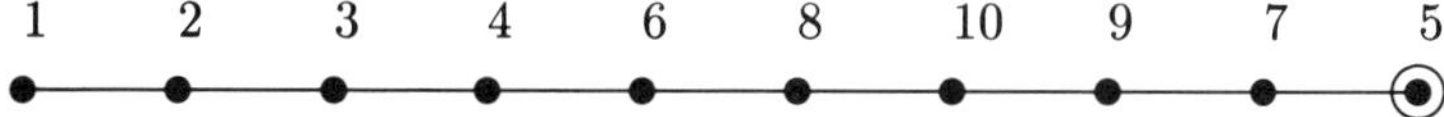

PROJECTIVES:

Nr.	1	2	3	4	5	6	7	8	9	10
CC	r	r	r	r	r	r	r	r	r	r
N&C	×	∘	×	∘	∘	×	×	∘	∘	×
$\chi_2 \otimes \chi_9$	1	2	1							
$\chi_2 \otimes \chi_{61}$			2	6		5		1		
$\chi_2 \otimes \chi_{65}$			1	6		8		3		
$\chi_2 \otimes \chi_{34}$			1	2		1				
$\chi_2 \otimes \chi_{114}$				2	1	16	2	108	46	139
$\chi_2 \otimes \chi_{59}$							1		1	
$\chi_2 \otimes \chi_{69}$								1	1	2

Group: M Prime: 23 Block: 1

Nr.	CAS-Nr.	Degree	CC	N&C
1	1	1	r	×
2	36	7731661927392812500	r	∘
3	41	28624326769272448614	42	∘
4	42	28624326769272448614	41	∘
5	46	64335692588991774720	r	×
6	59	456719917691248640000	60	∘
7	60	456719917691248640000	59	∘
8	65	1281000554262325081785	r	∘
9	83	1616491110022607929687	84	∘
9	84	1616491110022607929687	83	∘
10	96	3910090818374773783296	r	×
11	97	3926116519750656000000	r	×
12	101	6000207726850645023929	r	∘

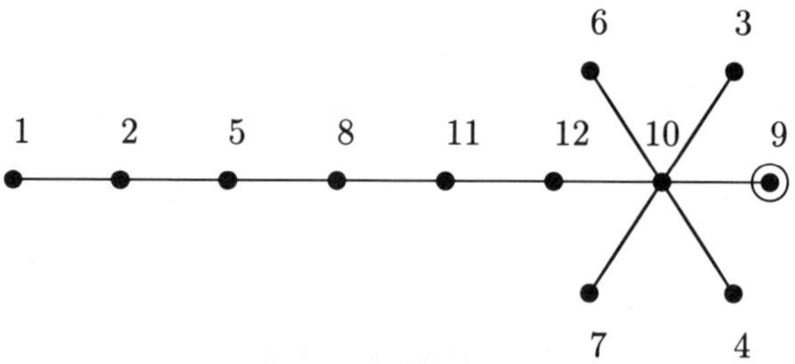

PROJECTIVES:

Nr.	1	2	3	4	5	6	7	8	9	10	11	12
CC	r	r	4	3	r	7	6	r	r	r	r	r
N&C	×	∘	∘	∘	×	∘	∘	∘	∘	×	×	∘
$\chi_5 \otimes \chi_5$	1	1										
$\chi_2 \otimes \chi_{19}$		1			1							
$\chi_3 \otimes \chi_{136}$				1		2	2		44	103	63	117
$\chi_2 \otimes \chi_{34}$					1			1				
$\chi_2 \otimes \chi_{126}$						1	1		1	4		1
$\chi_3 \otimes \chi_{35}$								1			1	
$\chi_2 \otimes \chi_{73}$											1	1

We have the following two trees consistent with the above set of projectives:

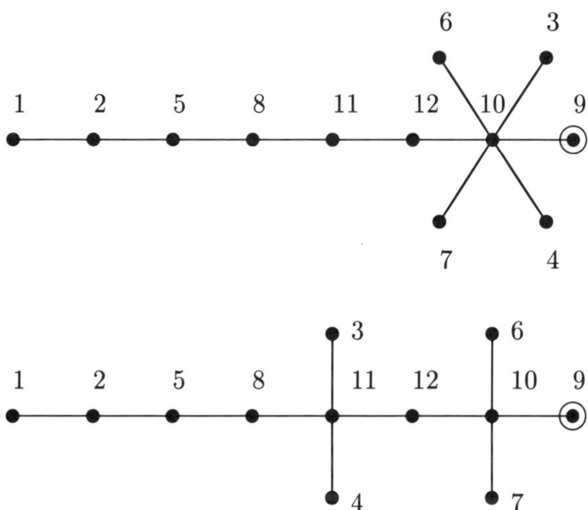

We only have to fix the complex conjugate pair χ_{41}, χ_{42} (nodes 3 and 4). The following tensor products solve the problem:

$$\chi_2 \otimes \chi_{41} = \chi_{157}$$

and

$$\chi_2 \otimes \chi_{42} = \chi_{157}$$

Let us assume that χ_2 (node 1 in Block 2) has Green correspondent $1B0$. Then it follows, because $1B0 \otimes 22Ax = 22Bx$ and χ_{157} is node 7 in Block 2, that χ_{41} has the Green correspondent $22A5$ or $22A7$. This contradicts the second tree. We remark that the irrationalities of the two pairs of complex conjugate characters are independent and also independent of the irrationalities of the complex conjugate characters in Block 4, so we do not have to worry about the planar embedding.

Group: M Prime: 23 Block: 2

Nr.	CAS-Nr.	Degree	CC	N&C
1	2	196883	r	×
2	15	2374124840062976	r	∘
3	18	15178147608537368	r	×
4	50	918438233727730974720	r	∘
5	85	1912590851134599453125000	86	∘
5	86	1912590851134599453125000	85	∘
6	113	2181694185821505680397072	r	×
7	157	56356433273146675005489152	r	∘
8	162	66550339514356152000000000	r	×
9	169	103354104243912727763091456	r	×
10	176	130287135266837289237316743	r	∘
11	183	163216709667196367710937500	r	∘
12	187	177966317773633111417870812	r	×

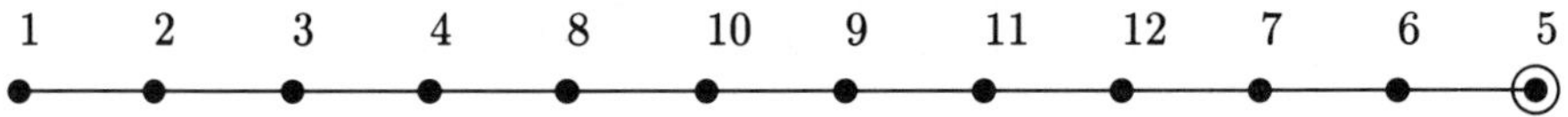

PROJECTIVES:

Nr.	1	2	3	4	5	6	7	8	9	10	11	12
CC	r	r	r	r	r	r	r	r	r	r	r	r
N&C	×	∘	×	∘	∘	×	∘	×	×	∘	∘	×
$\chi_2 \otimes \chi_5$	1	1										
$\chi_2 \otimes \chi_8$		1	1									
$\chi_2 \otimes \chi_{43}$			1	1								
$\chi_2 \otimes \chi_{62}$				1				2		1		
$\chi_2 \otimes \chi_{124}$					1	4	74	61	138	138	241	251
$\chi_2 \otimes \chi_{105}$						1	10	7	17	16	31	32
$\chi_2 \otimes \chi_{70}$							1	3	1	4		1
$\chi_3 \otimes \chi_{39}$							1	1		1		1
$\chi_2 \otimes \chi_{79}$								13	2	15	2	2
$\chi_2 \otimes \chi_{69}$								1	1	1	2	1
$\chi_4 \otimes \chi_{26}$											1	1

Group: M Prime: 23 Block: 3

Nr.	CAS-Nr.	Degree	CC	N&C
1	3	21296876	r	×
2	13	1044868466775133	r	∘
3	31	2858599095072164062 5	r	×
4	55	176813080258312695312 5	56	∘
4	56	176813080258312695312 5	55	∘
5	58	238298741750624242187 5	r	∘
6	95	351532203382732066094400	r	×
7	144	2750191760970910224718 7500	r	×
8	161	643261634275226242057031 25	r	∘
9	168	889438206202883432616723 93	r	∘
10	178	1361076441944737726132031 25	r	×
11	182	1615618649711711132875406 25	r	×
12	184	1722488523976517456534375 00	r	∘

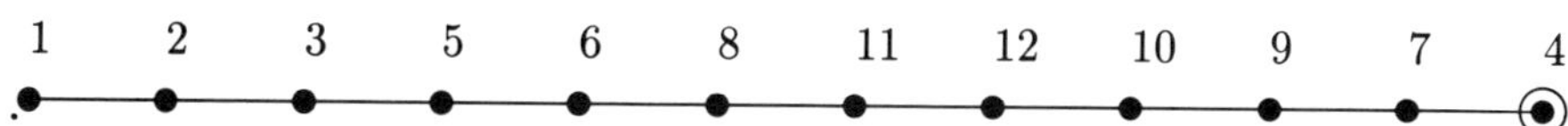

PROJECTIVES:

Nr.	1	2	3	4	5	6	7	8	9	10	11	12
CC	r	r	r	r	r	r	r	r	r	r	r	r
N&C	×	∘	×	∘	∘	×	×	∘	∘	×	×	∘
$\chi_2 \otimes \chi_6$	1	1										
$\chi_2 \otimes \chi_{32}$		1	2		2	1						
$\chi_2 \otimes \chi_{12}$		1	1									
$\chi_2 \otimes \chi_{62}$			4		9	6		1				
$\chi_2 \otimes \chi_{25}$			1		1							
$\chi_2 \otimes \chi_{156}$				1		1	328	806	1070	1683	2061	2196
$\chi_2 \otimes \chi_{79}$					3	5	3	3	7	4	1	
$\chi_2 \otimes \chi_{73}$						1	3	1	6	3		
$\chi_2 \otimes \chi_{105}$							3	9	10	21	29	34
$\chi_2 \otimes \chi_{74}$								1	1	2	3	3
$\chi_2 \otimes \chi_{69}$								1	1	1	2	1

Group: M Prime: 23 Block: 4

Nr.	CAS-Nr.	Degree	CC	N&C
1	4	842609326	r	×
2	10	125510727015275	r	∘
3	23	336041615485626050	r	×
4	37	1304153504203429687 50	r	∘
5	64	10145274012943412428800	r	×
6	81	14961479414922601090 2528	82	∘
7	82	14961479414922601090 2528	81	∘
8	128	75671515765424524257 81250	129	∘
8	129	75671515765424524257 81250	128	∘
9	134	94794957458053056531 25000	r	∘
10	142	226266213651605370999 27552	r	×
11	148	373107152115466240000 00000	r	×
12	153	426014748606395796698 96397	r	∘

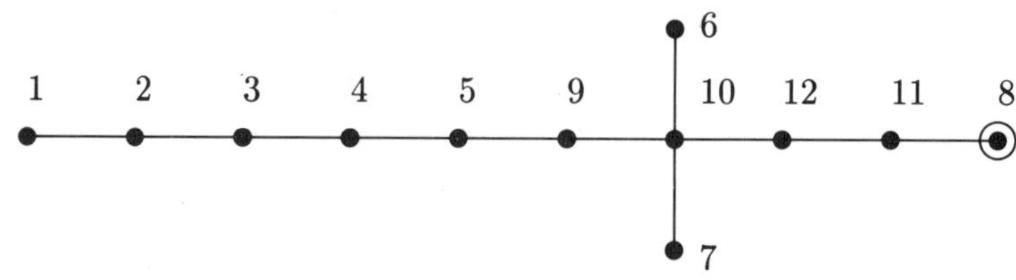

PROJECTIVES:

Nr.	1	2	3	4	5	6	7	8	9	10	11	12
CC	r	r	r	r	r	7	6	r	r	r	r	r
N&C	×	∘	×	∘	×	∘	∘	∘	∘	×	×	∘
$\chi_2 \otimes \chi_5$	1	1										
$\chi_2 \otimes \chi_{12}$		1	1									
$\chi_2 \otimes \chi_{43}$			2	5	4				1			
$\chi_2 \otimes \chi_{25}$			1	1								
$\chi_2 \otimes \chi_{32}$				1	1							
$\chi_2 \otimes \chi_{94}$					1				22	30	2	11
$\chi_2 \otimes \chi_{139}$						1	1	11	88	171	107	177
$\chi_2 \otimes \chi_{99}$								2		1	6	5

We are left with only the two possible trees below:

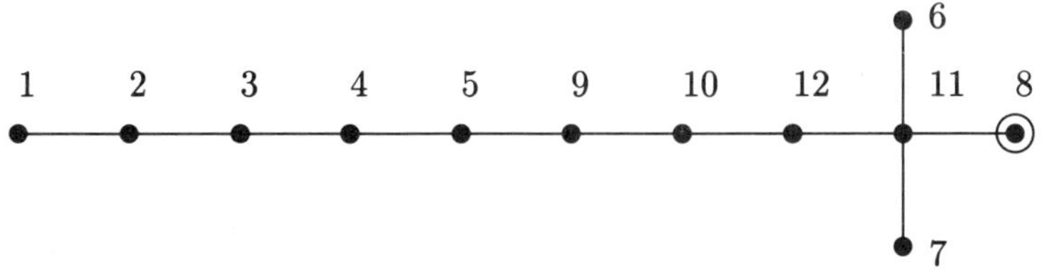

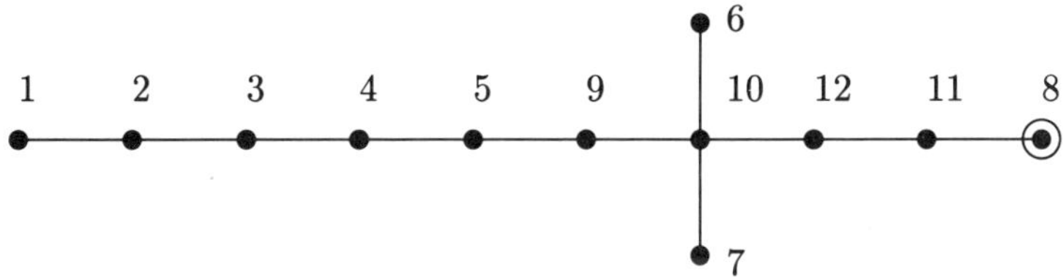

Let us assume that χ_4 (node 1) has Green correspondent $1D0$. The following tensor product holds in M:

$$\chi_4 \otimes \chi_4 \approx \chi_1$$

So in the normalizer of the 23-sylowsubgroup we get

$$1D0 \otimes 1D0 \approx 1A0$$

Here $\approx$ denotes the tensor product restricted to the principal block. The tensor products

$$\chi_4 \otimes \chi_{81}$$

and

$$\chi_4 \otimes \chi_{82}$$

do not contain the characters χ_{41} and χ_{42} (nodes 3 and 4 in the principal block). If the first tree is right, we get by Green correspondence that $1D0 \otimes 22D5$ contains $22A5$, so that χ_{41} should be in one of the above tensor products. This is a contradiction.

Group: M Prime: 23 Block: 5

Nr.	CAS-Nr.	Degree	CC	N&C
1	7	293553734298	r	×
2	9	36173193327999	r	∘
3	16	8980616927734375	17	∘
3	17	8980616927734375	16	∘
4	21	251098487132187500	r	×
5	51	1201241700908448332364	r	×
6	57	2351753641814605348320	r	∘
7	76	86551489469233273849000	r	×
8	92	277540481294528814140625	r	∘
9	137	9592584386918582979657728	r	∘
10	138	10023854998171489083984375	r	×
11	151	41762322738385820195625000	r	×
12	152	42001454087954515167503490	r	∘

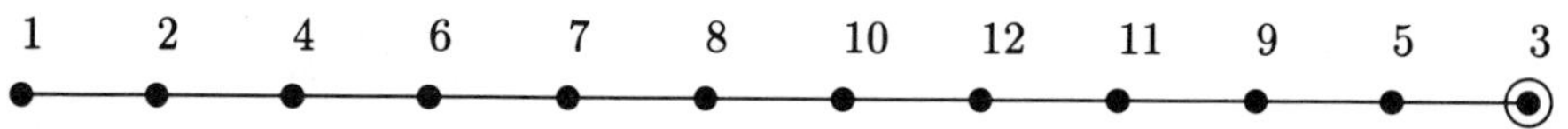

PROJECTIVES:

Nr.	1	2	3	4	5	6	7	8	9	10	11	12
CC	r	r	r	r	r	r	r	r	r	r	r	r
N&C	×	∘	∘	×	×	∘	×	∘	∘	×	×	∘
$\chi_2 \otimes \chi_5$	1	1										
$\chi_2 \otimes \chi_{14}$		1		1								
$\chi_5 \otimes \chi_{156}$			1		1320	2276	97035	320739	11249399	11810228	49081685	49417853
$\chi_2 \otimes \chi_{34}$				1		2	1					
$\chi_3 \otimes \chi_{35}$					1	1	1		2		1	
$\chi_2 \otimes \chi_{115}$					1		2	2	23	2	51	31
$\chi_2 \otimes \chi_{32}$							1	1				
$\chi_2 \otimes \chi_{156}$								1	95	129	479	512
$\chi_2 \otimes \chi_{71}$									1		1	
$\chi_2 \otimes \chi_{69}$										1		1
$\chi_3 \otimes \chi_{53}$											1	1

Group: M Prime: 31 Block: 1

Nr.	CAS-Nr.	Degree	CC	N&C
1	1	1	r	×
2	11	190292345709543	r	∘
3	16	8980616927734375	17	∘
4	17	8980616927734375	16	∘
5	22	290568421805921077	r	×
6	30	8754193822112578125	r	∘
7	65	12810005542623250817856	r	×
8	105	689763222744895005949242	106	∘
8	106	689763222744895005949242	105	∘
9	111	1599110387863558882812500	r	∘
10	123	5514132424881463208443904	r	×
11	127	7375892500409609408203125	r	∘
12	137	9592584386918582979657728	r	×
13	140	14930164283563048960000000	r	×
14	160	63750812845035828079008441	r	∘
15	167	86206621680977834911875000	r	∘
16	175	129572518017902934396764160	r	×

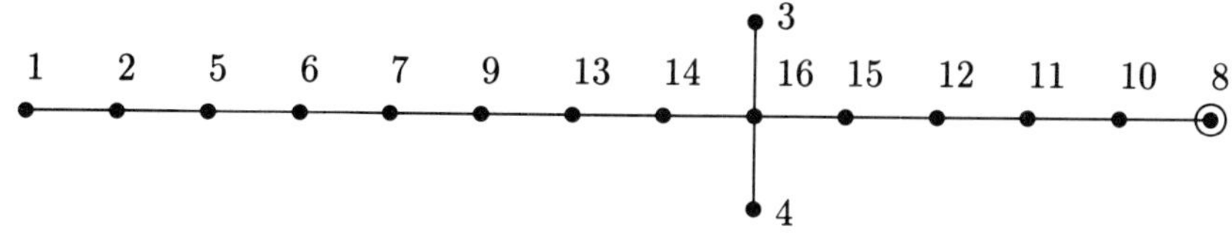

PROJECTIVES:

Nr.	1	2	3	4	5	6	7	8	9	10	11	12	13	14	15	16
CC	r	r	4	3	r	r	r	r	r	r	r	r	r	r	r	r
N&C	×	○	○	○	×	○	×	○	○	×	○	×	×	○	○	×
$\chi_3 \otimes \chi_3$	1	1														
$\chi_2 \otimes \chi_9$		1			1											
$\chi_3 \otimes \chi_{86}$				1				4	1	22	23	25	53	300	388	617
$\chi_2 \otimes \chi_{23}$					3	4	1									
$\chi_2 \otimes \chi_{57}$						2	6		6		1	1	2			
$\chi_2 \otimes \chi_{51}$												1			1	
$\chi_3 \otimes \chi_{59}$													1	7	6	12
$\chi_2 \otimes \chi_{74}$														1	1	2

We have the following four trees consistent with the above set of projectives. In the course of the proof for the second block we shall show that the second of them is correct.

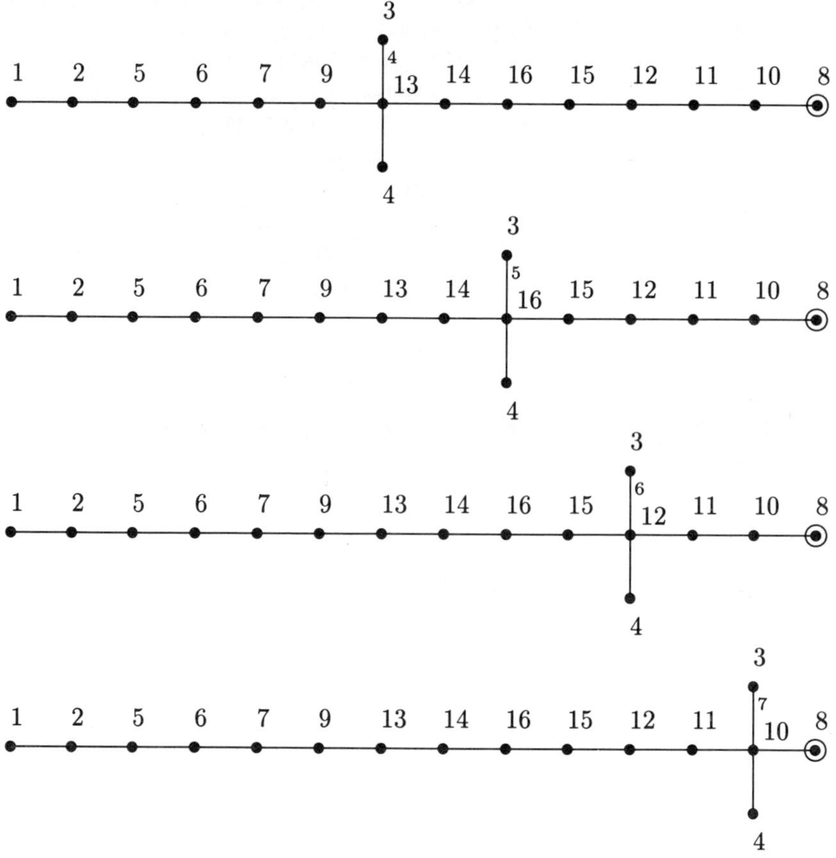

Group: M Prime: 31 Block: 2

Nr.	CAS-Nr.	Degree	CC	N&C
1	2	196883	r	×
2	7	293553734298	r	∘
3	13	1044868466775133	r	×
4	37	1304153504203429687500	r	∘
5	43	379913824694312370176	r	×
6	55	176813080258312695312 5	56	∘
7	56	1768130802583126953125	55	∘
8	73	6068376205205758732606 5	r	∘
9	107	68976672617955508099422 3	108	∘
9	108	689766726179555080994223	107	∘
10	114	2216343020913351966796875	r	×
11	138	10023854998171489083984375	r	×
12	148	37310715211546624000000000	r	∘
13	157	56356433273146675005489152	r	×
14	163	69084859008005036431224066	r	∘
15	190	203314261261157852274218750	r	∘
16	193	241866941438795926688759808	r	×

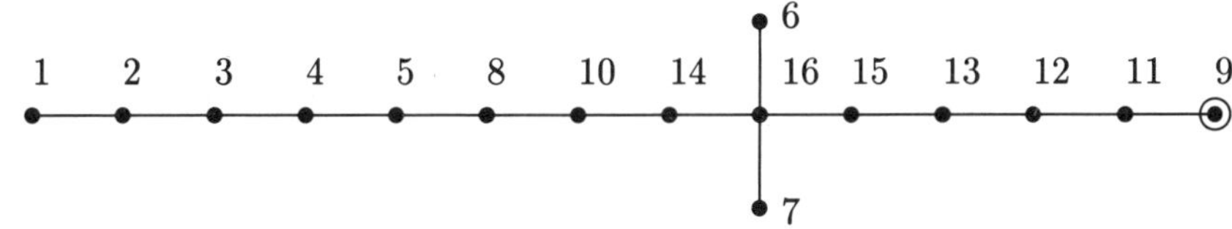

PROJECTIVES:

Nr.	1	2	3	4	5	6	7	8	9	10	11	12	13	14	15	16
CC	r	r	r	r	r	7	6	r	r	r	r	r	r	r	r	r
N&C	×	∘	×	∘	×	∘	∘	∘	∘	×	×	∘	×	∘	∘	×
$\chi_2 \otimes \chi_3$	1	1														
$\chi_2 \otimes \chi_9$		1	1													
$\chi_2 \otimes \chi_{32}$			1	2	1											
$\chi_2 \otimes \chi_{66}$				1	2			1		1				1		
$\chi_2 \otimes \chi_{34}$					1			1								
$\chi_2 \otimes \chi_{128}$						1			1		21	71	100	117	374	443
$\chi_2 \otimes \chi_{86}$							1				1	1	1	2	9	11
$\chi_2 \otimes \chi_{35}$								1		1						
$\chi_3 \otimes \chi_{47}$										1		1	1	3	3	5
$\chi_2 \otimes \chi_{83}$											1	2	1	2	7	9
$\chi_2 \otimes \chi_{81}$												1	2	2	8	9
$\chi_2 \otimes \chi_{71}$													1	1	3	3
$\chi_3 \otimes \chi_{39}$													1		2	1

The following three trees are consistent with the above set of projectives:

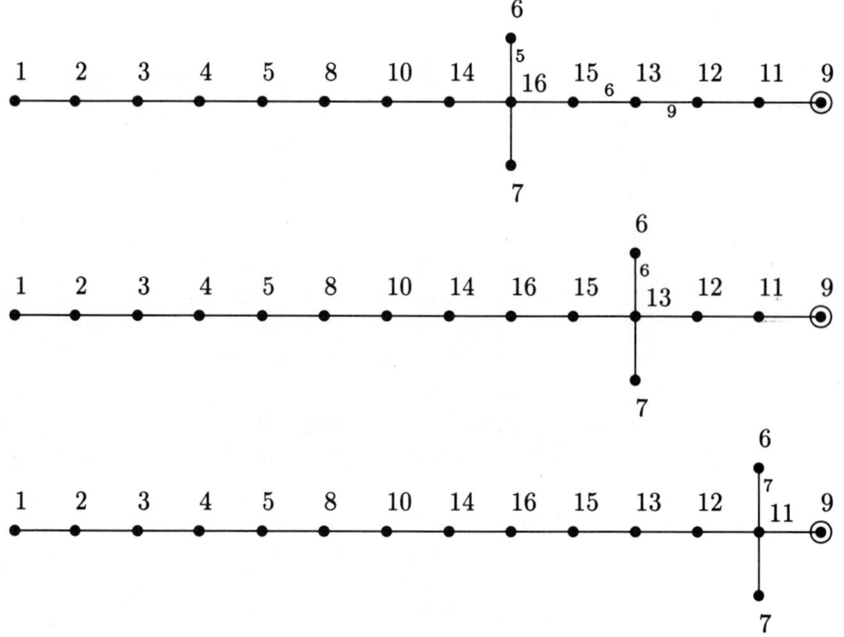

To exclude the last two possibilities, we use the following tensor products:

$$\begin{aligned} \chi_2 \otimes \chi_{16} &\approx_B \chi_{55} \\ \chi_7 \otimes \chi_{16} &\approx_B \chi_{193} \end{aligned}$$

In terms of nodes these tensor products are as follows:

$$\begin{aligned} 1_B \otimes 3_A &\approx_B 6_B \\ 2_B \otimes 3_A &\approx_B 16_B \end{aligned}$$

Of course the letters A and B denote the blocks 1 and 2 resp. Since $(1_B + 2_B)$ is projective, we also have that $(1_B + 2_B) \otimes 3_A \approx 6_B + 16_B$ is projective, i.e. node 6_B is joined to node 16_B. This means that the first possibility is correct.

Now let $1B0$ denote the Green correspondent of the leaf at node 1_B, and let $30Ax$ denote the Green correspondent of the edge joining node 3_A with the real stem in the principal block. Then $1B0 \otimes 30Ax = 30Bx$. It follows that x equals 5 and so the second possibility in the principal block is correct. Note that the two edges incident to node $13B$ have labels 6 and 9, i.e. the Green correspondents of the two trivial source modules with Brauer character 13_B are $1B6$ and $1B9$.

Group: M | Prime: 31 | Block: 3

Nr.	CAS-Nr.	Degree	CC	N&C
1	6	19360062527	r	×
2	8	3879214937598	r	∘
3	20	60359800576579350	r	×
4	26	3503434660075044981	27	∘
4	27	3503434660075044981	26	∘
5	36	77316619273928125000	r	∘
6	41	286243267692724486144	42	×
7	42	286243267692724486144	41	×
8	69	31714653744947491918600	r	×
9	97	392611651975065600000000	r	×
10	121	4926670174323484069683200	r	∘
11	122	5334046162969208352215625	r	∘
12	124	5514132424881463208443904	125	×
13	125	5514132424881463208443904	124	×
14	146	33684388830359981044531200	r	×
15	181	138988549876584520148320256	r	×
16	185	173865305251972140447265625	r	∘

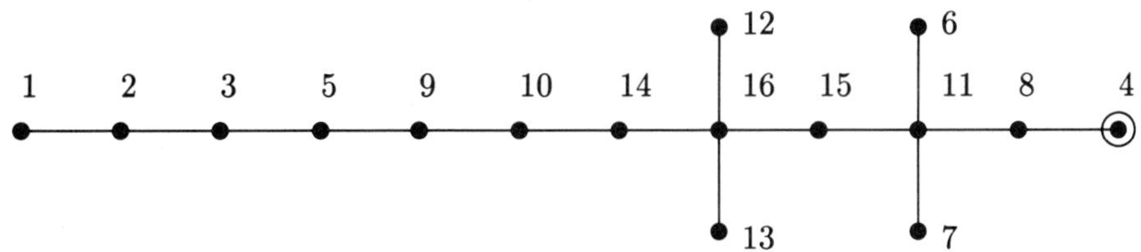

PROJECTIVES:

Nr.	1	2	3	4	5	6	7	8	9	10	11	12	13	14	15	16
CC	r	r	r	r	r	7	6	r	r	r	r	13	12	r	r	r
N&C	×	∘	×	∘	∘	×	×	×	×	∘	∘	×	×	×	×	∘
$\chi_2 \otimes \chi_3$	1	1														
$\chi_2 \otimes \chi_{18}$		1	1													
$\chi_2 \otimes \chi_{21}$			1		1											
$\chi_3 \otimes \chi_{164}$				1		1	1	63	607	9000	10515	11093	11093	65680	275247	344269
$\chi_2 \otimes \chi_{49}$					1				2	1						
$\chi_3 \otimes \chi_{135}$						1		9	63	1094	1359	1451	1451	8394	35568	44484
$\chi_2 \otimes \chi_{98}$								1	2	7	1			9	8	12
$\chi_2 \otimes \chi_{50}$										1				1		
$\chi_2 \otimes \chi_{87}$											1			1	4	4
$\chi_2 \otimes \chi_{83}$												1	1	1	9	12

The following three trees are consistent with the above set of projectives:

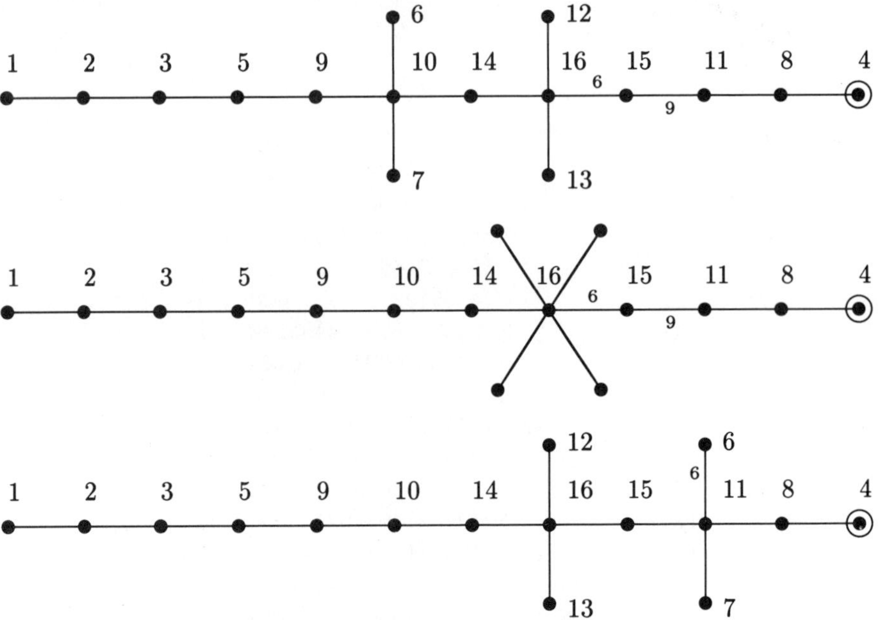

To rule out the first two we use the following tensor products:

$$\begin{array}{lll} \chi_2 \otimes \chi_6 & \approx_B & \chi_2 + \chi_7 + \chi_{13} \\ \chi_2 \otimes \chi_{41} & \approx_B & \chi_{157} \end{array}$$

In terms of nodes these tensor products are as follows:

$$\begin{array}{lll} 1_B \otimes 1_C & \approx_B & 1_B + 2_B + 3_B \\ 1_B \otimes 6_C & \approx_B & 13_B \end{array}$$

Of course the index C denotes the nodes of the third block. The first of these tensor products is used to find the component of $1B0 \otimes 1C0$ in the second block, where $1C0$ is the Green correspondent of the edge at node 1_C. Since both $1B0$ and $1C0$ are trivial source modules, so is the component of their product in Block 2. It follows that its Green correspondent has Brauer character 1_B or 3_B. In particular this component is indecomposable. Since it must also be self-dual, it has to be $1B0$, since there is no self-dual trivial source module with Brauer character 3_B. Now let $1Cx$ denote the Green correspondent of the edge joining node 6_C with the real stem. Then $1B0 \otimes 1Cx = 1Bx$. By the tensor product above, x must be 6 or 9, since these are the two labels at the edges incident to node 13_B. It follows that the last of the above trees is correct. Please note that the irrationalities of the two pairs of complex conjugate characters are independent, so we do not have to worry about the planar embedding.

Group: M Prime: 41 Block: 1

Nr.	CAS-Nr.	Degree	CC	N&C
1	1	1	r	×
2	2	196883	r	×
3	4	842609326	r	○
4	9	36173193327999	r	○
5	15	2374124840062976	r	×
6	16	8980616927734375	17	×
7	17	8980616927734375	16	×
8	19	39660520552077425	r	×
9	24	2500435234254428856	r	○
10	26	3503434660075044981	27	○
11	27	3503434660075044981	26	○
12	30	8754193822112578125	r	○
13	44	640558364167263622626	45	×
14	45	640558364167263622626	44	×
15	46	643356925889917747200	r	×
16	47	691170144025469730622	48	×
17	48	691170144025469730622	47	×
18	55	1768130802583126953125	56	×
19	56	1768130802583126953125	55	×
20	59	4567199176912486400000	60	×
21	60	4567199176912486400000	59	×
22	63	7226910362631220625000	r	×
23	66	19795913912408993711352	r	○
24	71	42940402913709544921875	72	○
25	72	42940402913709544921875	71	○
26	83	161649111002260792968750	84	○
27	84	161649111002260792968750	83	○
28	97	392611651975065600000000	r	○
29	99	597787522207315571077947	100	×
30	100	597787522207315571077947	99	×
31	102	626877403613887304040448	103	×
32	103	626877403613887304040448	102	×
33	107	689766726179555080994223	108	○
34	108	689766726179555080994223	107	○
35	128	7567151576542452425781250	129	○
36	129	7567151576542452425781250	128	○
37	130	7850934959207940600000000	r	×
38	131	8394037047155083487634450	r	×
39	159	62038057486792249132974080	r	○
40	161	64326163427522624205703125	r	○
41	173	124982156072747647257292800	r	×

The projectives below, in addition to

$$\chi_{16} \otimes (\chi_2 + \chi_4) \approx \chi_{55} + \chi_{128}$$

(in terms of nodes $6 \otimes (2+3) \approx 18 + 35$)) determine the following **subtree** (where we have disregarded the planar embedding):

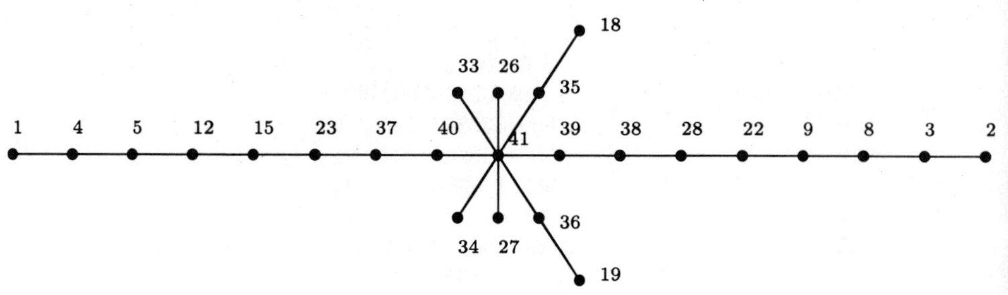

PROJECTIVES:

Nr.	1	2	3	4	5	6	7	8	9	10	11	12	13	14	15	16	17	18	19	20
CC	r	r	r	r	r	7	6	r	r	11	10	r	14	13	r	17	16	19	18	21
N&C	×	×	○	○	×	×	×	×	○	○	○	○	×	×	×	×	×	×	×	×
$\chi_3 \otimes \chi_3$	1	1	1	1																
$\chi_2 \otimes \chi_5$		1	1	1	1															
$\chi_2 \otimes \chi_3$		1	1																	
$\chi_2 \otimes \chi_{10}$			1	2	3			2	1			1								
$\chi_2 \otimes \chi_{18}$				2	3			3	3			2			1					
$\chi_2 \otimes \chi_{11}$				1	1			1	1											
$\chi_2 \otimes \chi_{25}$					1			2	3			2			2					
$\chi_{85} \otimes \chi_3$						1												1		
$\chi_2 \otimes \chi_{62}$									2			3			7					
$\chi_2 \otimes \chi_{64}$									1			2			4					
$\chi_2 \otimes \chi_{50}$									1											
$\chi_{164} \otimes \chi_3$										1	1		1	1				4	4	11
$\chi_2 \otimes \chi_{174}$													1	1						
$\chi_2 \otimes \chi_{115}$																1	1			
$\chi_{83} \otimes \chi_3$																		1		
$\chi_2 \otimes \chi_{85}$																		1		
$\chi_2 \otimes \chi_{126}$																				1
$\chi_{59} \otimes \chi_3$																				
$\chi_2 \otimes \chi_{137}$																				
$\chi_2 \otimes \chi_{73}$																				
$\chi_2 \otimes \chi_{132}$																				
$\chi_2 \otimes \chi_{122}$																				
$\chi_2 \otimes \chi_{105}$																				

PROJECTIVES (continued):

Nr.	22	23	24	25	26	27	28	29	30	31	32	33	34	35	36	37	38	39	40	41
CC	r	r	25	24	27	26	r	30	29	32	31	34	33	36	35	r	r	r	r	r
N&C	×	○	○	○	○	○	○	×	×	×	×	○	○	○	○	×	×	○	○	×
$\chi_3 \otimes \chi_3$																				
$\chi_2 \otimes \chi_5$																				
$\chi_2 \otimes \chi_3$																				
$\chi_2 \otimes \chi_{10}$																				
$\chi_2 \otimes \chi_{18}$																				
$\chi_2 \otimes \chi_{11}$																				
$\chi_2 \otimes \chi_{25}$	1	1																		
$\chi_{85} \otimes \chi_3$					3	3		4	4	2	2	4	4	50	49	27	24	310	324	682
$\chi_2 \otimes \chi_{62}$	9	4					9									1	2		1	
$\chi_2 \otimes \chi_{64}$	5	2					6										2		1	1
$\chi_2 \otimes \chi_{50}$	1	1														1				
$\chi_{164} \otimes \chi_3$	3	17	76	76	336	336	607	1189	1189	1236	1236	1406	1406	15145	15145	14918	15896	122051	126333	247237
$\chi_2 \otimes \chi_{174}$			1	1	7	7	5	22	22	20	20	21	21	249	249	224	226	1886	1969	3880
$\chi_2 \otimes \chi_{115}$			1	1			3									11	19	44	36	53
$\chi_{83} \otimes \chi_3$					3	3		3	3	2	2	4	4	43	43	19	19	266	272	589
$\chi_2 \otimes \chi_{85}$					1									2	1			3	3	9
$\chi_2 \otimes \chi_{126}$					1	1		1	1	1	1			19	19	9	10	112	116	243
$\chi_{59} \otimes \chi_3$														3	3			8	9	22
$\chi_2 \otimes \chi_{137}$		2	1	1			10			2	2			6	6	44	52	144	155	225
$\chi_2 \otimes \chi_{73}$		1					1									2	2	1	1	
$\chi_2 \otimes \chi_{132}$								2	2	1	1	1	1	17	17	20	17	130	143	266
$\chi_2 \otimes \chi_{122}$								1	1	1	1	1	1	11	11	10	8	78	85	165
$\chi_2 \otimes \chi_{105}$												2	1	1	1			11	9	25

Group: M　　Prime: 47　　Block: 1

Nr.	CAS-Nr.	Degree	CC	N&C
1	1	1	r	×
2	5	18538750076	r	∘
3	9	36173193327999	r	×
4	20	60359800576579350	r	∘
5	36	7731661927392812500 0	r	×
6	41	2862432676927244861 44	42	∘
7	42	286243267692724486144	41	∘
8	44	640558364167263622626	45	∘
9	45	640558364167263622626	44	∘
10	53	1480279477146615234375	54	∘
10	54	1480279477146615234375	53	∘
11	59	4567199176912486400000	60	×
12	60	4567199176912486400000	59	×
13	68	24670833602960142274950	r	∘
14	85	191259085113459945312500	86	×
15	86	191259085113459945312500	85	×
16	94	331150814995116217581480	r	×
17	119	4004308274823270400000000	r	×
18	124	5514132424881463208443904	125	×
19	125	5514132424881463208443904	124	×
20	137	9592584386918582979657728	r	∘
21	164	74612213529720383654779356	r	∘
22	169	103354104243912727763091456	r	×
23	181	138988549876584520148320256	r	×
24	185	173865305251972140447265625	r	∘

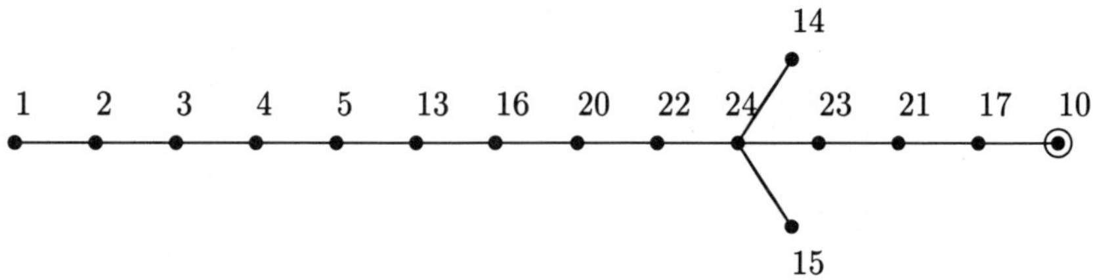

14

1 2 3 4 5 13 16 20 22 24 23 21 17 10

15

Warning! This is not the complete Brauer tree, but the maximal subtree we know. There are 12 possibilities for the location of the remaining pairs of complex conjugate characters. These are described below.

PROJECTIVES:

Nr.	1	2	3	4	5	6	7	8	9	10	11	12	13	14	15	16	17	18	19	20	21	22	23	24
CC	r	r	r	r	r	7	6	9	8	r	12	11	r	15	14	r	r	19	18	r	r	r	r	r
N&C	×	o	×	o	×	o	o	o	o	o	×	×	o	×	×	×	×	×	×	o	o	×	×	o
$\chi_2 \otimes \chi_2$	1	1																						
$\chi_{15} \otimes \chi_2$		1	2	2	1																			
$\chi_8 \otimes \chi_2$		1	2	1																				
$\chi_{21} \otimes \chi_2$			1	1	1								1											
$\chi_{29} \otimes \chi_2$				1	2								2			1								
$\chi_{157} \otimes \chi_2$						1	1							1	1	1	62	74	74	126	1040	1412	1900	2357
$\chi_{99} \otimes \chi_2$								1									1	1	1		12	15	23	28
$\chi_{126} \otimes \chi_2$											1	1		1	1		3	13	13	9	132	182	268	342
$\chi_{83} \otimes \chi_2$														1				1	1		4	4	9	12
$\chi_{55} \otimes \chi_2$														1										1
$\chi_{70} \otimes \chi_2$																1				2		1		

The projectives above determine the subtree we have given. They also show that node 11_A is joined to one of $\{20_A, 21_A, 24_A\}$, hence to the real stem. Node 18_A is linked to the real stem, too, since it is joined to one of $\{21_A, 24_A\}$. It follows from these observations that nodes 6_A and 8_A are leaves, i.e. are irreducible on reduction modulo 47.

To restrict the number of possibilities for the Brauer tree, we shall use the following tensor products:

$$\begin{array}{lcl} \chi_3 \otimes \chi_{16} & \approx_A & \chi_{85} \\ \chi_3 \otimes \chi_{26} & \approx_A & \chi_{164} \\ \chi_3 \otimes \chi_{41} & \approx_B & \chi_{158} + \chi_{175} + \chi_{176} \end{array}$$

In terms of nodes these tensor products are as follows:

$$\begin{array}{lcl} 1_B \otimes 4_B & \approx_A & 14_A \\ 1_B \otimes 7_B & \approx_A & 21_A \\ 1_B \otimes 6_A & \approx_B & 21_B + 23_B + 24_B \end{array}$$

We shall also use the following scalar products:

$$\begin{array}{rcl} (\chi_3 \otimes \chi_{59}, \chi_{71}) & = & 0 \\ (\chi_3 \otimes \chi_{59}, \chi_{72}) & = & 0 \\ (\chi_3 \otimes \chi_{59}, \chi_{105}) & = & 0 \\ (\chi_3 \otimes \chi_{59}, \chi_{106}) & = & 0 \\ (\chi_5 \otimes \chi_{41}, \chi_{85}) & = & 0 \\ (\chi_5 \otimes \chi_{41}, \chi_{86}) & = & 0 \end{array}$$

Again we give these scalar products in terms of nodes:

$$\begin{array}{rcl} (1_B \otimes 11_A, 10_B) & = & 0 \\ (1_B \otimes 11_A, 11_B) & = & 0 \\ (1_B \otimes 11_A, 13_B) & = & 0 \\ (1_B \otimes 11_A, 14_B) & = & 0 \\ (2_A \otimes 6_A, 14_A) & = & 0 \\ (2_A \otimes 6_A, 15_A) & = & 0 \end{array}$$

The projectives immediately show that there is no branch going off left of node 20_A. Thus if 11_A is joined to 20_A, its Green correspondent is $1A4$. Then the Green correspondent of $1_B \otimes 11_A$ is $1B0 \otimes 1A4 = 1B4$. The trivial source module with Green correspondent $1B4$ has Brauer character in $\{10_B, 11_B, 13_B, 14_B\}$ by the tree for Block 2. None of these nodes occurs in $1_B \otimes 11_A$ as can be seen from the scalar products given above. This contradiction shows that 11_A is not joined to 20_A.

We next show that 6_A is neither joined to 14_A nor to 15_A. This follows from the fact that 6_A is a leaf and that none of 14_A or 15_A is a constituent in the projective $6_A \otimes (1_A + 2_A)$. By the projectives above 8_A is joined to one of $\{17_A, 18_A, 19_A, 22_A, 23_A\}$. It is clear from this that 14_A is a leaf.

We proceed to show that there is no non-real branch coming off the tree at node 17_A. This follows if we can show that there must be an edge incident to 21_A with label 11. We clearly have $1B0 \otimes 1B0 \approx_A 1A0$ and so $1B0 \otimes 46B11 \approx_A 46A11$. The second tensor product now proves our claim.

Let $1Ax$ be the Green correspondent of the leaf 14_A. The formula $1B0 \otimes 1B7 = 1A7$ and the first tensor product show that x equals 7. It follows that to the left (on the nought side) of the pair 14_A, 15_A there are exactly two pairs of complex conjugate characters on the tree. The third tensor product implies that 6_A, 7_A must be one of these, since the Green correspondent of $1_B \otimes 6_A$ clearly has Brauer character 21_B.

The irrationalities of all pairs of complex conjugate characters in this block are independent (i.e. generate pairwise disjoint fields). So we do not have to worry about the numbering of the two members of a pair. To summarize, we have shown that 6_A is joined to one of $\{18_A, 22_A\}$, 8_A to one of $\{18_A, 22_A, 23_A\}$, and each of 11_A, 18_A to one of $\{21_A, 24_A\}$. Furthermore, there are exactly two pairs of complex conjugate characters located left of 14_A, 15_A, of which 6_A, 7_A is one. This leaves exactly 12 possible planar embedded trees.

Group: M Prime: 47 Block: 2

Nr.	CAS-Nr.	Degree	CC	N&C
1	3	21296876	r	×
2	6	19360062527	r	∘
3	15	2374124840062976	r	×
4	16	8980616927734375	17	×
5	17	8980616927734375	16	×
6	19	39660520552077425	r	∘
7	26	3503434660075044981	27	∘
8	27	3503434660075044981	26	∘
9	66	19795913912408993711352	r	×
10	71	42940402913709544921875	72	×
11	72	42940402913709544921875	71	×
12	74	70660346341309333984375	75	∘
12	75	70660346341309333984375	74	∘
13	105	689763222744895005949242	106	×
14	106	689763222744895005949242	105	×
15	110	1361549126105752982272875	r	∘
16	123	5514132424881463208443904	r	×
17	140	14930164283563048960000000	r	×
18	141	16109407269221032565630370	r	∘
19	151	41762322738385820195625000	r	×
20	156	51324350389558097414062500	r	∘
21	158	58437394633227526183321600	r	∘
22	161	64326163427522624205703125	r	×
23	175	129572518017902934396764160	r	×
24	176	130287135266837289237316743	r	∘

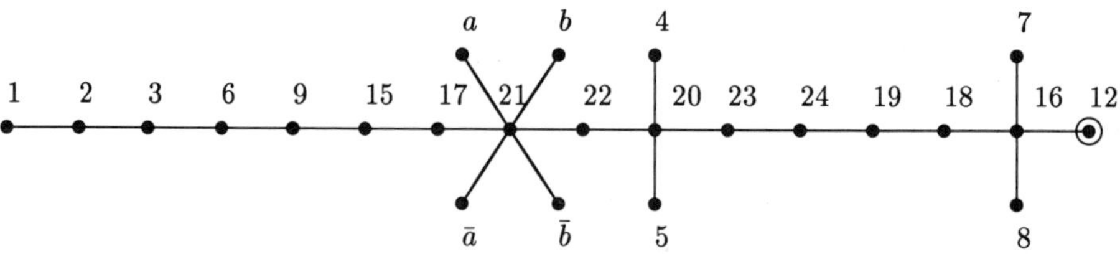

Here, $\{a, \bar{a}, b, \bar{b}\} = \{10, 11, 13, 14\}$. If we denote the Green correspondent of node 1_B by $1B0$, then the label on the edges around 21_B are $\{4, 5, 6, 18, 19, 20\}$ and the label on the edge joining node 4_B with the real stem is 7.

PROJECTIVES:

Nr.	1	2	3	4	5	6	7	8	9	10	11	12	13	14	15	16	17	18	19	20	21	22	23	24
CC	r	r	r	5	4	r	8	7	r	11	10	r	14	13	r	r	r	r	r	r	r	r	r	r
N&C	×	∘	×	×	×	∘	∘	∘	×	×	×	∘	×	×	∘	×	×	∘	×	∘	∘	×	×	∘
$\chi_2 \otimes \chi_2$	1	1																						
$\chi_5 \otimes \chi_2$		1	1																					
$\chi_{24} \otimes \chi_2$			2			3			1															
$\chi_{55} \otimes \chi_2$				1																1				
$\chi_{57} \otimes \chi_2$						1			4						5		2	1	1					
$\chi_{58} \otimes \chi_2$						1			4						4		2				1			
$\chi_{107} \otimes \chi_2$							1						2	1		1		1	3	9	8	9	19	16
$\chi_{26} \otimes (\chi_1 + \chi_5)$							1						1	1		1			1	3	5	4	9	8
$\chi_{62} \otimes \chi_2$									4						8	1	5	1	1		2	1		1
$\chi_{61} \otimes \chi_2$									3						6	1	4	2	2		1			1
$\chi_{73} \otimes \chi_2$									1						3	1	4	4	7		3	1	3	7
$\chi_{71} \otimes \chi_2$											1								1		1		2	3
$\chi_{126} \otimes \chi_2$												1			1	8	24	23	61	92	100	116	212	204
$\chi_{105} \otimes \chi_2$													1	1		1		1	3	9	8	9	19	16
$\chi_{87} \otimes \chi_2$															1		1	2	5	4	2	4	9	10
$\chi_{70} \otimes \chi_2$															1		1	1	3		1	1	2	4
$\chi_{83} \otimes \chi_2$																				3	1	3	3	2
$\chi_{69} \otimes \chi_2$																				1		1	1	1
$\chi_{74} \otimes \chi_2$																					1	1	2	2

Group: M Prime: 59 Block: 1

Nr.	CAS-Nr.	Degree	CC	N&C
1	1	1	r	×
2	5	18538750076	r	∘
3	10	125510727015275	r	×
4	16	8980616927734375	17	×
5	17	8980616927734375	16	×
6	23	336041615485626050	r	∘
7	26	3503434660075044981	27	×
8	27	3503434660075044981	26	×
9	38	15594307673918258285O	r	×
10	41	286243267692724486144	42	×
11	42	286243267692724486144	41	×
12	57	2351753641814605348320	r	∘
13	74	70660346341309333984375	75	∘
14	75	70660346341309333984375	74	∘
15	81	149614794149226010902528	82	∘
16	82	149614794149226010902528	81	∘
17	83	161649111002260792968750	84	∘
18	84	161649111002260792968750	83	∘
19	99	597787522207315571077947	100	∘
20	100	597787522207315571077947	99	∘
21	102	626877403613887304040448	103	∘
21	103	626877403613887304040448	102	∘
22	105	689763222744895005949242	106	∘
23	106	689763222744895005949242	105	∘
24	122	5334046162969208352215625	r	×
25	127	7375892500409609408203125	r	×
26	135	9592298143650890255171584	136	∘
27	136	9592298143650890255171584	135	∘
28	156	51324350389558097414062500	r	∘
29	160	63750812845035828079008441	r	∘
30	174	125517264890136048242396811	r	×

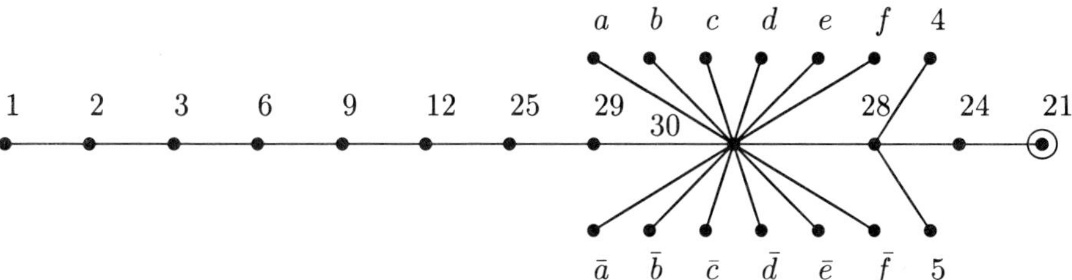

Warning ! This is not the complete Brauer tree since the nodes 7, 8 and 10, 11 are missing.

The notation

$$\{a, \bar{a}, b, \bar{b}, c, \bar{c}, d, \bar{d}, e, \bar{e}, f, \bar{f}\} = \{13, 14, 15, 16, 17, 18, 19, 20, 22, 23, 26, 27\}$$

is used in the above tree. We do not know the location of the nodes 7 and 10.

Observe that $(\chi_{10} \otimes \chi_i, \chi_j) = 0$ for $i, j \in \{16, 17, 26, 27, 41, 42\}$, $i \neq j$, $(i, j) \neq (41, 42)$ or $(42, 41)$. Since χ_{10} is node 3 and has a Green correspondent $1A1$, no two of the above characters can be incident to edges with subsequent labels. In particular, since χ_{16} (node 4) is joined to node 28, this can not be the case for χ_{26} nor χ_{41} (nodes 7 and 10). Also, not both of 7 and 10 are joined to node 29 at the same time.

It now follows from the projectives below that node 7 is joined to one of $\{22, 23, 26, 27, 29\}$.

We remark that in $\chi_5 \otimes \chi_{41}$ (in terms of nodes $2 \otimes 10$) none of χ_{74}, χ_{75}, χ_{81} nor χ_{82} (nodes 13, 14, 15 and 16) occurs. Since $\chi_1 + \chi_5$ is projective, it follows from this observation and the projectives given below that node 10 must be joined to one of $\{17, 18, 19, 20, 21, 22, 23, 26, 27, 29\}$. If we neglect the embedding and algebraic conjugacy (among the complex conjugate pairs) we thus have 17 possible trees left.

PROJECTIVES:

Nr.	1	2	3	4	5	6	7	8	9	10	11	12	13	14	15	16	17	18	19	20	21	22	23	24	25	26	27	28	29	30
CC	r	r	r	5	4	r	8	7	r	11	10	r	14	13	16	15	18	17	20	19	r	23	22	r	r	27	26	r	r	r
N&C	×	○	×	×	×	○	×	×	×	×	×	○	○	○	○	○	○	○	○	○	○	○	○	×	×	○	○	○	○	×
$\chi_2 \otimes \chi_2$	1	1																												
$\chi_4 \otimes \chi_2$		1	1																											
$\chi_{12} \otimes \chi_2$			1			1																								
$\chi_{56} \otimes \chi_2$					1																							1		
$\chi_{37} \otimes \chi_2$						1			2			1																		
$\chi_{108} \otimes \chi_2$								1														1	2	1		2	2	9	7	21
$\chi_{157} \otimes \chi_2$										1	1		1	1	2	2	1	1	10	10	8	10	10	77	97	130	130	719	872	1731
$\chi_{35} \otimes \chi_3$												1													1					
$\chi_{54} \otimes \chi_3$														1														2	2	5
$\chi_{81} \otimes \chi_2$																1												1	2	4
$\chi_{86} \otimes \chi_2$																		1								1	1	2	2	7
$\chi_{45} \otimes \chi_2$																				1										1
$\chi_{105} \otimes \chi_2$																						1	1	1		2	2	9	7	21
$\chi_{70} \otimes \chi_2$																									1				2	1

Group: M Prime: 71 Block: 1

Nr.	CAS-Nr.	Degree	CC	N&C
1	1	1	r	×
2	4	842609326	r	∘
3	8	3879214937598	r	×
4	12	222879856734249	r	∘
5	16	8980616927734375	17	∘
6	17	8980616927734375	16	∘
7	26	3503434660075044981	27	∘
8	27	3503434660075044981	26	∘
9	39	172399434201593354756	40	∘
9	40	172399434201593354756	39	∘
10	41	286243267692724486144	42	×
11	42	286243267692724486144	41	×
12	44	640558364167263622626	45	∘
13	45	640558364167263622626	44	∘
14	49	776097192277137500000	r	×
15	53	1480279477146615234375	54	∘
16	54	1480279477146615234375	53	∘
17	81	149614794149226010902528	82	∘
18	82	149614794149226010902528	81	∘
19	91	261575621299360905468750	r	×
20	101	600020772685064502392907	r	∘
21	102	626877403613887304040448	103	×
22	103	626877403613887304040448	102	×
23	105	689763222744895005949242	106	×
24	106	689763222744895005949242	105	×
25	123	5514132424881463208443904	r	∘
26	124	5514132424881463208443904	125	∘
27	125	5514132424881463208443904	124	∘
28	128	7567151576542452425781250	129	×
29	129	7567151576542452425781250	128	×
30	133	9416031858681585751556096	r	×
31	135	9592298143650890255171584	136	∘
32	136	9592298143650890255171584	135	∘
33	149	38471795739256565080575180	r	∘
34	152	42001454087954515167503490	r	×
35	175	129572518017902934396764160	r	∘
36	177	135226984222789977095703125	r	×

PROJECTIVES:

Nr.	1	2	3	4	5	6	7	8	9	10	11	12	13	14	15	16	17	18	19	20
CC	r	r	r	r	6	5	8	7	r	11	10	13	12	r	16	15	18	17	r	r
N&C	×	∘	×	∘	∘	∘	∘	∘	∘	×	×	∘	∘	×	∘	∘	∘	∘	×	∘
$\chi_2 \otimes \chi_2$	1	1																		
$\chi_6 \otimes \chi_2$		1	2	1																
$\chi_{31} \otimes \chi_2$				1										2						1
$\chi_{55} \otimes \chi_2$					1															
$\chi_{107} \otimes \chi_2$							1													
$\chi_{26} \otimes (\chi_1 + \chi_4)$							1													
$\chi_{157} \otimes \chi_2$										1	1						2	2	5	6
$\chi_{99} \otimes \chi_2$												1								
$\chi_{65} \otimes \chi_2$														1						4
$\chi_{166} \otimes \chi_2$															1	1	3	3	7	10
$\chi_{74} \otimes \chi_3$															1					1
$\chi_{81} \otimes \chi_2$																		1		
$\chi_{61} \otimes \chi_2$																				3
$\chi_{102} \otimes \chi_2$																				
$\chi_{105} \otimes \chi_2$																				
$\chi_{83} \otimes \chi_2$																				
$\chi_{85} \otimes \chi_2$																				
$\chi_{41} \otimes \chi_3$																				
$\chi_{71} \otimes \chi_2$																				

PROJECTIVES (continued):

Nr.	21	22	23	24	25	26	27	28	29	30	31	32	33	34	35	36
CC	22	21	24	23	r	27	26	29	28	r	32	31	r	r	r	r
N&C	×	×	×	×	∘	∘	∘	×	×	×	∘	∘	∘	×	∘	×
$\chi_2 \otimes \chi_2$																
$\chi_6 \otimes \chi_2$																
$\chi_{31} \otimes \chi_2$																
$\chi_{55} \otimes \chi_2$								1								
$\chi_{107} \otimes \chi_2$			2	1	1	2	2	1	1		2	2	5	5	19	24
$\chi_{26} \otimes (\chi_1 + \chi_4)$																1
$\chi_{157} \otimes \chi_2$	8	8	10	10	74	74	74	100	100	112	130	130	506	569	1784	1858
$\chi_{99} \otimes \chi_2$						1	1	2	2		1	1	3	5	18	17
$\chi_{65} \otimes \chi_2$										5			2	1	1	
$\chi_{166} \otimes \chi_2$	14	14	14	14	115	100	100	139	139	193	193	193	805	867	2687	2810
$\chi_{74} \otimes \chi_3$	1	1	1	1	9	10	10	15	15	15	18	18	64	78	236	240
$\chi_{81} \otimes \chi_2$													2	1	5	7
$\chi_{61} \otimes \chi_2$					1					4			1	1		
$\chi_{102} \otimes \chi_2$	1	1			1	1	1	1	1	1	2	2	6	6	19	21
$\chi_{105} \otimes \chi_2$			1	1	1	2	2	1	1		2	2	5	5	19	24
$\chi_{83} \otimes \chi_2$						1	1	2	2		1	1		1	3	2
$\chi_{85} \otimes \chi_2$						1	1	2	1		1	1	1	1	4	5
$\chi_{41} \otimes \chi_3$											1				1	2
$\chi_{71} \otimes \chi_2$													1		2	3

The projectives above determine the following **subtree** of the Brauer tree. Here, $\{a, \bar{a}, b, \bar{b}\} = \{7, 8, 31, 32\}$.

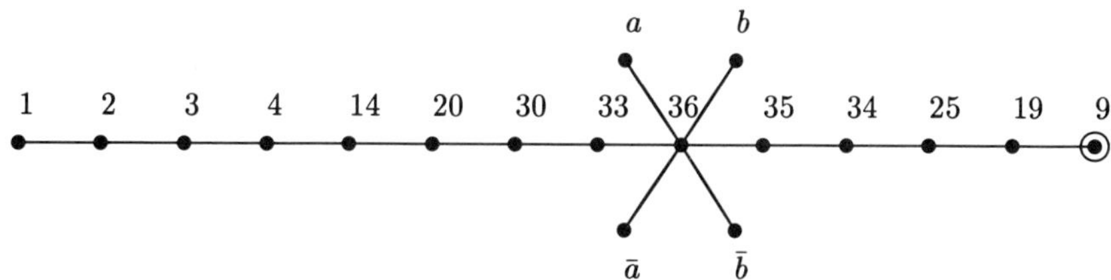

The projectives also determine the following branch of the tree, where $z \in \{31, 35\}$:

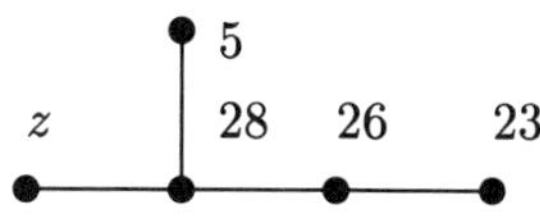

A

APPENDIX

A.1 The names

CAS-Name	ATLAS-Name
M11	M_{11}
2M12	$2.M_{12}$
J1	J_1
12M22	$12.M_{22}$
2J2	$2.J_2$
M23	M_{23}
2HS	2.HS
3J3	$3.J_3$
M24	M_{24}
3MCL	$3.M^cL$
HE	He
2RU	2.Ru
6SUZ	6.Suz

CAS-Name	ATLAS-Name
3ON	3. O'N
C3	Co_3
C2	Co_2
6F22	$6.Fi_{22}$
HA	HN
LY	Ly
TH	Th
F23	Fi_{23}
2C1	$2.Co_1$
J4	J_4
3F24′	$3.Fi_{24}'$
2BM	2.B
M	M

A.2 ATLAS vs. CAS order

The numbering of the nodes of a Brauer tree in our tables refers to the ordering of the characters in the CAS character tables. In a few cases this does not agree with the ordering of the ATLAS tables.

If there are several characters of the same degree, which are not algebraically conjugate, then their order on the ATLAS character table agrees with their order on the CAS table, unless indicated in the table below. The same applies to orbits of algebraically conjugate characters. Please note that a single character of the ATLAS compound table can split into more than one CAS character. This happens if it is a projective character belonging to a cocycle of order greater than 2. In these cases, of course, the CAS numbers are larger than the ATLAS numbers of the corresponding characters.

We have not attempted to adjust the ordering of the characters inside distinct orbits of algebraically conjugate characters whose values intersect in a non-trivial extension of the rational numbers. This matters only in cases where problems of algebraic conjugacy have been solved. The reader who is in doubt about the consistency of our tree with his or her character table should just imitate our proof to get a consistent tree.

Group	Degree	ATLAS-Nr.	CAS-Nr.
2M12	11	2	3
		3	2
2HS	1925	20	21
		21	20
6SUZ	20020	52	53
		53	52
	35100	{54, 55}	{56, 57}
	35100	{56, 57}	{54, 55}
	4368	118	160, 162
		119	159, 161
	27456	127	178, 180
		128	177, 179
3ON	495	33	37, 38
		34	35, 36
	5643	36	43, 44
		37	41, 42
C3	253	3	4
		4	3
	31625	22	23
		23	24
		24	22
	129536	33	34
		34	33
C2	31625	16	17
		17	16
	398475	38	39
		39	38
	664125	46	48
		47	47
		48	46
	1771000	52	53
		53	52

Group	Degree	ATLAS-Nr.	CAS-Nr.
6F22	50050	15 16	16 15
	75075	17 18	18 17
	450450	35 36	36 35
	1297296	102 103	103 102
	2555904	113 114	114 113
	405405	135 136	156, 158 155, 157
	1351350	150 151 152 153	187, 191 186, 190 188, 192 185, 189
	1965600	161 162	208, 210 207, 209
	314496	{176, 179} {177, 178}	{237, 238, 241, 242} {239, 240, 243, 244}
LY	45648306	45 46	46 45
F23	166559744	64 65	65 64
	264536064	76 77	77 76
2C1	469945476	94 95	95 94
J4	1183406741	43 44	44 43
3F24C	111223322200	74 75	75 74
	156321775827	88 89	89 88
2BM	665029816320000	92 93	93 92
	1370975689505565	106 107	107 106
	3555194033015625	127 128	128 127

A.3 Known decomposition matrices

The following table contains the sporadic simple groups or their covering groups for which all decomposition matrices are known.

Group	References
M11	James(1973)
2M12	James (1973); Humphreys (1980)
12M22	James (1973); Humphreys (1982b); Benson (1985)
M23	James (1973)
M24	James (1973); Parker (1987)
J1	Fong (1974); Feit (1984b)
2J2	Hiss and Lux (1988); Feit (1984b)
HS	Humphreys (1982a); Thackray (1981)

The next table contains those sporadic simple groups not already listed above, for which some decomposition matrices for blocks with noncyclic defect groups are known.

Group	Prime	References
MCL	2	Thackray (1981)
MCL	5	Hiss, Lux and Parker (1988); Woldar (1986)
C2	5	Lux (1988)
HE	7	Ryba (1988)

The next table contains those sporadic simple groups not already listed above, where alomst complete knowledge about some decomposition matrices for blocks of noncyclic defect exists.

Group	Prime	References
HE	3	Lux (1988)
HE	5	Lux (1988)
6SUZ	5	Lux (1988)
F22	5	Lux (1988)
2C1	7	Lux (1988)

As mentioned in the introduction, there are only ten Brauer trees left in the covering groups of the sporadic simple groups, which could not be determined up to algebraic conjugacy. The following table lists them all.

Group	Prime	Block Nr.
TH	19	1
2BM	13	6
2BM	19	2
2BM	31	2
2BM	47	1 and 2
M	41	1
M	47	1
M	59	1
M	71	1

Finally we give the trees which have been determined up to algebraic conjugacy independently by other people.

Group	Prime(s)	References
M11	all	James (1973)
2M12	all	James (1973); Humphreys (1980)
12M22	all	James (1973); Humphreys (1982b); Benson (1985)
M23	all	James (1973)
M24	all	James (1973)
J1	all	Feit (1984b); Parker (1987)
2J2	all	Feit (1984b); Parker (1987)
2HS	all	Thackray (1981); Feit (1984b); Parker (1987)
3J3	all	Feit (1984b)
3MCL	all	Feit (1984b); Thackray (1981); Parker (1987)
HE	7	Parker (1987)
2RU	7	Parker (1987)
6SUZ	7	Parker (1987)
HA	19	Parker (1987)
2C1	23	Parker (1987)
TH	31	Parker (1987)
M	31	Parker (1987)

BIBLIOGRAPHY

Alperin, J. L. (1986) *Local representation theory*, Cambridge studies in advanced mathematics 11, Cambridge University Press, Cambridge.

Benson, D. (1985) Brauer trees for $12M_{22}$, *J. Algebra* **95**, 398–408.

Benson, D. and Parker, R. (1984) The Green ring of a finite group, *J. Algebra* **87**, 290–331.

Brauer, R. (1941) Investigations on group characters, *Ann. of Math.* **42**, 936–58.

Burkhardt, R. (1979) Über die Zerlegungszahlen der Suzukigruppen, *J. Algebra* **59**, 421–33.

Burry, D. W. (1979) A strengthened theory of vertices and sources, *J. Algebra* **59**, 300–44.

Conway, J. H., Curtis, R. T., Norton, S. P., Parker, R. A. and Wilson, R. A. (1985) *Atlas of finite groups*, Oxford University Press, London/New York.

Dade, E. C. (1966) Blocks with cyclic defect groups, *Ann. of Math.* **84**, 20–48.

Deriziotis, D. I. and Michler, G. O. (1987) Character table and blocks of finite simple triality groups ${}^3D_4(q)$, *Trans. Amer. Math. Soc.* **303**, 39–70.

Feit, W. (1982) *The representation theory of finite groups*, North-Holland, Amsterdam.

Feit, W. (1984a) Possible Brauer trees, *Illinois J. of Math.* **28**, 43–56.

Feit, W. (1984b) Blocks with cyclic defect groups for some sporadic groups, in *Representation theory II, groups and orders*, Proceedings, Ottawa, Lecture Notes in Mathematics Vol. 1178, pp. 25–63, Springer Verlag, Berlin/New York/Heidelberg/Tokyo.

Fischer, B. (1986) Character tables of local subgroups of sporadic groups, (unpublished).

Fong, P. (1974) On decomposition numbers of J_1 and $R(q)$, in *Symposia Mathematica*, Vol. XIII, pp. 415–22, Academic Press, London.

Fong. P. (1989) Brauer trees in classical groups, preprint.

Fong, P. and Srinivasan, B. (1980) Blocks with cyclic defect groups in $GL(n,q)$, *Bull. Amer. Math. Soc.* **3**, 1041–4.

Fong, P. and Srinivasan, B. (1982) The blocks of finite general linear and unitary groups, *Invent. Math.* **69**, 109–53.

Fong, P. and Srinivasan, B. (1984) Brauer trees in $GL(n,q)$, *Math. Z.* **187**, 81–8.

Goldschmidt, D. M. (1980) *Lectures on Character Theory*, Publish or Perish, Berkeley.

Green, J. A. (1974a) *Vorlesungen über modulare Darstellungstheorie endlicher Gruppen*, Vorlesungen aus dem Mathematischen Institut Giessen, Heft 2.

Green, J. A. (1974b) Walking around the Brauer tree, *J. Austral. Math. Soc.* **17**, 197–213.

Hiss, G. (1986) The modular characters of the Tits simple group and its automorphism group, *Comm. Algebra* **14**, 125–54.

Hiss, G. and Lux, K. (1988) The Brauer characters of the Hall–Janko group, *Comm. Algebra* **16**, 357–98.

Hiss, G., Lux, K. and Parker, R. A. (1988) The 5-modular characters of the McLaughlin group, (unpublished).

Humphreys, J. F. (1980) The projective characters of the Mathieu group M_{12} and of its automorphism group, *Math. Proc. Camb. Phil. Soc.* **87**, 401–12.

Humphreys, J. F. (1982a) The modular characters of the Higman–Sims group, *Proc. Roy. Soc. Edinburgh* **92A**, 319–35.

Humphreys, J. F. (1982b) The projective characters of the Mathieu group M_{22}, *J. Algebra* **76**, 1–24.

Isaacs, I. M. (1976) *Character theory of finite groups*, Academic Press, New York/London.

James, G. D. (1973) The modular characters of the Mathieu groups, *J. Algebra* **27**, 57–111.

James, G. D. (1978) *The representation theory of the symmetric groups*, Lecture Notes in Mathematics, Vol. 682, Springer-Verlag, Berlin/New York.

Kawata, S. (1987) On the Loewy structure in the blocks of the Mathieu groups with cyclic defect groups, *Comm. Algebra* **15**, 1519–31.

Külshammer, B. (1987) (unpublished).

Kupisch, H. (1968) Projektive Moduln endlicher Gruppen mit zyklischer *p*-Sylow Gruppe, *J. Algebra* **10**, 1–7.

Landrock, P. (1983) *Finite group algebras and their modules*, London Mathematical Society Lecture Notes Series, Vol. 84, Cambridge University Press.

Lindsey, J. H. (1974) Groups with a T.I. cyclic Sylow subgroup, *J. Algebra* **30**, 181–235.

Lux, K. (1987) *Brauerbäume sporadisch einfacher Gruppen*, Dissertation, RWTH Aachen.

Lux, K. (1988) Zerlegungsmatrizen einiger sporadischer Gruppen, (unpublished).

Michler, G. (1975) Green correspondence between blocks with cyclic defect groups II, in *Representation of algebras*, Lecture Notes in Mathematics Vol. 488, pp. 210–35, Springer-Verlag, Berlin/Heidelberg/New York/Tokyo.

Neubüser, J., Pahlings, H., and Plesken, W. (1984) CAS; Design and use of a system for the handling of characters of finite groups, in *Computational group theory*, pp. 195–247, Academic Press.

Parker, R. A. (1987) *A collection of modular characters*, (unpublished).

Peacock, R. M. (1975) Blocks with a cyclic defect group, *J. Algebra* **37**, 74–103.

Ryba, A. J. .E. (1988) Calculation of the 7-modular characters of the Held group, *J. Algebra* **117**, 240–55.

Shamash, J. (1987) Blocks and Brauer trees for groups of type $G_2(q)$, in *Procceedings of the A.M.S. summer institute on representations of finite groups and related topics*, Arcata, CA.

Thackray, J. G. (1981) *Modular representations of finite groups*, Ph. D. thesis, Cambridge University.

Thompson, J. G. (1986) Bilinear forms in characteristic p and the Frobenius–Schur indicator, in *Group theory*, Beijing 1984, Lecture Notes in Mathematics Vol. 1185, pp. 221–30, Springer-Verlag, Berlin/Heidelberg/New York/Tokyo.

Ward, H. N. (1966) On Rees series of simple groups, *Trans. Amer. Math. Soc.* **121**, 62–89.

Willems, W. (1976) *Metrische Moduln über Gruppenringen*, Dissertation, Universität Mainz.

Woldar, A. J. (1986) On the 5-decomposition matrix for McLaughlin's sporadic simple group, *Comm. Algebra* **14**, 277–91.

INDEX